Electrical Power System Essentials

Electrical Power System Essentials

Pieter Schavemaker
Principal Consultant
The Netherlands

Lou van der Sluis
Professor emeritus Electrical Power Systems
Delft University of Technology
The Netherlands

Second Edition

This edition first published 2017
© 2017 John Wiley & Sons Ltd

First edition published 2008 by John Wiley & Sons Ltd.

Registered office
John Wiley & Sons Ltd, The Atrium, Southern Gate, Chichester, West Sussex, PO19 8SQ,
United Kingdom

For details of our global editorial offices, for customer services and for information about
how to apply for permission to reuse the copyright material in this book please see our website
at www.wiley.com.

Library of Congress Cataloging-in-Publication Data

Names: Schavemaker, Pieter. | Van der Sluis, Lou.
Title: Electrical power system essentials / Pieter Schavemaker and Lou van
    der Sluis.
Description: Second edition. | Chichester, West Sussex : John Wiley & Sons,
    Inc., 2017. | Includes bibliographical references and index.
Identifiers: LCCN 2016045881| ISBN 9781118803479 (cloth) | ISBN 9781118803462
    (epub)
Subjects: LCSH: Electric power systems. | Electric power distribution. |
    Electric power production.
Classification: LCC TK1001 .S3555 2017 | DDC 621.319/13−dc23 LC record available at
https://lccn.loc.gov/2016045881

A catalogue record for this book is available from the British Library.

Cover image: Reproduced by permission of TenneT TSO B.V.
Cover design by Wiley

Set in 10/12pt WarnockPro by SPi Global, Chennai, India

10 9 8 7 6 5 4 3 2 1

Printed and bound in Great Britain by TJ International Ltd, Padstow, Cornwall

Contents

Preface

In the field of power system analysis, an extensive amount of high-quality literature is available. Most of these textbooks follow more or less the same line and cover the same topics. This book differs from existing materials because the (steady-state) modeling of the power system components is covered in appendices. Therefore, the focus in the chapters itself is not on the modeling, but on the structure, functioning, and organization of the power system. The appendices contribute to the book by offering material that is not an integral part of the main text, but support and enhance it and as such are an integral part of the book. The book contains a large number of problems of which the extensive solutions are presented in a separate chapter.

The following is a short summary of the contents of the chapters and the appendices.

Chapter 1 (Introduction to Power System Analysis)

This first chapter describes the scope of the material and is an introduction to the steady-state analysis of power systems. Questions such as "why AC," "why 50 or 60 Hz," "why sinusoidally shaped AC," "why a three-phase system" are addressed. The basics for a steady-state analysis of balanced three-phase power systems are outlined, such as phasors, single-line diagrams, active power, reactive power, complex power, power factor, and per-unit normalization.

Chapter 2 (The Generation of Electric Energy)

The conversion from a primary source of energy to electrical energy is the topic of Chapter 2. The primary source of energy can be fossil fuels such as gas, oil, and coal or uranium, but can come from renewable sources as well: wind energy, hydropower, solar power, or geothermal power. In order to understand the nature of a thermal power plant, which is still the main source of power in the system, the principles of thermodynamics are briefly discussed. The final conversion from mechanical energy to electrical energy is achieved by the synchronous machine. The coupling of the machine with the grid and the actual power injection is analyzed.

Chapter 3 (The Transmission of Electric Energy)

The transmission and distribution network is formed by the overhead lines, the underground cables, the transformers, and the substations between the points of power injection and power consumption. Various substation concepts are presented, together with substation components and the protection installed. The transformers, overhead transmission lines, underground cables, gas-insulated transmission lines, protective relay operating principles, surge arresters, fuses, and circuit breakers are then considered in more detail. The transformer design, possible phase shift, and specific properties due to the magnetic core are highlighted. As overhead transmission lines are the most visible part of the power system, they are discussed from the point of view of what may be seen and why it is like that. The underground cables are also considered, contrasting them with overhead transmission. The chapter ends with the principles of HVDC transmission.

Chapter 4 (The Utilization of Electric Energy)

The power system is designed and arranged in such a way that demand may be fulfilled: consumers are supplied with the requested amount of active and reactive power at constant frequency and with a constant voltage. A load actually transforms the AC electrical energy into another form of energy. The focus in this chapter is on the various types of loads that transform the AC electrical energy into mechanical energy (synchronous and induction motors), light, heat, DC electrical energy (rectifiers), and chemical energy. After that, the individual loads in the system are clustered and classified as grid users according to three categories: residential loads (mostly single-phase loads), commercial and industrial loads (often three-phase loads), and electric railways (either DC or single-phase AC).

Chapter 5 (Power System Control)

Continuous control actions are necessary in the system for the control of the voltage, to maintain the balance between the amount of generated and consumed electricity, and to keep the system frequency at either 50 or 60 Hz. It is demonstrated that, in transmission networks, there is more or less a "decoupling" between the active power and the voltage angles on one side and the reactive power and voltage magnitudes on the other, which is the basis for the control. The power balance is maintained (primary control), and the system frequency deviation minimized (secondary control), by controlling the active power output of the generators. Voltage is controlled locally either at generator buses by adjusting the generator voltage control or at fixed points in the system where tap-changing transformers, capacitor banks, or other reactive power consumers/producers are connected. Flexible AC transmission systems (FACTS) devices are large power-electronic devices; they are operated in a shunt configuration for reactive power and voltage control, or they are connected in series to control the power flow.

Chapter 6 (Energy Management Systems)

In the control center, the transmission and distribution of electrical energy are monitored, coordinated, and controlled. The energy management system (EMS) is the interface between the operator and the actual power system. The supervisory control and data acquisition (SCADA) system collects real-time measured data from the system and presents it to the computer screen of the operator, and it sends control signals from the control center to the actual components in the network. The EMS is in fact an extension of the basic functionality of the SCADA system and includes tools for the analysis and the optimal operation of the power system. The state estimator serves as a "filter" for the collected measurement data; it determines the state of the power system that matches best with the available measurements. This is necessary input for other analysis programs in the EMS, such as the load flow or power flow and the optimal power flow. The load flow computation is one of the most important power system computations, giving us insight into the steady-state behavior of the power system. Therefore, besides the well-known Newton–Raphson load flow, a decoupled load flow and the DC load flow are also presented.

Chapter 7 (Electricity Markets)

At a broad conceptual level, there exists such a thing as a "common market model" that provides for both spot market trading coordinated by a grid/market operator and for bilateral contract arrangements scheduled through the same entity. The spot market is based on a two-sided auction model: both the supply and demand bids are sent to the power exchange. Market equilibrium occurs when the economic balance among all participants is satisfied and the benefits for society, called "the social welfare," are at their maximum value. The power system is a large interconnected system, so that multiple market areas are physically interconnected with each other: this facilitates the export of electricity from low-price areas to high-price areas.

Chapter 8 (Future Power Systems)

In this chapter some developments, originating from the complex technological, ecological, sociological, and political playing field and their possible consequences on the power system, are highlighted. A large-scale implementation of electricity generation based on renewable sources, for example, will cause structural changes in the existing distribution and transmission networks. Many of these units are decentralized generation units, rather small-scale units that are connected to the distribution networks often by means of a power-electronic interface. A transition from the current "vertically operated power system" into a "horizontally operated power system" in the future is not unlikely. Energy storage can be applied to level out large power fluctuations when the power is generated by renewable energy sources, driven by intermittent primary energy. The complexity of the system increases because of the use of FACTS devices, power-electronic interfaces, intermittent power production, and so on. Chaotic phenomena are likely to occur in the near future and large system blackouts will probably happen more often.

Appendix A (Maxwell's Laws)

Circuit theory can be regarded as describing a restricted class of solutions of Maxwell's equations. In this appendix, power series approximations will be applied to describe the electromagnetic field. It is shown that the zero- and first-order terms in these approximations (i.e., the quasi-static fields) form the basis for the lumped-circuit theory. By means of the second-order terms, the validity of the lumped-circuit theory at various frequencies can be estimated. It is the electrical size of the structure – its size in terms of the minimum wavelength of interest in the bandwidth over which the model must be valid – that dictates the sophistication and complexity of the required model. A criterion is derived that relates the dimensions of the electromagnetic structure to the smallest wavelength under consideration so that the validity of the lumped-element model can be verified.

Appendix B (Power Transformer Model)

Transformers essentially consist of two coils around an iron core. The iron core increases the magnetic coupling between the two coils and ensures that almost all the magnetic flux created by one coil links the other coil. The central item of this appendix is the mathematical description of the voltage–current relations of the transformer. First, the voltage–current relation of an ideal transformer, including the impedance transformation, is given. After that, a more general description of the transformer by means of magnetically coupled coils is derived. In the next step, the nonideal behavior of the transformer, comprising leakage flux and losses in the windings and in the iron core, is taken into account, and a transformer equivalent circuit is derived. The appendix ends with an overview of single-phase equivalent models of three-phase transformers.

Appendix C (Synchronous Machine Model)

A synchronous generator generates electricity by conversion of mechanical energy into electrical energy. The two basic parts of the synchronous machine are the rotor and the armature or stator. The iron rotor is equipped with a DC-excited winding, which acts as an electromagnet. When the rotor rotates and the rotor winding is excited, a rotating magnetic field is present in the air gap between the rotor and the armature. The armature has a three-phase winding in which the time-varying EMF is generated by the rotating magnetic field. For the analysis of the behavior of the synchronous machine in the power system, a qualitative description alone is not sufficient. The central item of this appendix is the mathematical description of the voltage–current relation of the synchronous generator. Based on the voltage–current relation, a circuit model is developed that is connected to an infinite bus to study the motor and generator behavior.

Appendix D (Induction Machine Model)

The induction machine is an alternating current machine that is very well suited to be used as a motor when it is directly supplied from the grid. The stator of the induction machine has a three-phase winding; the rotor is equipped with a short-circuited rotor winding. When the rotor speed is different from the speed of the rotating magnetic field generated by the stator windings, we describe the rotor speed as being asynchronous, in which case the short-circuited rotor windings are exposed to a varying magnetic field that induces an EMF and currents in the short-circuited rotor windings. The induced rotor currents and the rotating stator field result in an electromagnetic torque that attempts to pull the rotor in the direction of the rotating stator field. The central item of this appendix is the mathematical description of the voltage–current relation and the torque–current relations of the induction machine. Based on the voltage–current relation, a circuit model is developed.

Appendix E (The Representation of Lines and Cables)

When we speak of electricity, we think of current flowing through the conductors of overhead transmission lines and underground cables on its way from generator to load. This approach is valid because the physical dimensions of the power system are generally small compared to the wavelength of the currents and voltages in steady-state analysis. This enables us to apply Kirchhoff's voltage and current laws and use lumped elements in our modeling of overhead transmission lines and underground cables. We can distinguish four parameters for a transmission line: the series resistance (due to the resistivity of the conductor), the inductance (due to the magnetic field surrounding the conductors), the capacitance (due to the electric field between the conductors), and the shunt conductance (due to leakage currents in the insulation). Three different models are derived, which, depending on the line length, can be applied in power system analysis.

In the process of writing this book, we sometimes felt like working on a film script: we put the focus on selected topics and zoomed in or out whenever necessary, as there is always a delicate balance between the thing that you want to make clear and the depth of the explanation to reach this goal. We hope that we have reached our final goal and that this book provides you with a coherent and logical introduction to the interesting world of electrical power systems!

While writing this book we gratefully made use of the lecture notes that have been used over the years at the Delft University of Technology and the Eindhoven University of Technology in the Netherlands. The appendices on the modeling of the transformer, the synchronous machine, and the induction machine are based on the excellent Dutch textbook of Dr. Martin

Hoeijmakers on the conversion of electrical energy. We are very grateful for the careful reading of the manuscript by Prof. Emeritus Koos Schot, Robert van Amerongen, and Jan Heijdeman. We would like to thank Ton Kokkelink and Rene Beune, both from TenneT TSO B.V., for their valuable comments on Chapters 5 and 7, respectively. We appreciate the contribution to the problems and their solutions of Romain Thomas, and Dr. Laura Ramirez Elizondo.

The companion website for the book is http://www.wiley.com/go/ powersystem, where PowerPoint slides for classroom use can be downloaded.

Pieter H. Schavemaker
Lou van der Sluis
The Netherlands
Spring 2017

List of Abbreviations

AC	alternating current
ACE	area control error
ACSR	aluminum conductor steel reinforced
ATC	available transmission capacity
AVR	automatic voltage regulator
BES	battery energy storage
CAES	compressed air energy storage
CHP	combined heat and power
CO_2	carbon dioxide
CT	current transformer
DAM	day-ahead market
DC	direct current
DG	decentralized generation, distributed generation, dispersed generation
EMF	electromotive force
EMS	energy management system
ENTSO-E	European network of transmission system operators for electricity
FACTS	flexible AC transmission systems
GIL	gas-insulated transmission line
GTO	gate turnoff thyristor
HVDC	high-voltage DC
ID	intraday
IEC	International Electrotechnical Commission
IEEE	Institute of Electrical and Electronics Engineers
IGBT	insulated gate bipolar transistor
IPP	independent power producer
ISO	independent system operator
LCC	line commutated converter
LED	light-emitting diode
LFC	load frequency control

LL	line-to-line
LN	line-to-neutral
LTI	linear time-invariant
MCP	market clearing price
MCV	market clearing volume
NEC	net export curve
OTC	over the counter
pu	per unit
PV	photovoltaic
PWM	pulse-width modulation
PX	power exchange
RMS	root mean square
SCADA	supervisory control and data acquisition
SF_6	sulfur hexafluoride
SIPL	switching impulse protective level
SMES	superconducting magnetic energy storage
SSSC	static synchronous series compensator
STATCOM	static synchronous compensator
SVC	static var compensator
TCR	thyristor-controlled reactor
TCSC	thyristor-controlled series capacitor
TSC	thyristor-switched capacitor
TSO	transmission system operator
UCTE	Union for the Coordination of Transmission of Electricity
UPFC	unified power flow controller
VSC	voltage source converter
XPLE	cross-linked polyethylene

List of Symbols

Text Symbols

Bold uppercase text symbols generally refer to matrices, for example, A.
Bold lowercase text symbols generally refer to vectors, for example, x.

Various notations of a voltage:

$v, v(t)$	the sinusoidal time-varying quantity
V	the phasor representation of the sinusoidal time-varying quantity; the DC quantity
$|V|$	the effective or RMS value of the sinusoidal time-varying quantity; the length of the phasor representation of the sinusoidal time-varying quantity

The polarity of the voltage is indicated in circuit diagrams in one of the three following ways:

 DC voltage source; the long plate indicates the positive terminal, and the short plate the negative terminal

 AC voltage source; the plus sign indicates the positive terminal, and the minus sign the negative terminal

 arrow: it specifies the voltage between two terminals/points in the circuit diagram; the arrowhead indicates the positive terminal, and the tail the negative terminal

Graphical Symbols

Graphical symbols in a circuit diagram:

⌇⌇⌇	inductance
—∥—	capacitance
—◁◁◁—	resistance
—▭—	impedance, admittance, general load
—⊟—	fuse
⌇∥⌇	transformer
⌇ ⌇	magnetically coupled coils
≐	DC voltage source
(∼)	AC voltage source
(↑)	current source
—▶⊢	diode
—▶⊢	power-electronic switching device (e.g., thyristor, GTO)
⊥	earth, neutral, reference

Graphical symbols in a single-line or one-line diagram:

———	transmission link, line, cable
✕	circuit breaker
+	disconnector
▮	busbar, node
⟶	load

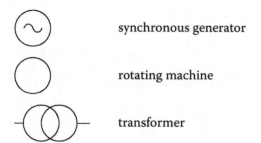

synchronous generator

rotating machine

transformer

1

Introduction to Power System Analysis

1.1 Introduction

As electricity comes out of the alternating current (AC) outlet every day, and has already been doing so for more than 100 years, it may nowadays be regarded as a commodity. It is a versatile and clean source of energy; it is fairly cheap and "always available." In the Netherlands, for instance, an average household encountered only 20 minutes' interruption to their supply in the year 2014 [1] out of a total of 8760 hours, resulting in an availability of 99.996195%!

Society's dependence on this commodity has become critical and the social impact of a failing power system is beyond imagination:

- Cars would not be refueled as gas station pumps are driven by electricity.
- The sliding doors of shops and shopping malls would not be able to open or close and people would therefore be locked out or in.
- Electrified rail systems, such as subways and trains, would come to a standstill.
- Traffic lights would not work.
- Refrigerators would stop.
- Heating/cooling installations would fail.
- Cash dispensers would be offline.
- Computers would serve us no longer.
- Water supplies would stop or run out.

Many more examples may be given, but the message is clear: electric power systems are the backbone of modern society (see Figure 1.1), and chaos would result if the electricity supply failed for an extended period.

Our society needs engineers who know how to design, build, and operate an electrical power system. So let us discover what lies beyond the AC outlet and enter the challenging world of power system analysis.

Electrical Power System Essentials, Second Edition. Pieter Schavemaker and Lou van der Sluis.
© 2017 John Wiley & Sons Ltd. Published 2017 by John Wiley & Sons Ltd.

Figure 1.1 The Earth's city lights, indicating the most urbanized areas. The Visible Earth, NASA.

1.2 Scope of the Material

Power system analysis is a broad subject, too broad to cover in a single textbook. The authors confine themselves to an overview of the structure of the power system (from generation via transmission and distribution to customers) and only take into account its steady-state behavior. This means that only the power frequency (50 or 60 Hz) is considered. An interesting aspect of power systems is that the modeling of the system depends on the time scale under review. Accordingly, the models for the power system components that are used in this book have a limited validity; they are only valid in the steady-state situation and for the analysis of low-frequency phenomena. In general, the time scales we are interested in are as follows:

- Years, months, weeks, days, hours, minutes, and seconds for steady-state analysis at power frequency (50 or 60 Hz)
  This is the time scale on which this book focuses. Steady-state analysis covers a variety of topics such as planning, design, economic optimization, load flow/power flow computations, fault calculations, state estimation, protection, stability, and control.
- Milliseconds for dynamic analysis (kHz)
  Understanding the dynamic behavior of electric networks and their components is important in predicting whether the system, or a part of the system, remains in a stable state after a disturbance. The ability of a power system to maintain stability depends heavily on the controls in the system to dampen the electromechanical oscillations of the synchronous generators.

- Microseconds for transient analysis (MHz)

  Transient analysis is of importance when we want to gain insight into the effect of switching actions, for example, when connecting or disconnecting loads or switching off faulty sections, or into the effect of atmospheric disturbances, such as lightning strokes, and the accompanying overvoltages and overcurrents in the system and its components.

Although the power system itself remains unchanged when different time scales are considered, components in the power system should be modeled in accordance with the appropriate time frame. An example to illustrate this is the modeling of an overhead transmission line. For steady-state computations at power frequency, the wavelength of the sinusoidal voltages and currents is 6000 km (in the case of 50 Hz):

$$\lambda = \frac{v}{f} = \frac{3 \times 10^5}{50} = 6000 \,\text{km} \tag{1.1}$$

λ the wavelength [km]
v the speed of light ≈ 300000 [km/s]
f the frequency [Hz $= 1/s$]

Thus, the transmission line is, so to speak, of "electrically small" dimensions compared to the wavelength of the voltage. The Maxwell equations can therefore be approximated by a quasi-static approach, and the transmission line can accurately be modeled by lumped elements (see also Appendix A). Kirchhoff's laws may fruitfully be used to compute the voltages and currents. When the effects of a lightning stroke have to be analyzed, frequencies of 1 MHz and higher occur and the typical wavelength of the voltage and current waves is 300 m or less. In this case the transmission line is far from being "electrically small," and it is not allowed to use the lumped-element representation anymore. The distributed nature of the transmission line has to be taken into account, and we have to calculate with traveling waves.

Despite the fact that we mainly use lumped-element models in our book, it is important to realize that the energy is mainly stored in the electromagnetic fields surrounding the conductors rather than in the conductors themselves as is shown in Figure 1.2. The Poynting vector, being the outer product of the electric field intensity vector and the magnetic field intensity vector, indicates the direction and intensity of the electromagnetic power flow [2, 3]:

$$S = E \times H \tag{1.2}$$

S the Poynting vector [W/m^2]
E the electric field intensity vector [V/m]
H the magnetic field intensity vector [A/m]

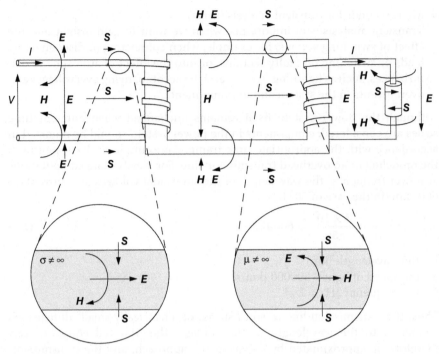

Figure 1.2 Transmission line–transformer–transmission line–load: the energy is stored in the electromagnetic field.

Due to the finite conductivity of the conductor material and the finite permeability of the transformer core material, a small electric field component is present inside the conductor and a small magnetic field component results in the transformer core:

$$E = \frac{J}{\sigma} \tag{1.3}$$

J the current density vector [A/m^2]
σ the conductivity [S/m]

$$H = \frac{B}{\mu} \tag{1.4}$$

B the magnetic flux density vector [T = A H/m^2]
μ the permeability [H/m]

This leads to small Poynting vectors pointing toward the conductor and the transformer core: the losses in the transmission line and the transformer are fed from the electromagnetic field, as is the power consumed by the load.

1.3 General Characteristics of Power Systems

Most of the power systems are 50 or 60 Hz three-phase AC systems. The voltage levels used are quite diverse. In the following sections, we explain why these choices have been made.

1.3.1 AC versus DC Systems

The choice for AC systems over DC systems can be brought back to the "battle" between Nikola Tesla (1856–1943) and Thomas Alva Edison (1847–1931). Edison managed to let a light bulb burn for 20 hours in the year 1879. He used a 100 V DC voltage and this was one of the main drawbacks of the system. At that time a DC voltage could not be transformed to another voltage level, and the transportation of electricity at the low voltage level of 100 V over relatively short distances already requires very thick copper conductors to keep the voltage drop within limits; this makes the system rather expensive. Nevertheless, it took quite some time before AC became the standard. The reason for this was that Edison, besides being a brilliant inventor, was also a talented and cunning businessman as will become clear from the following anecdote. Edison tried to conquer the market and made many efforts to have the DC adopted as the universal standard. But behind the scenes he also tried hard to have AC adopted for a special application: the electric chair. After having accomplished this, Edison intimidated the general public into choosing DC by claiming that AC was highly dangerous, the electric chair being the proof of this! Eventually AC became the standard because transformers can quite easily transform the voltage from lower to higher voltage levels and vice versa.

Nowadays, power-electronic devices make it possible to convert AC to DC, DC to AC, and DC to DC with a high rate of efficiency, and the obstacle of altering the voltage level in DC systems has disappeared. What determines, in that case, the choice between AC and DC systems? Of course, financial investments do play an important role here. The incremental costs of DC transmission over a certain distance are less than the incremental costs of AC, because in a DC system two conductors are needed whereas three-phase AC requires three conductors. On the other hand, the power-electronic converters for the conversion of AC to DC at one side, and from DC to AC at the other side, of the DC transmission line are more expensive than the AC transmission terminals. If the transmission distance is sufficiently long, the savings on the conductors overcome the cost of the converters, as shown in Figure 1.3, and DC transmission is, from a capital investment point of view, an alternative to AC.

The following are a few of the examples of high-voltage DC (HVDC) applications.

- Long submarine crossings. For example, the Baltic cable between the Scandinavian countries and Germany and the 600 km cable connection between Norway and the Netherlands (the NorNed Cable Project).

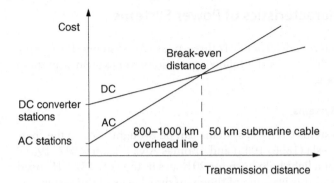

Figure 1.3 Break-even distance for HVDC [4].

- Asynchronous interconnection to interconnect networks that operate at different frequencies. For example, the HVDC intertie connection between the 50 Hz, 500 kV Argentinean system and the 60 Hz, 525 kV Brazilian system.
- Asynchronous interconnection to interconnect networks that operate at the same frequency but cannot be connected by means of AC due to stability reasons or operational differences. For example, the Scandinavian system is asynchronously connected to the western continental European system; the same applies for the US Eastern Interconnection and the US Western Interconnection.

Also in our domestic environment DC systems are present as the majority of our electronic equipment works internally with a DC voltage: personal computers, hi-fi equipment, video, DVD players, the television, and so on.

Shape of the alternating voltage

When an alternating voltage is considered, several types of alternating voltage are possible, such as sinusoidal, block, or triangular-shaped voltages, as depicted in Figure 1.4. For power systems, the sinusoidal alternating voltage is the right one to choose. By approximation, the power system can be considered to be a linear time-invariant (LTI) dynamic system. The elementary operations in such a system are multiplication with a constant number and addition and subtraction of quantities and delay in time (phase shift). When we perform these operations on a sinusoidal signal of constant frequency, another

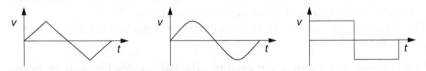

Figure 1.4 Alternating voltages: triangular, sinusoidal, and block.

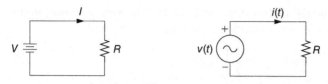

Figure 1.5 The definition of RMS values of sinusoidal quantities.

sinusoidal signal with the same frequency is the result. The same applies for differentiation and integration. The signal, after manipulation, may differ in amplitude or may be out of phase with the original signal, but the frequency and the shape of the signal have not been affected. This is not the case when the other alternating voltage shapes are used.

In other words, a sinusoidal excitation of a linear system results in a sinusoidal response. Therefore, all the voltages and currents in the power system are sinusoidal and have the same frequency so that the components in the system can be designed for this wave shape.

Sinusoidal alternating voltage

When we talk about an alternating sinusoidal voltage (or current), we generally refer to the so-called RMS root mean square (RMS) value or effective value of the voltage (or current). This RMS or effective value of a sinusoidal alternating voltage (or current) is the equivalent value of the corresponding direct voltage (or current) that dissipates the same amount of power in a given resistor during one time period of the alternating voltage (or current). We derive this equality for the DC and AC circuit shown in Figure 1.5.

The power dissipated in the resistance in the DC circuit is

$$P = \frac{V^2}{R} = I^2 R \tag{1.5}$$

When we write the voltage and the current in the AC circuit as

$$v(t) = \sqrt{2}|V| \sin(\omega t) \quad \text{and} \quad i(t) = \sqrt{2}|I| \sin(\omega t) \tag{1.6}$$

ω the angular frequency ($\omega = 2\pi f$) [rad/s]

the instantaneous power dissipated in the AC circuit is

$$p(t) = \frac{v^2(t)}{R} = i^2(t)R \tag{1.7}$$

and for the average power this results in

$$P = \frac{1}{T}\int_0^T \frac{v^2}{R}dt = \frac{1}{T}\int_0^T i^2 R dt \tag{1.8}$$

T the period of the sine wave ($T = 1/f = 2\pi/\omega$) [s]

When the power in the DC circuit (Equation 1.5) and the average power in the AC circuit (Equation 1.8) are demanded to be equal, we can write

$$V^2 = \frac{1}{T}\int_0^T v^2 dt \quad \text{and} \quad I^2 = \frac{1}{T}\int_0^T i^2 dt \tag{1.9}$$

and substitution of the equations for the alternating voltage and current (Equation 1.6) gives us

$$V = \sqrt{\frac{1}{T}\int_0^T v^2 dt} = \sqrt{2}|V|\sqrt{\frac{1}{T}\int_0^T \sin^2(\omega t)dt} = \sqrt{2}|V|\sqrt{\frac{1}{2}} = |V|$$

$$I = \sqrt{\frac{1}{T}\int_0^T i^2 dt} = \sqrt{2}|I|\sqrt{\frac{1}{T}\int_0^T \sin^2(\omega t)dt} = \sqrt{2}|I|\sqrt{\frac{1}{2}} = |I| \tag{1.10}$$

|V| the RMS or effective value of the alternating voltage
|I| the RMS or effective value of the alternating current

So summing up, the RMS or effective value of a sinusoidal alternating voltage (or current) is equal to the value of the equivalent direct voltage (or current) that dissipates the same amount of power in a given resistor during one time period of the alternating voltage (or current).

The expression RMS is related to the previously derived expressions:

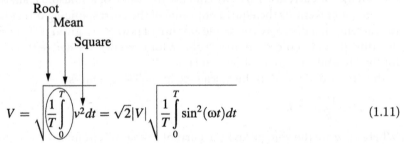

$$V = \sqrt{\frac{1}{T}\int_0^T v^2 dt} = \sqrt{2}|V|\sqrt{\frac{1}{T}\int_0^T \sin^2(\omega t)dt} \tag{1.11}$$

The term below the square root in Equation 1.11 is in fact the mean of $\sin^2(\omega t)$, as indicated in Figure 1.6.

From the equations of the sinusoidal voltage and current (Equation 1.6), we see the relation between the RMS value and the peak value:

$$\sqrt{2}|V| = \hat{V} \quad \text{and} \quad \sqrt{2}|I| = \hat{I} \tag{1.12}$$

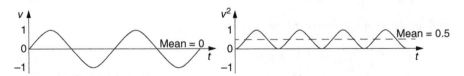

Figure 1.6 Mean value of a squared sine.

It is the RMS value of the sinusoidal voltage and current that is read by the common type of voltmeters and ammeters.

Example 1.1 *RMS and peak value*

When we speak of a voltage of 230 V, the RMS value of the voltage amounts to 230 V, while the peak value of the sinusoidal voltage is $\sqrt{2} \cdot 230 = 325$ V.

1.3.2 50 and 60 Hz Frequency

The choice of the frequency is not as arbitrary as one might think. Between 1885 and 1900, a diversity of frequencies was used in the United States: 140, 133⅓, 125, 83⅓, 66⅔, 60, 50, 40, 33⅓, 30, 25, and 16⅔ Hz [5–8]. Each frequency had its own field of application. The power frequency finally came out at 60 Hz in North America, Brazil, and Japan and at 50 Hz in most of the other countries. Nowadays, 16⅔ (Europe) and 25 Hz (North America) are in use for railway applications, and 400 Hz is a popular frequency on board of ships, airplanes, and oil rigs.

A too low frequency, such as 10 or 20 Hz, is useless for domestic lighting as the human eye records this as flicker. On the other hand, the frequency cannot be too high as:

- The hysteresis losses in the transformer core increase in proportion to the frequency while the eddy current losses increase in quadratic proportion to the frequency.
- The capacitive reactance of cables and transmission lines increases $(X = -1/\omega C)$.
- The inductive reactance, and the related voltage drop, increases $(X = \omega L)$.
- The electromagnetic interference with the radio traffic will grow.

Yet there is also an advantage in using a higher power system frequency – the power-to-weight ratio of transformers, motors, and generators is higher. In other words, the components can be smaller, while the power output is the same. The formula of Esson gives a generalized expression for the power of an electrical machine:

$$P = K \cdot D^2 \cdot l \cdot n \tag{1.13}$$

K the "output coefficient" [J/m³], which depends on the type of machine, the type of cooling, and the magnetic material used
D the diameter of the armature [m]
l the axial length of the armature [m]
n the rotational speed of the machine [1/s]

From Equation 1.13 we see that when we increase the rotational speed, by choosing a higher system frequency, the dimension of the machine can be smaller for the same output power.

Another example is the transformer. The relation between the applied voltage and the resulting flux is given by the following equations:

$$v_1(t) = \sqrt{2}|V_1| \cos(\omega t) = N_1 \frac{d\Phi}{dt}$$

$$\Phi(t) = C + \frac{\sqrt{2}|V_1|}{\omega N_1} \sin(\omega t) = \sqrt{2}|\Phi| \sin(\omega t) \tag{1.14}$$

$$|\Phi| = |B|A$$

N_1 the number of turns of the primary transformer winding
Φ the magnetic flux [Wb = V s]
C the integration constant [Wb]; zero in steady-state conditions
B the magnetic flux density [T = Wb/m^2]
A the cross-sectional area of the iron transformer core [m^2]

We see that when the applied voltage remains the same, a higher system frequency ($\omega = 2\pi f$) results in a lower effective value of the magnetic flux ($|\Phi|$) so that we can use a smaller cross-sectional area for the iron core when we keep the magnetic flux density constant.

When there is freedom to choose the system frequency, a higher frequency can be very advantageous, especially in the case that weight and volume play a role, for example, on board of airplanes and ships.

1.3.3 Balanced Three-Phase Systems

The transmission and distribution systems are three-phase systems. In this book we restrict ourselves to balanced three-phase power systems. In the case of a balanced three-phase system, the sinusoidal voltages are of equal magnitude in all three phases and shifted in phase by 120°, as shown in Figure 1.7:

$$v_a = \sqrt{2}|V| \cos(\omega t)$$

$$v_b = \sqrt{2}|V| \cos\left(\omega t - \frac{2\pi}{3}\right) \tag{1.15}$$

$$v_c = \sqrt{2}|V| \cos\left(\omega t - \frac{4\pi}{3}\right)$$

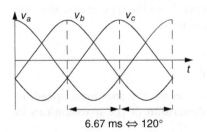

6.67 ms ⟺ 120°

Figure 1.7 Phase voltages in a balanced three-phase power system (50 Hz).

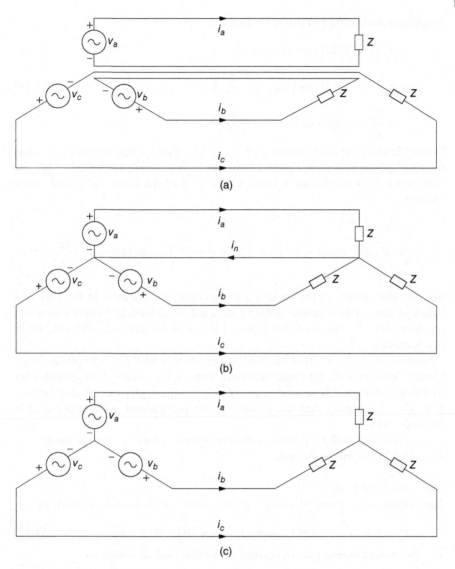

Figure 1.8 A balanced three-phase power system.

Furthermore, the network has identical impedances in each phase and the loads are identical. We can imagine the three-phase system consisting of three separate single-phase systems, as depicted in Figure 1.8 (a). Since the voltages in the three phases are equal in magnitude and 120° shifted in phase and since the impedances in the three phases are equal, the currents will also be equal in magnitude and shifted in phase by 120°:

$$i_a = \frac{v_a}{Z} = \sqrt{2}|I| \cos(\omega t - \varphi)$$

$$i_b = \frac{v_b}{Z} = \sqrt{2}|I| \cos\left(\omega t - \varphi - \frac{2\pi}{3}\right) \qquad (1.16)$$

$$i_c = \frac{v_c}{Z} = \sqrt{2}|I| \cos\left(\omega t - \varphi - \frac{4\pi}{3}\right)$$

When the three return conductors are combined as a single return conductor, the three-phase system in Figure 1.8 (b) can be drawn. The current through the common return conductor is equal to the sum of the three individual phase currents:

$$i_n = i_a + i_b + i_c$$

$$= \sqrt{2}|I| \left[\cos(\omega t - \varphi) + \cos\left(\omega t - \varphi - \frac{2\pi}{3}\right) + \cos\left(\omega t - \varphi - \frac{4\pi}{3}\right)\right] = 0$$

$$(1.17)$$

Because the current in the common return conductor is zero, in the case of a balanced three-phase power system with a balanced load, this common return conductor can be removed (see Figure 1.8 (c)), and a network with only three conductors results.

Because of the fact that in the power system many single-phase loads (domestic users and also some industrial ones) are connected, the utilities try to divide the single-phase loads equally over the three phases (see also Section 4.3). The assumption that the power system is balanced is therefore a valid approximation.

The reason to build a power system as a three-phase system is twofold as explained in the following text.

Power considerations
The voltage and current of a single-phase inductive load can be written as

$$v(t) = \sqrt{2}|V| \cos(\omega t) \quad \text{and} \quad i(t) = \sqrt{2}|I| \cos(\omega t - \varphi) \qquad (1.18)$$

and the instantaneous power, consumed by this load, amounts to

$$p(t) = 2|V||I| \cos(\omega t) \cos(\omega t - \varphi)$$

$$= |V||I|[\cos(\varphi) + \cos(2\omega t - \varphi)] \qquad (1.19)$$

From this expression we learn that the single-phase instantaneous power is not constant, but varies in time with double the power frequency ($2\omega t$). This is rather unpleasant, especially when the electrical energy is used for electrical motors and traction applications, because this results in a pulsating torque on the axis of the machine.

The voltages and currents of a balanced three-phase inductive load can be written as

$$v_a = \sqrt{2}|V|\cos(\omega t) \qquad\qquad i_a = \sqrt{2}|I|\cos(\omega t - \varphi)$$
$$v_b = \sqrt{2}|V|\cos\left(\omega t - \frac{2\pi}{3}\right) \qquad i_b = \sqrt{2}|I|\cos\left(\omega t - \varphi - \frac{2\pi}{3}\right) \qquad (1.20)$$
$$v_c = \sqrt{2}|V|\cos\left(\omega t - \frac{4\pi}{3}\right) \qquad i_c = \sqrt{2}|I|\cos\left(\omega t - \varphi - \frac{4\pi}{3}\right)$$

and the instantaneous power, consumed by this load, amounts to

$$p(t) = v_a i_a + v_b i_b + v_c i_c$$
$$= |V||I|[\cos(\varphi) + \cos(2\omega t - \varphi)] + |V||I|\left[\cos(\varphi) + \cos\left(2\omega t - \varphi - \frac{4\pi}{3}\right)\right]$$
$$+ |V||I|\left[\cos(\varphi) + \cos\left(2\omega t - \varphi - \frac{2\pi}{3}\right)\right] \qquad (1.21)$$
$$= 3|V||I|\cos(\varphi)$$

In a balanced three-phase power system, the instantaneous power is constant! This is in fact valid for every balanced power system with more than three phases. For an n-phase system, we can write the following general expression for the instantaneous power:

$$p(t) = n|V||I|\cos(\varphi) + |V||I|\sum_{k=1}^{n}\cos\left(2\omega t - \varphi - 2k\cdot\frac{2\pi}{n}\right) = n|V||I|\cos(\varphi)$$
$$+ |V||I|\cos(2\omega t - \varphi)\sum_{k=1}^{n}\cos\left(2k\cdot\frac{2\pi}{n}\right) \qquad (1.22)$$
$$+ |V||I|\sin(2\omega t - \varphi)\sum_{k=1}^{n}\sin\left(2k\cdot\frac{2\pi}{n}\right)$$

A close observation of the terms behind the summation signs reveals

$$\sum_{k=1}^{n}\cos\left(2k\cdot\frac{2\pi}{n}\right) = 0 \quad \forall n \geq 3$$
$$\qquad\qquad (1.23)$$
$$\sum_{k=1}^{n}\sin\left(2k\cdot\frac{2\pi}{n}\right) = 0 \quad \forall n \geq 1$$

Thus, the instantaneous power, as given in Equation 1.22, is constant for every number of phases greater than or equal to three. Then why do we apply a three-phase system and not a four- or five-phase system? This is because each phase requires its own conductor, and the balanced three-phase system is the system with the smallest number of phase conductors capable of delivering constant instantaneous power.

The power supplied by the balanced three-phase system equals three times the average power supplied by one of the three single-phase systems of which the system is built up, while only one extra conductor is required. As the cosine term with the double frequency in Equation 1.19 has an average value of zero, the average power supplied by the single-phase system amounts to

$$P_{1\phi} = |V||I| \cos(\varphi) \tag{1.24}$$

The average power supplied by the balanced three-phase system equals the instantaneous power (Equation 1.21):

$$P_{3\phi} = p(t) = 3|V||I| \cos(\varphi) = 3P_{1\phi} \tag{1.25}$$

The three-phase system depicted in Figure 1.8 (c) transports the same amount of power as the three-phase system built of three individual single-phase systems, as shown in Figure 1.8 (a), but with only half the number of conductors!

Rotating magnetic field

A three-phase system is able to produce a rotating magnetic field, as is visualized in Figure 1.9. This is a very important property as all AC machines

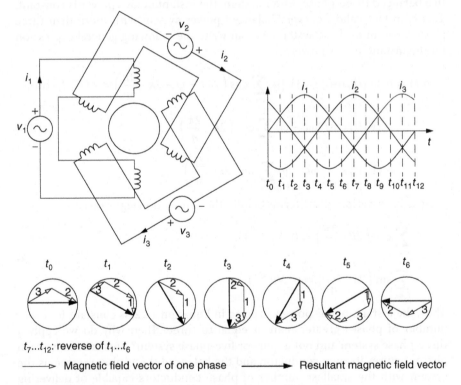

$t_7...t_{12}$: reverse of $t_1...t_6$

⟶ Magnetic field vector of one phase   ⟶ Resultant magnetic field vector

Figure 1.9 Magnetic field generated by a three-phase coil system [9].

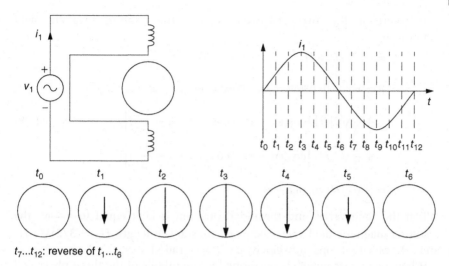

$t_7...t_{12}$: reverse of $t_1...t_6$

Figure 1.10 Magnetic field generated by a single-phase coil system.

operate on this principle. In a three-phase coil system, the resulting magnetic field vector rotates with a constant magnitude, while in a single-phase coil system the magnitude of the magnetic field vector varies in one direction (see Figure 1.10). The expressions for the currents in the three-phase coil system are

$$i_1 = \sqrt{2}|I| \sin(\omega t)$$
$$i_2 = \sqrt{2}|I| \sin\left(\omega t - \frac{2\pi}{3}\right) \quad (1.26)$$
$$i_3 = \sqrt{2}|I| \sin\left(\omega t - \frac{4\pi}{3}\right)$$

and the resulting magnetic fields can be written as

$$H_1 = \sqrt{2}|H| \sin(\omega t) \sin(x)$$
$$H_2 = \sqrt{2}|H| \sin\left(\omega t - \frac{2\pi}{3}\right) \sin\left(x - \frac{2\pi}{3}\right) \quad (1.27)$$
$$H_3 = \sqrt{2}|H| \sin\left(\omega t - \frac{4\pi}{3}\right) \sin\left(x - \frac{4\pi}{3}\right)$$

t the time [s]; $t = 0 \leftrightarrow H_1 = 0 \; \forall x$
x the circumferential position [rad]; $x = 0 \leftrightarrow H_1 = 0 \; \forall t$

Therefore, in the three-phase coil system, the resulting magnetic field amounts to

$$H_r = H_1 + H_2 + H_3$$

$$= \frac{1}{2}\sqrt{2}|H|[\cos(\omega t - x) - \cos(\omega t + x)]$$

$$+ \frac{1}{2}\sqrt{2}|H| \left[\cos(\omega t - x) - \cos\left(\omega t + x - \frac{4\pi}{3}\right)\right] \qquad (1.28)$$

$$+ \frac{1}{2}\sqrt{2}|H| \left[\cos(\omega t - x) - \cos\left(\omega t + x - \frac{2\pi}{3}\right)\right]$$

$$= \frac{3}{2}\sqrt{2}|H| \cos(\omega t - x)$$

When the time varies and the radial position is kept equal to $x = \omega t$, the resulting magnetic field vector remains constant in length $H_r = (3/2)\sqrt{2}|H|$ and rotates with a constant velocity $dx/dt = \omega$ [rad/s].

When a compass needle is positioned in the middle of the three-phase coil system, the needle keeps pace with the rotating field, which is a crude equivalent of the synchronous motor. When a copper cylinder is placed in the center of the three-phase coil system, the rotating field drags the cylinder around with it, and we have a primitive equivalent of the induction motor.

Example 1.2 *"Two-phase" system*
It is possible to create a system with two-phase windings that delivers constant instantaneous power and that produces a rotating magnetic field. To achieve this, the voltages in the two phases should be 90° out of phase. The voltages and currents of a balanced two-phase inductive load can be written as

$$v_a = \sqrt{2}|V|\cos(\omega t) \qquad i_a = \sqrt{2}|I|\cos(\omega t - \varphi)$$

$$v_b = \sqrt{2}|V|\cos\left(\omega t - \frac{\pi}{2}\right) \quad i_b = \sqrt{2}|I|\cos\left(\omega t - \varphi - \frac{\pi}{2}\right) \qquad (1.29)$$

and the instantaneous power, consumed by this load, amounts to

$$p(t) = v_a i_a + v_b i_b$$

$$= |V||I|[\cos(\varphi) + \cos(2\omega t - \varphi)]$$

$$+ |V||I|[\cos(\varphi) + \cos(2\omega t - \varphi - \pi)] \qquad (1.30)$$

$$= 2|V||I|\cos(\varphi)$$

The resulting magnetic field vector of the two-phase coil system rotates with a constant amplitude as visualized in Figure 1.11.

A drawback of such a two-phase system is that the current through the return conductor is not equal to zero:

$$i_n = i_a + i_b \neq 0 \qquad (1.31)$$

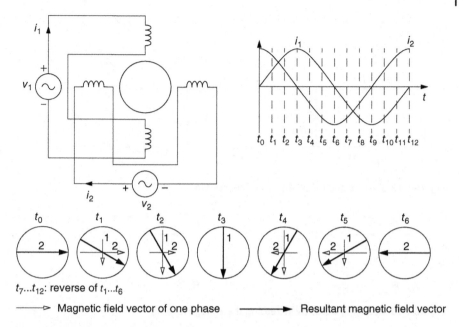

$t_7...t_{12}$: reverse of $t_1...t_6$

→▷ Magnetic field vector of one phase ⟶ Resultant magnetic field vector

Figure 1.11 Magnetic field generated by a two-phase coil system.

Accordingly, this system requires as many conductors (three) as a balanced three-phase system, but only a power $p(t) = 2|V||I| \cos(\varphi)$ is transported (compared with $p(t) = 3|V||I| \cos(\varphi)$ in the three-phase system).

1.3.4 Voltage Levels

The range of voltage ratings finds its origin in the use of carbon arc lamps in the early days of the power system. These lamps, the source of electric lighting before the incandescent lamp was invented, worked with a DC voltage of 55 V. In those days the systems were laid out as three-wire systems, with conductors at a potential of −55, 0, and 55 V as shown in Figure 1.12 (b). In this configuration a higher voltage of 110 V was available as well. By using such a three-conductor system, one could save considerably on copper. In the system with two conductors (Figure 1.12 (a)), the losses equal $I^2 \cdot 2R$, with a resistance per conductor of R and a total copper weight of 100%. In the case of the three-conductor system, the losses equal $(I^2/4) \cdot 2r$ (note that the current in the middle conductor equals zero). In the case of equal losses in the two systems, the resistance per conductor in the three-conductor system equals $r = 4R$. Therefore, the copper weight per conductor is reduced to 25%, which brings the total required copper weight to $3/2 \cdot 25\% = 37.5\%$ of the copper needed for the two-wire layout.

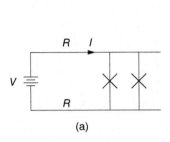
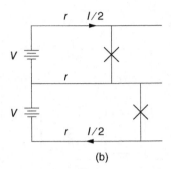

Figure 1.12 Two- (a) and three- (b) conductor system.

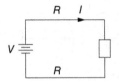
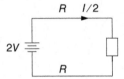

Figure 1.13 Reducing losses by increasing the voltage level.

The choice for a higher voltage level of 220 V (and in Europe later on for 230 V) was made in order to reduce losses, as is illustrated in Figure 1.13. Both voltage sources deliver a power of $VI = 2V \cdot I/2$. The losses in the circuit with supply voltage V equal $I^2 \cdot 2R$, while the losses in the circuit with double this supply voltage amount to $(I^2/4) \cdot 2R$, which means a loss reduction of 75%. When the choice was made for the alternating voltage system, the same voltage levels were maintained.

The voltage ratings in a power system can be divided into three levels:

- The generation level: 10–25 kV. The power is generated at a relatively low voltage level to keep the high-voltage insulation of the generator armature within limits.
- The transmission level: 110–420 kV and higher (in the former Soviet Union even the 1200 kV level is in operation).
- The distribution level: 10–72.5 kV.

The power is supplied to the customer at a variety of voltage levels; heavy-industrial consumers can be connected from 150 to 10 kV, while households are connected to the 0.4 kV voltage level. The changeover between voltage levels is made by transformers. The voltage ratings used in the Dutch system are shown in Figure 1.14.

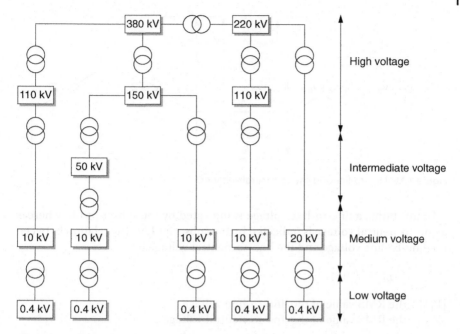

Figure 1.14 Voltage levels and transformation steps in the Dutch power system; *this voltage level can be 20 kV as well.

Line-to-line and line-to-neutral voltages

The voltage ratings of three-phase systems are usually expressed as line-to-line voltages instead of line-to-neutral voltages, as is illustrated in Figure 1.15. When the line-to-neutral voltages in the phases a and b are

$$v_{an} = \sqrt{2}|V| \cos(\omega t)$$
$$v_{bn} = \sqrt{2}|V| \cos\left(\omega t - \frac{2\pi}{3}\right),$$
(1.32)

the line-to-line voltage between the phases a and b can be calculated as follows:

$$v_{ab} = v_{an} - v_{bn} = \sqrt{2}|V| \left[\cos(\omega t) - \cos\left(\omega t - \frac{2\pi}{3}\right)\right]$$
$$= \sqrt{2}|V| \left[2 \cdot \cos\left(\frac{\pi}{6}\right) \cdot \cos\left(\omega t + \frac{\pi}{6}\right)\right] = \sqrt{3}\left[\sqrt{2}|V| \cos\left(\omega t + \frac{\pi}{6}\right)\right]$$
(1.33)

The amplitude of the line-to-line voltage is $\sqrt{3}$ times the amplitude of the line-to-neutral voltage ($|V_{ab}| = \sqrt{3}|V_{an}|$), and the line-to-line voltage v_{ab} leads the line-to-neutral voltage v_{an} by 30 electrical degrees.

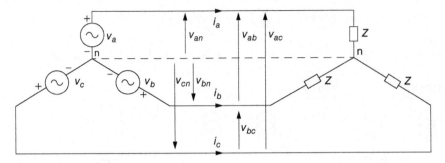

Figure 1.15 Line-to-line and line-to-neutral voltages.

In this book, a line-to-line voltage is indicated by the subscript LL, whereas a line-to-neutral voltage is indicated by the subscript LN. The relation between the line-to-line voltage and the line-to-neutral voltage is

$$|V_{LL}| = \sqrt{3}|V_{LN}|$$ (1.34)

$|V_{LL}|$ the RMS value of the line-to-line voltage
$|V_{LN}|$ the RMS value of the line-to-neutral voltage

Example 1.3   *Line-to-line and line-to-neutral voltage*
The highest voltage level of the Dutch transmission network is 380 kV. This is the RMS value of the line-to-line voltage. The RMS value of the corresponding line-to-neutral voltage equals $380\,\text{kV}/(\sqrt{3}) \approx 220\,\text{kV}$. The lowest voltage used in the Netherlands is 0.4 kV (line-to-line voltage). The corresponding line-to-neutral voltage amounts to $400/(\sqrt{3}) \approx 230\,\text{V}$.

1.4 Phasors

Power system calculations in the steady-state situation are considerably simplified by introducing the phasor. A phasor is an arrow in the complex plane that has a one-to-one relation with a sinusoidal signal as can be seen from Figure 1.16. When the sine/cosine in Figure 1.16 has a frequency of 50 Hz, a radius with a length $\sqrt{2}$ times the length of the phasor rotates counterclockwise in the complex plane with a frequency of 50 Hz.

Consider the following general sinusoidal voltage and current expressions:

$$v(t) = \sqrt{2}|V|\cos(\omega t) \quad \text{and} \quad i(t) = \sqrt{2}|I|\cos(\omega t - \varphi)$$ (1.35)

In order to express these quantities as phasors, we apply Euler's identity:

$$e^{j\varphi} = \cos(\varphi) + j\sin(\varphi)$$ (1.36)

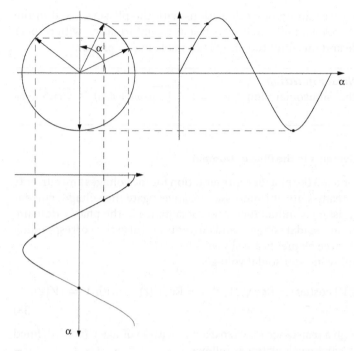

Figure 1.16 Relation between a counterclockwise rotating radius and a sinusoidal signal.

and the sinusoidal voltage and current can be written as

$$v(t) = \mathrm{Re}\{\sqrt{2}|V|e^{j\omega t}\} = \mathrm{Re}\{\sqrt{2}Ve^{j\omega t}\} \quad \text{with} \quad V = |V| = |V|\angle 0$$

$$i(t) = \mathrm{Re}\{\sqrt{2}|I|e^{j(\omega t-\varphi)}\} = \mathrm{Re}\{\sqrt{2}Ie^{j\omega t}\} \quad \text{with} \quad I = |I|e^{-j\varphi} = |I|\angle -\varphi$$

$$(1.37)$$

Re   the operator that takes the real part of a complex quantity
V the voltage phasor
I the current phasor

The phasor represents a sinusoidal signal: the length of the phasor equals the effective or RMS value of the signal, and the angle of the phasor matches the phase shift of the signal with respect to a reference (an ideal cosine (or sine)). We see that the frequency component is absent when we use the phasor notation.

The voltages produced by the synchronous generators in the power system are 50 Hz (or 60 Hz) sinusoidal voltages. As the power system is supposed to be a linear system (see Section 1.3.1) in steady state, the voltage out of the AC outlet at home is also a sinusoidal voltage with a frequency of 50 Hz; only the amplitude of the voltage differs and a phase shift may have occurred. Therefore, in steady-state calculations, the frequency gives no extra information and can be omitted so that we can do our calculations with the phasors, fixed in the complex plane. No relevant information is lost as the information with respect to the phase angle and the amplitude is still available in this phasor.

Example 1.4 *Phasor notation*

The phasor of the sinusoidal signal $v(t) = 141.4\cos(\omega t + \pi/6)$ is written as $V = 100\angle 30°$.

1.4.1 Network Elements in the Phasor Domain

After having introduced the phasor representation for the voltages and currents in the sinusoidal steady state, it is necessary to investigate the voltage–current relations of the resistance, inductance, and capacitance in the phasor domain. In Figure 1.17, the sinusoidal voltages and currents and also the corresponding phasors for those three elements are shown.

Consider the following sinusoidal voltage:

$$v(t) = \sqrt{2}|V|\cos(\omega t) = \mathrm{Re}\{\sqrt{2}Ve^{j\omega t}\} = \mathrm{Re}\{v'(t)\} \quad \text{with } V = |V|\angle 0°$$

(1.38)

The current through a resistance R, when excited with a voltage $v'(t)$ (as defined in Equation 1.38), can be calculated as follows:

$$i'(t) = \frac{v'(t)}{R} = \sqrt{2}\cdot\frac{V}{R}\cdot e^{j\omega t} = \sqrt{2}Ie^{j\omega t}$$

(1.39)

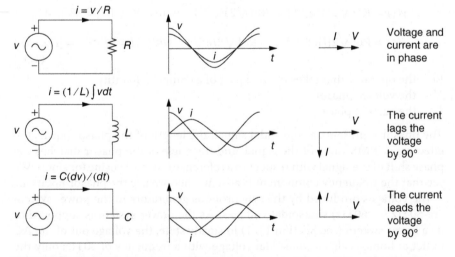

Figure 1.17 Relation between the sinusoidal voltage and current and the corresponding phasors for a resistance, inductance, and capacitance.

Therefore, the voltage–current relation of a resistance in the phasor domain is

$$I = \frac{V}{R} \tag{1.40}$$

When we take the real part of Equation 1.39, we get the expression of the time-varying sinusoidal current:

$$i(t) = \text{Re}\{i'(t)\} = \sqrt{2} \cdot \frac{|V|}{R} \cdot \cos(\omega t) \tag{1.41}$$

The current through a capacitor C, when connected to a voltage source $v'(t)$ (as defined in Equation 1.38), can be obtained as follows:

$$i'(t) = C\frac{dv'}{dt} = \sqrt{2} \cdot j\omega CV \cdot e^{j\omega t} = \sqrt{2}Ie^{j\omega t} \tag{1.42}$$

Thus, the voltage–current relation of a capacitor in the phasor domain is

$$I = j\omega CV \tag{1.43}$$

Taking the real part of Equation 1.42 leads to the expression of the time-varying sinusoidal current:

$$i(t) = \text{Re}\{i'(t)\} = \sqrt{2} \cdot \omega C|V| \cdot -\sin(\omega t) = \sqrt{2} \cdot \omega C|V| \cdot \cos\left(\omega t + \frac{\pi}{2}\right) \tag{1.44}$$

The current through an inductor L, driven by a voltage $v'(t)$ (as defined in Equation 1.38), is

$$i'(t) = \frac{1}{L}\int v'\,dt = \sqrt{2} \cdot \frac{V}{j\omega L} \cdot e^{j\omega t} = \sqrt{2}Ie^{j\omega t} \tag{1.45}$$

and the voltage–current relation of an inductor in the phasor domain is

$$I = \frac{V}{j\omega L} \tag{1.46}$$

The real part of Equation 1.45 gives us the expression for the time-varying sinusoidal current:

$$i(t) = \text{Re}\{i'(t)\} = \sqrt{2} \cdot \frac{|V|}{\omega L} \cdot \sin(\omega t) = \sqrt{2} \cdot \frac{|V|}{\omega L} \cdot \cos\left(\omega t - \frac{\pi}{2}\right) \tag{1.47}$$

The voltage–current relations are summarized in Table 1.1.

Table 1.1 Voltage–current relations.

Element	Time Domain	Phasor Domain
Resistance	$v = iR$	$V = IR$
Capacitance	$i = C(dv)/(dt)$	$I = j\omega CV$
Inductance	$v = L(di)/(dt)$	$V = j\omega LI$

In the phasor domain, a general expression for the impedance can be written as

$$Z = \frac{V}{I} = R + jX \qquad (1.48)$$

Z the impedance [Ω]
R the resistance [Ω]
X the reactance [Ω]

When X has a positive sign, the energy-storage element is an inductor ($jX = j\omega L$). When X has a negative sign, the energy-storage element is a capacitor ($jX = 1/(j\omega C) = -j/(\omega C)$).

When X equals zero, there is no energy storage and the impedance is purely resistive. This is the case when we have no capacitors or inductors or when the frequency is zero and we deal with DC.

In a similar way, a general expression for the admittance can be written as

$$Y = \frac{I}{V} = G + jB \qquad (1.49)$$

Y the admittance [S]
G the conductance [S]
B the susceptance [S]

1.4.2 Calculations in the Phasor Domain

When we set aside the "deeper meaning" of the phasor (being the representation of a sinusoidal signal), a phasor is nothing more than a complex number represented by a vector in the complex plane. Therefore, the mathematical rules for vector calculus can be applied for phasors.

A vector can be described by its length and its angle (polar coordinates) or represented by its real and imaginary part (rectangular or Cartesian coordinates). The relation between these two forms of notation is shown in Figure 1.18. Basically, it makes no difference whether a phasor is expressed by

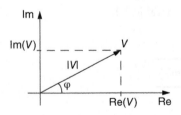

$$V = |V|\angle\varphi = \mathrm{Re}(V) + j\mathrm{Im}(V)$$
$$|V| = \sqrt{\mathrm{Re}(V)^2 + \mathrm{Im}(V)^2}$$
$$\varphi = \tan^{-1}\left(\frac{\mathrm{Im}(V)}{\mathrm{Re}(V)}\right)$$
$$\mathrm{Re}(V) = |V|\cos(\varphi)$$
$$\mathrm{Im}(V) = |V|\sin(\varphi)$$

Figure 1.18 The phasor as a vector in the complex plane.

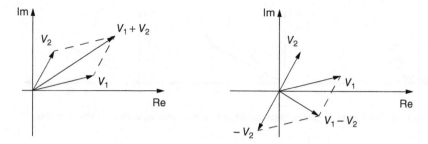

Figure 1.19 Two basic operations on vectors: addition and subtraction.

the Cartesian notation or the polar notation; however, for certain calculations it can be advantageous to use one or the other. For addition and subtraction of two phasors, the Cartesian notation is in general the easiest to apply as shown graphically in Figure 1.19:

$$V_1 = a + jb \quad \text{and} \quad V_2 = c + jd \tag{1.50}$$

$$V_1 + V_2 = (a + c) + j(b + d) \quad \text{addition}$$
$$V_1 - V_2 = (a - c) + j(b - d) \quad \text{subtraction} \tag{1.51}$$
$$V_1^* = a - jb \qquad\qquad\qquad \text{complex conjugate}$$

For multiplication and division of phasors, the polar notation is more handy to use:

$$V = |V|\angle\alpha \quad \text{and} \quad I = |I|\angle\beta \tag{1.52}$$

$$VI = |V||I|\angle(\alpha + \beta) \quad \text{multiplication}$$

$$\frac{V}{I} = \frac{|V|}{|I|}\angle(\alpha - \beta) \quad \text{division} \tag{1.53}$$

$$V^* = |V|\angle -\alpha \qquad \text{complex conjugate}$$

Other complex quantities, such as the complex power (which is introduced in Section 1.6.2) or impedance or admittance, can be drawn as a vector in the complex plane too as is shown in Figure 1.20, and the same mathematical rules apply for those quantities as well. These quantities, however, cannot be interpreted as phasors, as they do not have the same mathematical background.

The familiar complex operator j (i in the mathematical literature; j is common practice in the electrotechnical world in order to avoid confusion between the current and the complex operator) is in fact a vector too:

$$j = e^{j90} = 0 + j1 = 1\angle 90° \tag{1.54}$$

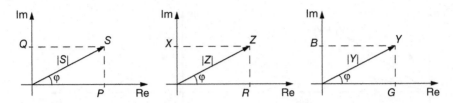

Figure 1.20 The complex power, impedance, and admittance as vectors in the complex plane.

Multiplication of a vector in the complex plane with the j operator causes a counterclockwise rotation of the vector with 90° while the length of the vector is unchanged. Another multiplication with j leads to a rotation over 180°:

$$j^2 = e^{j180} = -1 + j0 = 1\angle 180° \tag{1.55}$$

In electrical power engineering another complex operator is commonly used, the so-called a operator:

$$a = e^{j120} = -0.5 + j0.866 = 1\angle 120° \tag{1.56}$$

A vector multiplied in the complex plane with the a operator rotates counterclockwise by 120° while the length of the vector remains unchanged. A repeated multiplication with the a operator results in

$$a^2 = e^{j240} = -0.5 - j0.866 = 1\angle 240°$$
$$a^3 = e^{j360} = 1 + j0 = 1\angle 0° \tag{1.57}$$

From Equations 1.56 and 1.57, it is evident that

$$1 + a + a^2 = 0 \tag{1.58}$$

This is shown graphically in Figure 1.21.

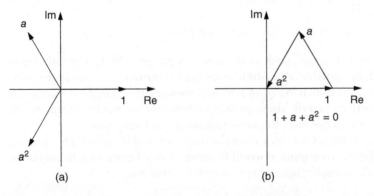

Figure 1.21 Vector diagrams of various powers of the a operator.

Figure 1.22 The relation between the line-to-line and the line-to-neutral voltage.

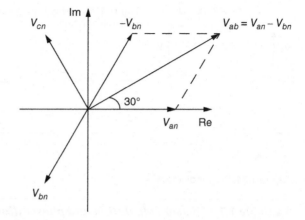

Example 1.5 Line-to-line and line-to-neutral voltage

When voltage phasors are applied, the relation between the line-to-neutral and the line-to-line voltage ($|V_{LL}| = \sqrt{3}|V_{LN}|$), as found in Section 1.3.4, can be calculated as follows. The phasor representation of the line-to-neutral voltages in the phases a and b, defined earlier in Equation 1.32, are $V_{an} = |V|\angle 0°$ and $V_{bn} = |V|\angle 240°$, respectively (see Figure 1.22). The line-to-line voltage between the phases a and b is

$$V_{ab} = V_{an} - V_{bn} = \frac{3}{2}|V| + j\frac{1}{2}\sqrt{3}|V|$$

$$= \sqrt{3}|V|\angle 30° = \sqrt{3}V_{an}\angle 30° \qquad (1.59)$$

This is simply the phasor representation of the line-to-line voltage in the time domain as denoted in Equation 1.33. The mathematical relation can be verified graphically from Figure 1.22.

Example 1.6 Balanced three-phase voltage

The voltage phasors in a balanced three-phase system can be written as (see Figure 1.22)

$$V_{an} = |V|\angle 0° = |V|$$

$$V_{bn} = |V|\angle 240° = a^2|V| \qquad (1.60)$$

$$V_{cn} = |V|\angle 120° = a|V|$$

Therefore, the sum of these voltage phasors equals zero (see also Equation 1.58 and Figure 1.21 (b)):

$$V_{an} + V_{bn} + V_{cn} = |V|(1 + a^2 + a) = 0 \qquad (1.61)$$

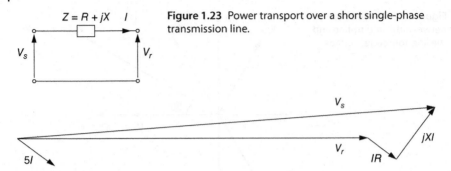

Figure 1.23 Power transport over a short single-phase transmission line.

Figure 1.24 Phasor diagram.

Example 1.7 *Phasor calculation and phasor diagram*

Given is a short single-phase transmission line, which can be modeled as a series impedance $Z = 4 + j7\,\Omega$ (see also Appendix E.4), with a (line-to-neutral) voltage at the receiving end (subscript r) of $10/\sqrt{3}$ kV. The current amounts to $|I| = 150\,$A. At the receiving end of the line, an inductive load is present, and the current phasor lags the voltage phasor by 36.9°. This situation is depicted in Figure 1.23. We want to calculate the voltage at the sending end of the line (subscript s).

The impedance of the transmission line can be written in polar coordinates $Z = 8\angle 60°\,\Omega$. When the voltage at the receiving end of the line is used as a reference, the corresponding phasor is $V_r = 5773.5\angle 0°$ V, and the current phasor is $I = 150\angle -36.9°$ A. The voltage at the sending end of the transmission line can be calculated in the following way:

$$V_s = V_r + ZI = 5773.5\angle 0° + 8\angle 60° \cdot 150\angle -36.9°$$
$$= 5773.5 + 1200\angle 23.1° = 6877.3 + j470.8 \tag{1.62}$$
$$= 6.9\angle 3.9° \text{ kV}$$

When the voltage and current phasors are drawn in a single diagram, the phasor diagram in Figure 1.24 results. The voltage drop across the transmission line $(V_s = V_r + ZI)$ is now visualized. For a better visibility, the current has been drawn at a five times larger scale: $5I$ instead of I. The voltage phasor IR is the product of the current phasor I and the resistance R: only the length of the current vector changes, whereas the angle remains the same. The voltage phasor jXI is the product of the current phasor I, the reactance value X, and the complex operator j. The length of the current vector alters (XI) and the vector rotates counterclockwise by 90° (jXI).

1.5 Equivalent Line-to-neutral Diagrams

When solving balanced three-phase systems, one can work with a single-phase equivalent of the three-phase system. In fact, the consecutive steps made

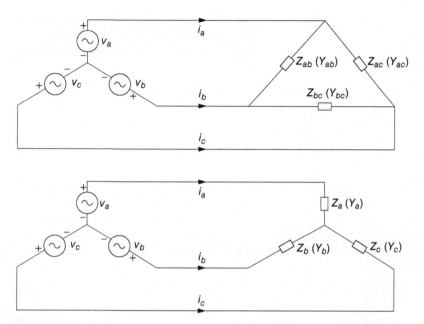

Figure 1.25 Conversion of a delta-connected load to a wye-connected load.

to arrive at the balanced three-phase system consisting of three conductors in Figure 1.8 are reversed: the three-phase system is split up into three single-phase networks, of which only one needs to be analyzed. When the voltages and currents are known in this single phase, one can simply obtain the expressions for the voltages and currents in the other two phases by rotating the corresponding phasors with 120° and 240°.

When the three-phase network contains delta-connected elements, they have to be converted to their equivalent wye connections first, as shown in Figure 1.25. The delta–wye transformation formulas for both impedances and admittances are given in Table 1.2.

Table 1.2 Delta–wye transformation.

Impedance	Admittance
$Z_a = \dfrac{Z_{ab}Z_{ac}}{Z_{ab} + Z_{ac} + Z_{bc}}$	$Y_a = \dfrac{Y_{ab}Y_{ac} + Y_{ab}Y_{bc} + Y_{ac}Y_{bc}}{Y_{bc}}$
$Z_b = \dfrac{Z_{ab}Z_{bc}}{Z_{ab} + Z_{ac} + Z_{bc}}$	$Y_b = \dfrac{Y_{ab}Y_{ac} + Y_{ab}Y_{bc} + Y_{ac}Y_{bc}}{Y_{ac}}$
$Z_c = \dfrac{Z_{ac}Z_{bc}}{Z_{ab} + Z_{ac} + Z_{bc}}$	$Y_c = \dfrac{Y_{ab}Y_{ac} + Y_{ab}Y_{bc} + Y_{ac}Y_{bc}}{Y_{ab}}$

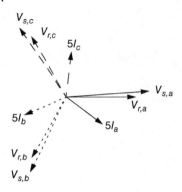

Figure 1.26 Phasors in Figure 1.24, obtained from single-phase computations (solid), are rotated counterclockwise with 120 (dashed) and 240 (dotted) degrees.

As we assume the system to be balanced (i.e., $Z_{ab} = Z_{ac} = Z_{bc}$ and $Y_{ab} = Y_{ac} = Y_{bc}$), the two following delta–wye transformation formulas can be derived:

$$Z_Y = \frac{Z_\Delta}{3} \tag{1.63}$$

$$Y_Y = 3Y_\Delta \tag{1.64}$$

Example 1.8   *Equivalent line-to-neutral diagram and single-phase computation*

The loaded single-phase short transmission line of Example 1.7 (p. 28), with a (line-to-neutral) voltage at the receiving end of $10/\sqrt{3}\,\text{kV}$, can be interpreted as an equivalent line-to-neutral diagram of a balanced three-phase short transmission line with a line-to-line voltage at the receiving end of $10\,\text{kV}$, which supplies a balanced wye-connected inductive load. As the voltages and currents are known in one phase (e.g., phase a), the voltages and currents in the other two phases (phases c and b) can be obtained by rotating the corresponding phasors with $120°$ and $240°$ counterclockwise as shown in Figure 1.26.

1.6   Power in Single-phase Circuits

In Section 1.3.3, we learned that the single-phase instantaneous power is a function of time and therefore not constant. In this section, we examine the power concept of a single-phase circuit more thoroughly, and we determine the relations between the voltage and current phasors and the power.

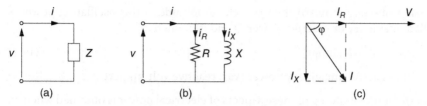

Figure 1.27 An inductive load split up into a resistor in parallel with an inductor.

1.6.1 Active and Reactive Power

The sinusoidal expressions for voltage and current of a general single-phase load, as shown in Figure 1.27 (a), are

$$v(t) = \sqrt{2}|V| \cos(\omega t) \quad \text{and} \quad i(t) = \sqrt{2}|I| \cos(\omega t - \varphi) \tag{1.65}$$

φ the phase shift between the voltage and the current: φ is positive for a current lagging the voltage and negative for a leading current

The instantaneous power consumed by the impedance Z amounts to

$$p(t) = 2|V||I| \cos(\omega t) \cos(\omega t - \varphi)$$
$$= |V||I| \cos(\varphi)[1 + \cos(2\omega t)] + |V||I| \sin(\varphi) \sin(2\omega t) \tag{1.66}$$
$$= P[1 + \cos(2\omega t)] + Q \sin(2\omega t)$$

The first term in Equation 1.66 ($P[1 + \cos(2\omega t)]$) describes an unidirectional component of the instantaneous power with an average value P, which is called the average power and is also addressed as real or active power. So

$$P = |V||I| \cos(\varphi) \tag{1.67}$$

P the active/real/average power [W]
$\cos(\varphi)$ the power factor (the cosine of the phase shift between the voltage and current (or, in other words, the cosine of the phase angle between the voltage and current phasor (see Figure 1.27 (c))))

The second term in Equation 1.66 ($Q \sin(2\omega t)$) is alternately positive and negative and has an average value of zero. This term describes a bidirectional, that is, oscillating, component of the instantaneous power. When this term has a positive sign, the power flow is toward the load; when the sign is negative, the power flows from the load back to the source of supply. Because the average value of this oscillating power component equals zero, it gives on average no transfer of energy toward the load. The amplitude of this oscillating power is called imaginary or reactive power Q and is defined as

$$Q = |V||I| \sin(\varphi) \tag{1.68}$$

Q the reactive/imaginary power [var; reactive volt-amperes]

A better understanding of these aspects of electrical power is obtained when an inductive load is modeled as a resistor in parallel with an inductor as visualized in Figure 1.27 (b). The current can be resolved into two components: one that is in phase with the voltage (the current through the resistor) and one that is 90° out of phase (the current through the inductor). As the length of the current phasor in phase with the voltage equals $|I_R| = |I| \cos(\varphi)$ (see the phasor diagram in Figure 1.27 (c)), the instantaneous current component in phase with the voltage can be written as

$$i_R = \sqrt{2}|I_R| \cos(\omega t) = \sqrt{2}|I| \cos(\varphi) \cos(\omega t) \tag{1.69}$$

Similarly, the length of the current phasor lagging the voltage with 90° equals $|I_X| = |I| \sin(\varphi)$, and the instantaneous current component lagging the voltage with 90° can be written as

$$i_X = \sqrt{2}|I_X| \sin(\omega t) = \sqrt{2}|I| \sin(\varphi) \sin(\omega t) \tag{1.70}$$

The instantaneous power consumed by the resistor (see Figure 1.28 (b)) equals

$$\begin{aligned} vi_R &= 2|V||I| \cos(\varphi)\cos^2(\omega t) \\ &= |V||I| \cos(\varphi)[1 + \cos(2\omega t)] \\ &= P[1 + \cos(2\omega t)] \end{aligned} \tag{1.71}$$

The instantaneous power toward the inductor (see Figure 1.28 (c)) equals

$$\begin{aligned} vi_X &= 2|V||I| \sin(\varphi) \sin(\omega t) \cos(\omega t) \\ &= |V||I| \sin(\varphi) \sin(2\omega t) \\ &= Q \sin(2\omega t) \end{aligned} \tag{1.72}$$

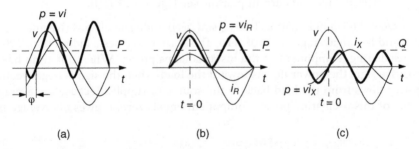

Figure 1.28 Voltage, current, and instantaneous power of an inductive load.

Figure 1.29 A simple series (a) and parallel (b) circuit.

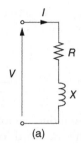
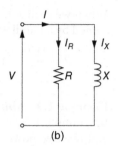

(a)   (b)

When we add both instantaneous power components (see Figure 1.28 (a)), we recognize the same expression as found earlier in Equation 1.66:

$$p = vi = vi_R + vi_X \tag{1.73}$$

Therefore, of the instantaneous power p, consumed by an element at any instant, a part (vi_R) is utilized for permanent consumption, such as conversion into heat. This part always has a positive value, that is, it cannot be returned to the rest of the circuit. The remainder (vi_X) is used to establish either a magnetic or electric field, that is, it is taken up and returned to the circuit with the rhythm of double the power frequency.

In a simple series circuit, as shown in Figure 1.29 (a), easier expressions for the active and reactive power can be derived. When the series impedance is written as $Z = R + jX$ and $|V| = |Z||I|$ is substituted into the equations for the active and reactive power, the following equations are the result:

$$P = |V||I|\cos(\varphi) = |I|^2|Z|\cos(\varphi) \tag{1.74}$$

$$Q = |V||I|\sin(\varphi) = |I|^2|Z|\sin(\varphi) \tag{1.75}$$

When we recognize that $R = |Z|\cos(\varphi)$ and $X = |Z|\sin(\varphi)$ (see Figure 1.20), the active and reactive power consumed by a series impedance can be written as

$$P = |I|^2 R \quad \text{and} \quad Q = |I|^2 X \tag{1.76}$$

In a simple parallel circuit, as shown in Figure 1.29 (b), easier expressions for the active and reactive power can also be derived. Deconstructing the current phasor into two components with a length $|I_R| = |I|\cos(\varphi)$ and $|I_X| = |I|\sin(\varphi)$ (see the phasor diagram in Figure 1.27 (c)) results in the following equations for the active and reactive power:

$$P = |V||I|\cos(\varphi) = |V||I_R| \tag{1.77}$$

$$Q = |V||I|\sin(\varphi) = |V||I_X| \tag{1.78}$$

When we realize that $|V| = |I_R|R = |I_X|X$, the active and reactive power consumed by a parallel impedance can be written as

$$P = \frac{|V|^2}{R} \quad \text{and} \quad Q = \frac{|V|^2}{X} \tag{1.79}$$

Example 1.9 *Single-phase power*

In Example 1.7 (p. 28), the single-phase inductive load consumes both active and reactive power (an explanation why the word "consumes" has been used here is given in Section 1.6.2):

$$P = |V_r||I| \cos(\varphi) = \frac{10 \times 10^3}{\sqrt{3}} \cdot 150 \cdot \cos(36.9) = 692.5\,\text{kW} \tag{1.80}$$

$$Q = |V_r||I| \sin(\varphi) = \frac{10 \times 10^3}{\sqrt{3}} \cdot 150 \cdot \sin(36.9) = 520\,\text{kvar} \tag{1.81}$$

The inductive load can be represented as a resistance in parallel with an inductance. The component values can be computed from the following relations (the frequency is 50 Hz):

$$Q = \frac{|V_r|^2}{X} \rightarrow 520 \times 10^3 = \left(\frac{10 \times 10^3}{\sqrt{3}}\right)^2 /(2\pi f L) \tag{1.82}$$

$$P = \frac{|V_r|^2}{R} \rightarrow 692.5 \times 10^3 = \left(\frac{10 \times 10^3}{\sqrt{3}}\right)^2 /R \tag{1.83}$$

This results in the following values: $L = 0.2\,\text{H}$ and $R = 48.1\,\Omega$.

The active power loss in the transmission line can be computed as follows:

$$P = |I|^2 R = (150)^2 \cdot 4 = 90\,\text{kW} \tag{1.84}$$

The reactive power consumed by the transmission line amounts to

$$Q = |I|^2 X = (150)^2 \cdot 7 = 157.5\,\text{kvar} \tag{1.85}$$

1.6.2 Complex Power

We now examine whether the expressions that we derived for the active and reactive power can be obtained from some kind of multiplication of the voltage phasor ($V = |V|\angle\alpha$) and the current phasor ($I = |I|\angle\beta$) as shown in Figure 1.30. The angle between the voltage and the current phasor is defined as $\varphi = \alpha - \beta$. Direct multiplication of both phasors gives the following expression:

$$\begin{aligned} VI &= |V||I|\angle(\alpha + \beta) \\ &= |V||I|[\cos(\alpha + \beta) + j\sin(\alpha + \beta)] \end{aligned} \tag{1.86}$$

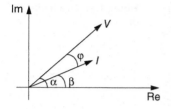 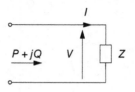

Figure 1.30 Phasor diagram of a single-phase load.

It is obvious that the equations of the active and reactive power do not result. Two other voltage and current phasor multiplications are shown in the following:

$$
\begin{aligned}
V^*I &= |V||I|\angle(-\alpha + \beta) \\
&= |V||I|[\cos(-\alpha + \beta) + j\sin(-\alpha + \beta)] \\
&= |V||I|[\cos(\varphi) - j\sin(\varphi)] \\
&= P - jQ
\end{aligned}
\tag{1.87}
$$

$$
\begin{aligned}
VI^* &= |V||I|\angle(\alpha - \beta) \\
&= |V||I|[\cos(\alpha - \beta) + j\sin(\alpha - \beta)] \\
&= |V||I|[\cos(\varphi) + j\sin(\varphi)] \\
&= P + jQ
\end{aligned}
\tag{1.88}
$$

In both equations, an active and reactive power component is present, but how about the sign of the reactive power?

We adopt the sign convention recommended by the International Electrotechnical Commission (IEC). A capacitor supplies reactive power, whereas an inductor consumes reactive power. Or in other words, the reactive power absorbed by an inductive load has a positive sign, and the reactive power absorbed by a capacitive load a negative sign. In the case that $\alpha > \beta$ (see Figure 1.30), the current lags the voltage. Therefore, the load is inductive and, in line with the IEC convention, consumes reactive power. To obtain the proper sign for the reactive power, it is necessary to calculate VI^*.

The mode of operation of an element, in terms of active and reactive power, can be seen from a quadrant diagram as shown in Figure 1.31. The non-reference phasor points to the quadrant of operation. As an example, let us consider the situation shown in the quadrant diagram in Figure 1.31. The current lags behind the voltage. Therefore, the load is inductive and consumes both active $(+P)$ and reactive power $(+Q)$. The load can be represented by a resistance in series with an inductance as shown in quadrant 1. Note that the quadrants 2 and 3 require an active element, that is, an underexcited generator and an overexcited generator (see also Section 2.5).

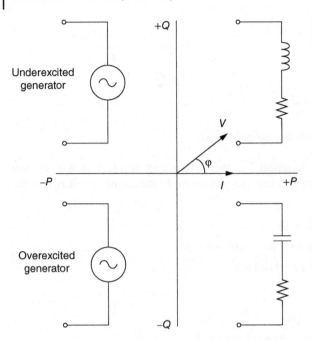

Figure 1.31 Quadrant diagram.

The quantity, obtained after the multiplication of the voltage phasor and the complex conjugated of the current phasor, is called the complex power S:

$$S = VI^* = P + jQ \qquad (1.89)$$

S the complex power [VA]

The complex power S is a complex quantity and can be expressed in both polar coordinates and rectangular or Cartesian coordinates, as shown in Figure 1.32:

$$
\begin{aligned}
S &= VI^* \\
&= |V||I|\angle\varphi \\
&= |S|\angle\varphi \quad \text{(polar)} \\
&= P + jQ \quad \text{(Cartesian)}
\end{aligned}
\qquad (1.90)
$$

φ the phase shift between the voltage and the current [rad]
$|S|$ the apparent power [VA]

The apparent power $|S|$ is defined as (see also Figure 1.32)

$$|S| = |V||I| = \sqrt{P^2 + Q^2} \qquad (1.91)$$

The apparent power is a useful practical quantity for specifying the rating of electrical apparatus when the maximum voltage and maximum current are fixed, and the phase angle is not considered.

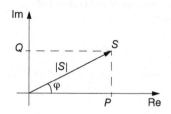

$$S = |S|\angle\varphi = P + jQ$$

$$|S| = \sqrt{P^2 + Q^2}$$

$$\varphi = \tan^{-1}\left(\frac{Q}{P}\right)$$

$$P = |S|\cos(\varphi)$$

$$Q = |S|\sin(\varphi)$$

Figure 1.32 The complex power (consumed by an inductive load).

Table 1.3 Power definitions.

Symbols	Terminology	Units
p	Instantaneous power	W
S	Complex power	VA
$|S|$	Apparent power	VA
P	Active power	W
	Real power	
	Average power	
Q	Reactive power	var
	Imaginary power	

The different symbols that we use to address electrical power are summarized in Table 1.3. There is no "deeper meaning" behind the different units (W, VA, var) that are used to express the different types of power values. (It is in fact no different from the impedance terms $Z = R + jX$; we do not use three different kinds of ohm!) But it is convenient that we are able to read from the unit which type of power is addressed.

Example 1.10   *Complex power*
Consider Example 1.7 (p. 28). In Example 1.9 (p. 34), the active and reactive power consumed by the single-phase inductive load were calculated:

$$P = |V_r||I|\cos(\varphi) = \frac{10 \times 10^3}{\sqrt{3}} \cdot 150 \cdot \cos(36.9) = 692.5\,\text{kW} \tag{1.92}$$

$$Q = |V_r||I|\sin(\varphi) = \frac{10 \times 10^3}{\sqrt{3}} \cdot 150 \cdot \sin(36.9) = 520\,\text{kvar} \tag{1.93}$$

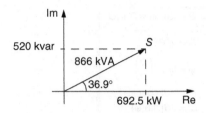

Figure 1.33 The complex power consumed by the inductive load.

Therefore, the complex power consumed by the load amounts to

$$S = P + jQ = 692.5 + j520 \, \text{kVA}$$
$$= |S|\angle\varphi = 866\angle36.9° \, \text{kVA} \tag{1.94}$$

This relation is displayed in Figure 1.33. The complex power can also be calculated in a different way:

$$S = V_r I^* = 5773.5\angle0° \cdot 150\angle36.9° = 866\angle36.9° \, \text{kVA} \tag{1.95}$$

Taking the real and imaginary parts of this complex power results in the values for the active and reactive power that we found earlier.

1.6.3 Power Factor

In previous sections, the active power has been defined as

$$P = |V||I|\cos(\varphi) = |S|\cos(\varphi) \tag{1.96}$$

The term $\cos(\varphi)$ is called the power factor, being the cosine of the phase shift between the voltage and current, that is, the cosine of the phase angle between the voltage and current phasor. In fact, the power factor is that part of the apparent power that is related to the mean energy flow, like mechanical energy in the case of a machine or heat in the case of a resistor.

The power factor can be computed by using several (equivalent) formulas. They can be obtained easily by inspection of the power diagram in Figure 1.32:

$$\cos(\varphi) = \frac{P}{|S|} = \frac{P}{\sqrt{P^2 + Q^2}}$$
$$= \cos\left(\tan^{-1}\left(\frac{Q}{P}\right)\right) \tag{1.97}$$

The power factor gives the information to calculate the active power from the apparent power. When we want to calculate the reactive power that is consumed, extra information is needed in order to specify whether the consumed reactive power is positive or negative. This information is expressed in words. When the current lags the voltage, the power factor is said to be lagging. When the current leads the voltage, the power factor is said to be leading.

It is evident from Equation 1.96 that a value of the power factor ($\cos(\varphi)$) close to 1 is beneficial. When a certain amount of active power is consumed at a fixed voltage, a power factor lower than 1 leads to increased currents and higher ohmic losses in the power system. From the utility point of view, it is therefore desirable that the large (industrial) loads have a power factor close to 1. Unfortunately, most of the heavy-industrial loads are inductive (electrical drives, for instance) and have rather low power factors. As a consequence, power-factor corrections should be made by these larger industries to improve their power factor. The power factors for households and small commercial users are usually closer to 1.

Example 1.11   *Power factor*
Consider Example 1.7 (p. 28). The power factor is defined as the cosine of the phase angle between the voltage and current phasor. Thus, the power factor of the inductive load is

$$\cos(\varphi) = \cos(36.9) = 0.8 \tag{1.98}$$

Because the current lags the voltage, the power factor is said to be 0.8 lagging. This value can be calculated from the consumed power as well (see also Example 1.10 (p. 37)):

$$\cos(\varphi) = \frac{P}{|S|} = \frac{P}{\sqrt{P^2 + Q^2}} = \frac{692.5 \times 10^3}{866 \times 10^3} = 0.8 \tag{1.99}$$

$$\cos(\varphi) = \cos\left(\tan^{-1} \left(\frac{Q}{P} \right) \right) = \cos\left(\tan^{-1} \left(\frac{520 \times 10^3}{692.5 \times 10^3} \right) \right) = 0.8 \tag{1.100}$$

Example 1.12   *Power-factor improvement*
Consider Example 1.7 (p. 28) and the previous example. The power factor of the inductive load amounts to 0.8 and is to be increased to the value of 1. This can be done by connecting a capacitor in parallel to the load. The value of the capacitance is such that the capacitor supplies the amount of reactive power that is consumed by the inductive load. When we assume that the voltage at the load terminals remains fixed, the required capacitance can be calculated as follows:

$$Q = \frac{|V_r|^2}{-X} \rightarrow 520 \times 10^3 = \left(\frac{10 \times 10^3}{\sqrt{3}} \right)^2 (2\pi f C) \tag{1.101}$$

The capacitance value amounts to $C = 50\,\mu\text{F}$.

After the power-factor correction, $\cos(\varphi) = 1$, the current equals

$$|I| = \frac{P}{|V_r| \cos(\varphi)} = \frac{692.5 \times 10^3}{(10 \times 10^3/(\sqrt{3})) \cdot 1} = 120 \, \text{A} \qquad (1.102)$$

Now, the power loss in the short transmission line is

$$P_{\text{loss}} = |I|^2 R = (120)^2 \cdot 4 = 57.6 \, \text{kW} \qquad (1.103)$$

which means that considerable savings, by a loss reduction from 90 kW at a $\cos(\varphi) = 0.8$ to 57.6 kW at a $\cos(\varphi) = 1$, have been established.

1.7 Power in Three-phase Circuits

The power (consumed by a load or produced by a generator) in a three-phase network can be found easily by adding up the power for each of the three phases. For balanced three-phase systems, the three-phase complex power is three times the complex power of the single-phase equivalent network. Therefore, the following equations hold for the complex, apparent, active, and reactive power:

$$S_{3\phi} = 3S_{1\phi} = 3V_{\text{LN}}I^* \qquad (1.104)$$

$$|S_{3\phi}| = 3|S_{1\phi}| = 3|V_{\text{LN}}||I| = \sqrt{3}|V_{\text{LL}}||I| \qquad (1.105)$$

$$P_{3\phi} = 3P_{1\phi} = 3|V_{\text{LN}}||I| \cos(\varphi) = \sqrt{3}|V_{\text{LL}}||I| \cos(\varphi) \qquad (1.106)$$

$$Q_{3\phi} = 3Q_{1\phi} = 3|V_{\text{LN}}||I| \sin(\varphi) = \sqrt{3}|V_{\text{LL}}||I| \sin(\varphi) \qquad (1.107)$$

V_{LL} the line-to-line voltage phasor
V_{LN} the line-to-neutral voltage phasor
I the current phasor

Example 1.13   *Three-phase power*
The loaded single-phase short transmission line in Example 1.7 (p. 28), with a (line-to-neutral) voltage at the receiving end of $|V_{r,\text{LN}}| = 10/\sqrt{3}$ kV, can be interpreted as an equivalent line-to-neutral diagram of a balanced three-phase short transmission line with a line-to-line voltage at the receiving end of $|V_{r,\text{LL}}| = 10$ kV, which is loaded by a balanced wye-connected inductive load. The three-phase active power consumed by the load can be calculated in the following (equivalent) ways (see also Example 1.9 (p. 34)):

$$P_{3\phi} = 3P_{1\phi} = 3 \cdot 692.5 \times 10^3 = 2077.5 \, \text{kW} \qquad (1.108)$$

$$P_{3\phi} = 3|V_{r,\text{LN}}||I| \cos(\varphi) \qquad \qquad (1.109)$$

$$= 3 \cdot \frac{10 \times 10^3}{\sqrt{3}} \cdot 150 \cdot \cos(36.9) = 2077.5 \, \text{kW}$$

$$P_{3\phi} = \sqrt{3}|V_{r,\mathrm{LL}}||I|\cos(\varphi)$$
$$= \sqrt{3} \cdot 10 \times 10^3 \cdot 150 \cdot \cos(36.9) = 2077.5 \text{ kW} \tag{1.110}$$

1.8 Per-unit Normalization

In power engineering a normalization of numerical values is common practice. This so-called per-unit (pu) normalization is similar to calculating with percentages. In the percentage system, the product of two quantities expressed as percentages must be divided by 100 to obtain the result in percentage. This drawback is circumvented in the pu system. In the pu system, 100% corresponds to 1 pu. The pu value of a certain quantity is defined as

$$\text{Per-unit value} = \frac{\text{actual value}}{\text{base value}} \tag{1.111}$$

In power system analysis, the following quantities are of interest and frequently normalized: voltages, currents, impedances, and powers. These quantities are interrelated and this means that a selection of two base values fixes immediately the base values for the other two. Therefore, selecting a base for the pu calculation requires a selection of two base values only, and the other two base values can be calculated. Taking into account the constant voltage nature of the power system and the fact that the equipment is rated in volt-amperes, the base voltage and (apparent) power are usually the quantities to specify the base values, as shown in Table 1.4.

Example 1.14 *Base quantities and per-unit calculation*
Consider Example 1.7 (p. 28). The calculation of the voltage at the sending end of the short transmission line is repeated here in per-unit quantities. The base

Table 1.4 Base quantities in the per-unit system.

Base Quantity	Single Phase (Line-to-Neutral Voltage, Single-Phase Power)	Three Phase (Line-to-Line Voltage, Three-Phase Power)
Voltage [V]	$|V_b|$ (selected)	$|V_b|$ (selected)
(Apparent) Power [VA]	$|S_b|$ (selected)	$|S_b|$ (selected)
Current [A]	$|I_b| = \dfrac{|S_b|}{|V_b|}$	$|I_b| = \dfrac{|S_b|}{\sqrt{3}|V_b|}$
Impedance [Ω]	$|Z_b| = \dfrac{|V_b|}{|I_b|} = \dfrac{|V_b|^2}{|S_b|}$	$|Z_b| = \dfrac{|V_b|/\sqrt{3}}{|I_b|} = \dfrac{|V_b|^2}{|S_b|}$

voltage and base (apparent) power are selected: $|V_b| = |V_{r,\text{LN}}| = 10/\sqrt{3}\,\text{kV}$ (line-to-neutral) and $|S_b| = 1\,\text{MVA}$ (single-phase power). Now the base current and base impedance can be calculated:

$$|I_b| = \frac{|S_b|}{|V_b|} = \frac{1 \times 10^6}{10 \times 10^3/\sqrt{3}} = 173.2\,\text{A} \tag{1.112}$$

$$|Z_b| = \frac{|V_b|^2}{|S_b|} = \frac{(10 \times 10^3/\sqrt{3})^2}{1 \times 10^6} = 33.3\,\Omega \tag{1.113}$$

The network quantities expressed in pu are obtained by dividing the actual value by the corresponding pu value: $|V_{r,\text{LN}}| = 1\,\text{pu}$, $I = 0.87\angle{-36.9°}\,\text{pu}$, $Z = 0.24\angle 60°\,\text{pu}$. Now, the voltage in pu at the sending end of the transmission line can be computed:

$$
\begin{aligned}
V_{s,\text{LN}} &= V_{r,\text{LN}} + ZI = 1\angle 0° + 0.24\angle 60° \cdot 0.87\angle{-36.9°} \\
&= 1 + 0.21\angle 23.1° \tag{1.114} \\
&= 1.19 + j0.08 = 1.19\angle 3.9°\,\text{pu}
\end{aligned}
$$

The value in volts is found by multiplication of the pu value with the base voltage value:

$$V_{s,\text{LN}} = \frac{10 \times 10^3}{\sqrt{3}} \cdot 1.19\angle 3.9° = 6.9\angle 3.9°\,\text{kV} \tag{1.115}$$

This is the same value as we found earlier in Example 1.7 (p. 28).

There are several reasons why the pu normalization is used. First of all, it is more comfortable to work with a voltage value of, for example, 1.00895 pu than with a value of 151343 V (in this case, 1 pu corresponds to a base value of 150 kV): the actual deviation in the voltage value can be more easily observed. Another advantage is that a pu quantity contains "more information." Consider, for instance, a voltage drop of 1500 V along a transmission line. This information is rather insignificant since we have no information about the nominal voltage rating of the transmission line. A voltage drop of 1500 V on a 10 kV transmission line is extraordinary, while on a 150 kV transmission line it is an acceptable value. When we have a pu value, for example, a voltage drop of 0.015 pu, all relevant information for a useful interpretation is present: the voltage drop amounts to 1.5% of the base voltage level. Another advantage of the pu calculation appears when we deal with systems with transformers. When the base quantities are selected properly, the transformers disappear from our line diagrams if we suppose them to be ideal. In Figure 1.34, an ideal transformer is shown, and the relations between voltage and current at the primary side

Figure 1.34 An ideal (single-phase) transformer.

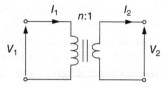

(subscript 1) and secondary side (subscript 2) are (see also Appendix B.2)

$$|V_2| = \frac{|V_1|}{n} \tag{1.116}$$

$$|I_2| = n|I_1| \tag{1.117}$$

n the turns ratio: the number of windings at the primary side divided by the number of windings at the secondary side ($n = N_1/N_2$)

To achieve the previously mentioned advantage, the base (apparent) power ($|S_b|$) is selected and must be the same at both the primary and the secondary sides of the transformer.

Furthermore, the base voltages at the primary and secondary sides of the transformer should be selected such that they have the same ratio as the turns ratio of the transformer windings:

$$|V_b|_2 = \frac{|V_b|_1}{n} \tag{1.118}$$

In that case, the base currents at the primary and secondary sides are related as follows:

$$|I_b|_2 = \frac{|S_b|}{|V_b|_2} = n \cdot \frac{|S_b|}{|V_b|_1} = n|I_b|_1 \tag{1.119}$$

Dividing the voltage relation of the ideal transformer (Equation 1.116) by the base voltage $|V_b|_2$ and substituting the relation $|V_b|_2 = |V_b|_1/n$ results in

$$\frac{|V_2|}{|V_b|_2} = \frac{|V_1|}{n|V_b|_2} = \frac{|V_1|}{|V_b|_1} \tag{1.120}$$

In other words, the pu voltage at the secondary side of the ideal transformer equals the pu voltage at the primary side. The same procedure can be followed for the current. Dividing the current relation of the ideal transformer (Equation 1.117) by the base current $|I_b|_2$ and substituting the relation $|I_b|_2 = n|I_b|_1$ results in

$$\frac{|I_2|}{|I_b|_2} = n \cdot \frac{|I_1|}{|I_b|_2} = \frac{|I_1|}{|I_b|_1} \tag{1.121}$$

In other words, the pu current at the secondary side of the ideal transformer equals the pu current at the primary side. As both the pu voltage and the pu current are equal at the primary and secondary sides of the ideal transformer, the electrotechnical symbol of the ideal transformer does not serve any purpose in the circuit diagram and can be left out.

Example 1.15 *Per-unit calculation and ideal transformers*
Consider the single-phase power system of Figure 1.35. In the system are two ideal transformers with ratings:

$$A - B: \ 10\,\text{MVA}, \ 13.8\,\text{kV}/138\,\text{kV}$$
$$B - C: \ 10\,\text{MVA}, \ 138\,\text{kV}/69\,\text{kV}$$

As base power the value for the apparent power is chosen: $|S_b| = 10\,\text{MVA}$. The base voltage in circuit B is chosen to be equal to $|V_b|_B = 138\,\text{kV}$. The base voltages in the circuits A and C are derived from the base voltage in circuit B and the turns ratio of the transformers:

$$|V_b|_A = \frac{1}{10} \cdot |V_b|_B = 13.8\,\text{kV}$$
$$|V_b|_C = \frac{1}{2} \cdot |V_b|_B = 69\,\text{kV} \tag{1.122}$$

Now, the base impedances in the three circuits can be calculated as

$$|Z_b|_A = \frac{|V_b|_A^2}{|S_b|} = 19\,\Omega$$

$$|Z_b|_B = \frac{|V_b|_B^2}{|S_b|} = 1904\,\Omega \tag{1.123}$$

$$|Z_b|_C = \frac{|V_b|_C^2}{|S_b|} = 476\,\Omega$$

The resistive load in circuit C, expressed in pu, is

$$R_C = \frac{100}{|Z_b|_C} = \frac{100}{476} = 0.21\,\text{pu} \tag{1.124}$$

When the resistance is referred to circuit B (see also Appendix B.2) and expressed in pu, the following value is the result:

$$R_B = 2^2 \cdot 100 = 400\,\Omega \quad R_B = \frac{400}{|Z_b|_B} = \frac{400}{1904} = 0.21\,\text{pu} \tag{1.125}$$

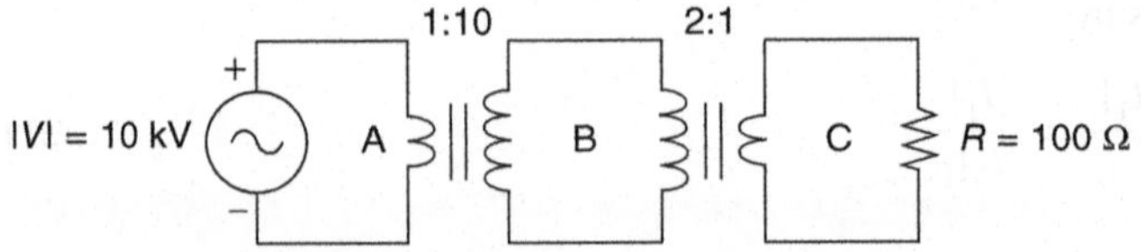

Figure 1.35 Single-phase circuit with ideal transformers.

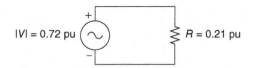

Figure 1.36 Per-unit diagram of the single-phase circuit in Figure 1.35.

When the resistance is referred to circuit A and expressed in pu, the following value results:

$$R_A = 0.1^2 \cdot 2^2 \cdot 100 = 4\ \Omega \quad R_A = \frac{4}{|Z_b|_A} = \frac{4}{19} = 0.21\,\text{pu} \quad (1.126)$$

Because of the well-chosen base quantities, the pu resistance of the load, when referred to another part of the system, is the same as the pu resistance of the load in its original position. It is evident that the single-phase circuit of Figure 1.35, when expressed in pu, turns into the simplified circuit, which is shown in Figure 1.36, in which all the ideal transformers have disappeared.

1.9 Power System Structure

The graphical layout of the three-phase power system is often displayed as a single-line equivalent or as a one-line diagram. Such a one-line diagram gives only an overview of the topology of the power system. The components are identified by means of standardized symbols and not by models built up from lumped-circuit elements (as is the case with the equivalent line-to-neutral diagram as introduced in Section 1.5). The symbols used in such a one-line diagram are shown in the list of symbols. Sometimes lumped-circuit elements appear in a "one-line diagram." An example of such a "one-line diagram" is shown in Figure 5.28 (Section 5.5.3): it shows a device used in the power system for sophisticated control actions. The lumped elements in the diagram show what is actually inside the device.

A one-line diagram that schematically displays part of the structure of a power system is shown in Figure 1.37. The one-line diagram shows a clear vertical structure: a relatively small number of large power stations supply the transmission network (380 and 150 kV). Besides some large industrial consumers that are connected to the higher voltage levels (150 kV), most of the power is transported and distributed to the consumption centers located at the lowest voltage levels (10 and 0.4 kV). Nowadays, more and more decentralized generation, that is, small-scale generators connected to the lower voltage levels (like the generator connected to the 10 kV bus shown in Figure 1.37), is being integrated in the system (see also Section 8.3). Examples of such decentralized generators are windmills, solar panels, and combined heat-power units (producing steam for industrial processes and electricity as a by-product).

Our book closely follows the vertical structure of the power system. In Chapter 2 the generation of electric energy is examined. In Chapter 3 the

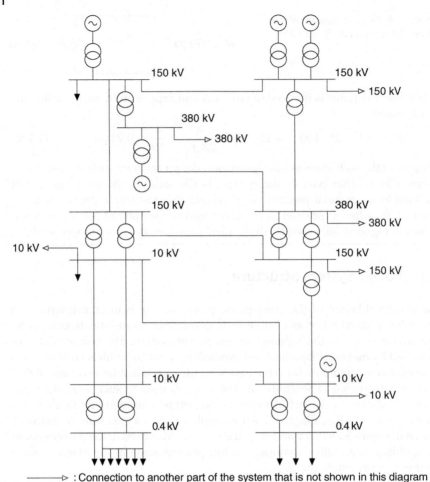

: Connection to another part of the system that is not shown in this diagram

Figure 1.37 One-line diagram of a section of a power system [10].

transmission and distribution are highlighted. The utilization of electricity is described in Chapter 4. Without controls, the power system cannot function, and the controls implemented are presented in Chapter 5. In the control center, the generation, transmission, and distribution of electrical energy are monitored, coordinated, and controlled by means of the energy management system (EMS). The EMS is the interface between the operator and the power system, and its fundamentals are outlined in Chapter 6. The electricity market, where all the previously treated items meet each other in a commercial sense, is introduced in Chapter 7. The book concludes by questioning the vertical structure of the power system in Chapter 8, when some thoughts on future power systems are offered. The relation between Maxwell's laws and lumped-element modeling is the topic of the first appendix. The lumped-element models for the power transformer, the synchronous machine, the induction machine, and the overhead lines and underground cables are derived in the other appendices.

Problems

1.1 For a specific node in a $50\,\text{Hz}$ single-phase circuit, the voltage equals $v(t) = 325.27 \sin(\omega t + \varphi)$ V, and the injected current is $i(t) = 141.42 \sin(\omega t - \varphi)$ A, where $\varphi = \pi/6$ rad.
  a. Calculate for $v(t)$ and $i(t)$ the peak value and the RMS value.
  b. Express the voltage and current as phasors.
  c. Is the circuit capacitive or inductive? Explain your answer.

1.2 In a three-phase circuit, the voltages of the phases a and b, with respect to the neutral n, are $V_{an} = 140\angle 45°$ V and $V_{bn} = 90\angle{-15°}$ V. Calculate V_{ba}.

1.3 In a balanced three-phase system, with a phase sequence abc, the Y-connected impedances are $Z = 10\angle 30°\,\Omega$. If $V_{bc} = 400\angle 90°$ V, calculate
  a. V_{cn}
  b. I_{cn}
  c. S consumed by the impedances

1.4 Two ideal voltage sources, $E_1 = 100\angle 0°$ V and $E_2 = 100\angle 30°$, are connected through an impedance $Z = j5\,\Omega$. For both voltage sources the generator convention is used, which means that the power delivered by the voltage sources is positive.
  a. Calculate the currents I_1 and I_2 delivered by both sources.
  b. Calculate the active power and the reactive power consumed by both sources.
  c. Which of the two sources is the generator?
  d. Calculate the losses.

1.5 In a balanced three-phase system, the power injected at node m equals $S = 100 + j60$ MVA, while the line-to-line voltage equals $380\angle 0°$ kV.
  a. Determine the power factor of the supplied power.
  b. Give an expression for the current phasor magnitude.

1.6 In a balanced three-phase system, the voltage between phases a and b is $V_{ab} = 173.2\angle 0°$ V. The Y-connected load is $Z = 10\angle 0°$ Ω. The phase sequence is *abc*.
   a. Calculate all phase-to-neutral voltages.
   b. Calculate all phase currents.

1.7 A three-phase Y-connected load consumes 250 kW, with a power factor of 0.8 lagging from a 400 V line. In parallel with this load is a three-phase capacitor bank connected, which delivers 60 kvar.
   a. Calculate the total phase current (combined load and capacitor bank).
   b. What is the resulting power factor?

1.8 The equivalent circuit to represent a load on a 110 kV transmission line consists of three Y-connected impedances of $Z = 80 + j30$ Ω. Assuming an RMS line-to-line voltage of 100 kV,
   a. What are the phase currents drawn by the load?
   b. What is the active and reactive power drawn by the load?

1.9 A three-phase transmission line can be represented by a phase impedance of $Z = 5 + j60$ Ω/phase. The complex power, measured at the sending end of the line, is $S_1 = 210 + j30$ MVA. The line has a fixed line-to-line voltage of 220 kV at the sending end.
   a. Calculate the voltage and the complex power at the receiving end of the line.
   b. What are the transmission losses?

1.10 Redo Problem 1.9 and use per-unit values. The base quantities are
   - $|S_b| = 300$ MVA (three phase)
   - $|V_b| = 130$ kV (line-to-neutral)

1.11 In a base system $|S_{b,1}|$ and $|V_{b,1}|$, the base impedance equals $|Z_{b,1}|$. $|Z_{b,2}|$ is the base impedance in a base system using $|S_{b,2}|$ and $|V_{b,2}|$. Show that the following equation holds:

$$|Z_{b,2}| = |Z_{b,1}| \frac{|S_{b,1}| \, |V_{b,2}|^2}{|S_{b,2}| \, |V_{b,1}|^2}$$

References

1 Netbeheer Nederland: *Betrouwbaarheid van elektriciteitsnetten in Neder-land*, Netbeheer Nederland, 2014.
2 Herrmann, F., and Schmid, G.B.: 'The Poynting vector field and the energy flow within a transformer', *American Journal of Physics*, Vol. **54**, Issue 6, June 1986, pp. 528–31.
3 Newcomb, W.A.: 'Where is the Poynting vector in an ideal trans-former?', *American Journal of Physics*, Vol. **52**, Issue 8, August 1984, pp. 723–4.
4 Hammons, T.J., Woodford, D., Loughtan, J., Chamia, M., Donahoe, J., et al.: 'Role of HVDC transmission in future energy development', *IEEE Power Engineering Review*, February 2000, pp. 10–25.
5 Furfari, F.A.: 'The evolution of power-line frequencies 133 1/3 to 25 Hz', *IEEE Industry Applications Magazine*, Vol. **6**, Issue 5, September/October 2000, pp. 12–14.
6 Lamme, B.G.: 'The technical story of the frequencies', *AIEE Transactions*, Vol. **37**, 1918, pp. 65–89.
7 Mixon, P.: 'Technical origins of 60 Hz as the standard AC frequency in North America', *IEEE Power Engineering Review*, Vol **19**, March 1999, pp. 35–37.
8 Owen, E.L.: 'The origins of 60 Hz as a power frequency', *IEEE Industry Applications Magazine*, Vol. **3**, Issue 6, November/December 1997, pp. 8–14.
9 Laithwaite, Eric R., and Freris, Leon L.: *Electric Energy: Its Generation, Transmission and Use*, McGraw-Hill, Maidenhead, 1980, ISBN 0-07-084109-8.
10 Hoeijmakers, Martin J.: *Elektrische omzettingen*, Delft University Press, Delft, 2003, ISBN 90-407-2455-5.

2

The Generation of Electric Energy

2.1   Introduction

The power system generates, transports, and distributes electric energy economically and reliably to the consumers, with the constraint that both the voltage and frequency are kept constant, within narrow margins, at the load side.
Power quality is a major issue these days, a nearly perfect sine wave of constant
frequency and amplitude and always available. Electrical engineering started
basically with electric power engineering at the turn of the nineteenth century when the revolution in electrical engineering took place. In a rather short
period of time, the transformer was invented, electric motors and generators
were designed, and the step from DC to AC transmission was made. Society
was completely changed, first by lighting, rapidly followed by the versatile application of electrical power. In this early period, independent operating power
companies used different voltage levels and operated their system at various
frequencies. Electrical engineers were among the first to realize that international standardization would become necessary in the modern world.

The expansion of the world demand for energy has been met by the
ever-increasing exploitation of fossil fuels, oil, natural gas, and coal. In
1972 *The Limits to Growth*, a report for the Club of Rome Project on the
Predicament of Mankind, was published [1]. The book became a best seller.
A mathematical model of the world was introduced with five basic variables:
population, capital, food, natural resources, and pollution. Different scenarios
were presented – from a modeling point of view rather simple scenarios – but
the bottom line for the exponentially growing world population picture is
how to feed the people and how to power the society. The constraints on
expansion of demand through government legislation have not worked, and
world energy prices have not increased very much. More oil wells and natural
gas reserves have been located and have moved the problem of the natural
resources beyond our horizon. However, the ongoing expansion of energy
demand will also be constrained by ecological considerations such as limits to
available sites for power stations, heat and water disposal, water availability,

Electrical Power System Essentials, Second Edition. Pieter Schavemaker and Lou van der Sluis.
© 2017 John Wiley & Sons Ltd. Published 2017 by John Wiley & Sons Ltd.

air pollution, and possible effects on our climate. The Kyoto agreement in 1997 to reduce CO_2 emission and the greenhouse effect nowadays influence the political decisions made by the governments of the developed countries. The result is an increase in the application of wind and solar power as a source of renewable energy.

The backbone of the electric power system is a number of generating stations, distributed over a territory and electrically operating in parallel. The primary source of energy can be water stored behind a dam (hydroelectric power); fossil fuels such as gas, oil, and coal (conventional thermal power); or uranium (nuclear thermal power) (see Table 2.1). In France the bulk of the generated electricity comes from nuclear power plants, while in Norway hydroelectric power is the source of energy. Electricity is not only generated locally but also transported to and imported from neighboring countries. A group of countries with long-term trading contracts and a common power flow and frequency control form a power pool.

The term "renewable energy" covers energy from a broad spectrum of more or less self-renewing energy sources such as sunlight, wind, flowing water, the Earth's internal heat, and biomass (such as energy crops and agricultural and industrial waste). U.S. power plants used renewable energy sources, including water, wind, biomass wood and waste, geothermal, and solar, to generate about 13% of the electricity produced in the United States during 2015. The largest share of electricity generated by renewable sources in 2015 came from hydroelectric power (46%), followed by wind (35%), biomass wood (8%), solar (5%), biomass waste (3%), and geothermal (3%) [2].

Similar to fossil fuels, renewable energy resources are not uniformly distributed throughout the world, but every region has some renewable energy resource. And when we consider that renewable energy systems generate little or no waste or pollutants that contribute to acid rain, urban smog, or excess carbon dioxide and other gases in the atmosphere, renewable energy technologies offer the promise of clean, abundant energy from self-renewing resources such as the sun, wind, Earth, and plants. The biomass, however, does

Table 2.1 Net generation capacity in some European countries as of 31 December 2015 [3].

Capacity [GW]	France	The Netherlands	Germany	Austria	Switzerland	Italy
Hydro power	25.4	–	9.6	13.6	13.8	22.2
Nuclear power	63.2	0.5	10.8	–	3.3	–
Conventional thermal	22.6	25.6	78.5	7.5	0.5	69.7
Renewables	18.2	5.7	89.3	2.8	1.1	32.3
Total	**129.4**	**31.8**	**188.2**	**23.9**	**18.7**	**124.2**

release carbon dioxide when it is converted to energy, but because the biomass absorbs carbon dioxide as it grows, the entire process of growing, using, and regrowing the biomass results in very low to zero carbon-dioxide emissions.

2.2 Thermal Power Plants

Thermal power plants, of which a photo is shown in Figure 2.1, convert the primary energy from coal, oil, gas, and nuclear into electrical energy. The first step in this conversion process is to transform the chemical energy of fossil fuels into thermal energy, either by combustion (in the case of coal, oil, and gas) or by fission (nuclear). Thermal energy is used to produce steam at high temperatures and pressures. The steam expands adiabatically (i.e. without heat extraction or injection) in a steam turbine, and the thermal energy of the steam is converted into mechanical energy when the steam passes the turbine blades and causes the turbine to rotate. The shaft of the turbine is connected with the shaft of a synchronous generator, and by electrical induction the mechanical energy is converted into electrical energy.

2.2.1 The Principles of Thermodynamics

The first law of thermodynamics is the law of conservation of energy: energy cannot be created and cannot be destroyed. We can only convert one form of energy into another form of energy. For a gas, the first law of thermodynamics

Figure 2.1 A thermal power plant. Reproduced with permission of TenneT TSO B.V.

can be written in mathematical form as

$$dQ = dU + dW = dU + p\,dV \tag{2.1}$$

Q the heat added to the system [J]
U the internal energy [J]
W the work done by the system [J]
p the gas pressure [Pa]
V the gas volume [m^3]

This formula is read as "the heat added to a system equals the change in the internal energy of the system plus the work done by the system."

The internal energy of an ideal gas is the kinetic energy of the molecules of the gas and is related to the absolute temperature T:

$$U = \frac{3}{2}nkT \tag{2.2}$$

n the number of molecules
k the universal Boltzmann gas constant [$k = 1.38 \times 10^{-23}$ J/K]
T the temperature [K]

In many types of engines, work is done by a gas expanding against a movable piston as in a steam engine or in an automobile engine. In a thermal power plant, the steam expands in a steam turbine: the high gas temperature and pressure cause the gas to expand rapidly between the turbine blades, drive the turbine, and do the work.

The states of a gas can be represented on a pV diagram. Each point on the pV diagram indicates a particular state of the gas. In Figure 2.2, following path A, the gas is heated at constant pressure until its volume is V_2, and then the gas is cooled down at constant volume until its pressure is p_2. The work done along path A is $p_1 (V_2 - V_1)$ for the horizontal part of the path and zero for the constant-volume part. Following path B, the work done is $p_2 (V_2 - V_1)$, which is much less than the work done along path A, because the work done by the

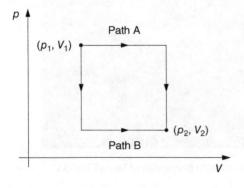

Figure 2.2 pV diagram of an ideal gas with two states (p_1, V_1) and (p_2, V_2).

gas is equal to the area between the pV curve and the V-axis:

$$W = \oint p\,dV \tag{2.3}$$

A trajectory in a pV diagram where the pressure is kept constant is called an isobar; a trajectory where the temperature remains constant is called an isotherm.

The second law of thermodynamics is formulated by Lord Kelvin as follows: "It is impossible to remove thermal energy from a system at a single temperature and convert it to mechanical work without changing the system or surrounding in some other way." We can also take the formulation by Carnot, which says that there is no machine possible that takes the heat of a body with a certain temperature and that is able to convert this heat totally in work, that is, there is always a need for a second body or heat reservoir at a lower temperature that can absorb a certain amount of heat. The law of conservation of energy gives us

$$Q_1 = W + Q_2 \tag{2.4}$$

The efficiency of such a heat engine is by definition the ratio of the work done to the heat absorbed from the hot reservoir:

$$\eta = \frac{W}{Q} = \frac{Q_1 - Q_2}{Q_1} = 1 - \frac{Q_2}{Q_1} \tag{2.5}$$

In Figure 2.3 the schematic representation of a heat engine is shown that removes heat from a hot reservoir with temperature T_1, does work W, and gives off heat to a cold reservoir at a temperature T_2.

For a machine without losses operating in a cycle, a so-called reversible machine, it can be proven that the efficiency of such a machine only depends on the temperatures of the hot and cold reservoir:

$$\eta_C = 1 - \frac{T_2}{T_1} \tag{2.6}$$

Figure 2.3 Schematic representation of a heat engine working between a hot reservoir (T_1) and a cold reservoir (T_2).

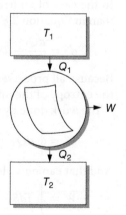

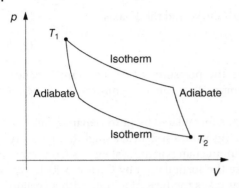

Figure 2.4 The *pV* diagram of the Carnot cycle.

This is called the Carnot efficiency and the cycle of the reversible machine is called the Carnot cycle. The Carnot cycle for an ideal gas follows two isothermal trajectories and two adiabatic trajectories. An adiabatic trajectory is a trajectory where no heat flows into or out of the system. The Carnot cycle, being a process that operates with the greatest possible efficiency, is a reversible process. The *pV* diagram of the Carnot cycle is shown in Figure 2.4. The work done per cycle is the surface enclosed by the isothermal and adiabatic trajectories:

$$W = \oint p\,dV \tag{2.7}$$

The entropy (S) is a measure of the disorder of a system. Like the pressure p, volume V, temperature T, and internal energy U, the entropy S is a function of the state of a system. It is the change in entropy dS of a system that is important. When dQ is the heat that must be added to the system in a reversible process, the change in entropy is defined as

$$dS = \frac{dQ}{T} \quad \text{or} \quad Q = \int_1^2 T\,dS \tag{2.8}$$

In the case of an irreversible process, the increase in entropy is always larger than in Equation 2.8, as a positive term dS_{irr} is added:

$$dS = \frac{dQ}{T} + dS_{irr} \tag{2.9}$$

Because all processes we encounter in our daily life are irreversible processes, the entropy of the universe increases!

The work done by a general process cycle is

$$W = \oint dQ \tag{2.10}$$

And that can now also be written as

$$W = \oint T\,dS - \oint T\,dS_{irr} \tag{2.11}$$

Because T and dS_{irr} are always positive, this equation tells us that the work W, and with it the efficiency η, is maximal for a reversible process for which the efficiency is the Carnot efficiency: $\eta = \eta_C$. The efficiency for an irreversible process can be written as

$$\eta_{irr} = \left(1 - \frac{T_2}{T_1}\right) - \frac{1}{Q_1}\oint TdS_{irr} \qquad (2.12)$$

The process of the steam cycle in a thermal power plant is not ideal because, for instance, steam is not an ideal gas (the pressure increase is, e.g., not adiabatic), and therefore the entropy increases. It was the Scotsman William John Macquorn Rankine who first gave an adequate description of the processes of the steam cycle in a thermal power plant, illustrated in Figure 2.5, which we call the Rankine cycle. The principle shape of the Rankine cycle is depicted in the entropy–temperature diagram in Figure 2.6. The Rankine cycle has the following trajectories:

1–2 The boiler-feed pump delivers the water to a huge boiler. The water temperature increases slightly because the pump has to operate against the steam pressure in the boiler tubes.

2–3 The provision of heat makes the water temperature rise to the boiling point.

3–4 The water boils and steam is produced at a constant boiling temperature, which depends on the pressure.

4–5 In the superheater, composed of a series of tubes surrounded by flames, the steam is raised about 200 °C in temperature. This ensures that the steam is dry and it raises the overall efficiency of the power plant. During the rise in temperature, the pressure remains constant, so that the trajectory 2–3–4–5 is an isobar.

5–6 The steam expands adiabatically in the steam turbine, and the thermal energy converts into mechanical energy. The steam temperature drops from T_{max} till T_0.

6–1 The low-temperature steam condenses to water in the condenser. The steam flows over cooling pipes and through those cooling pipes flows cold water from an outside source, such as a river or a lake. The heat is carried away and the condensing steam creates a vacuum.

The total energy added, $Q = \int_1^5 TdS$, is given by the surface area S_1–1–2–3–4–5–S_2–S_1. The energy that leaves the system with the cooling water is equal to the surface area 6–1–S_1–S_2–6. The efficiency of the Rankine cycle is

$$\eta = \frac{\text{surface area } 1-2-3-4-5-6-1}{\text{surface area } S_1-2-3-4-5-S_2-S_1} \qquad (2.13)$$

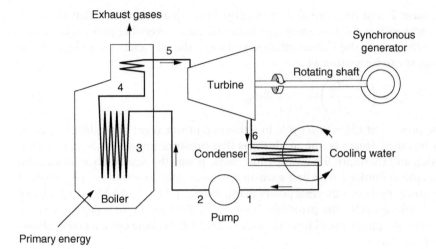

Figure 2.5 The principal components of a thermal power plant with the Rankine cycle.

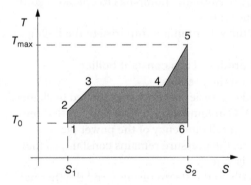

Figure 2.6 The Rankine cycle.

2.3 Nuclear Power Plants

The generation of electricity with nuclear power plants has made a rapid development since the commissioning of the first fission reactor in 1942. The rapid growth was hampered in April 1986 because of the Chernobyl (USSR) accident. Public opinion reacted against the use of nuclear energy, in particular in the industrialized countries. Large-scale application of renewable energy sources was expected to solve the energy problem and bring the world an environmentally friendly source of energy, while the generation of electricity through nuclear technology was widely out of favor internationally.

The nuclear power plants that are in operation have fission reactors that split uranium nuclei, and during that splitting process, mass is lost and converted into energy in accordance with Einstein's relation $E = mc^2$. Even when the losses in mass are very small, the released energy is enormous. The energy of the fission process is used to heat water and to produce steam as in conventional gas or coal-fired thermal power plants. Since the Second World War, another nuclear technology has also come into the picture: nuclear fusion. It is called fusion because it is based on fusing light nuclei, such as hydrogen isotopes, to release energy. The process is similar to that which powers the sun and other stars. Nuclear fusion is the holy grail that could solve the world's energy problem. Fusion power not only offers the potential of an almost limitless source of energy for future generations but also presents some formidable scientific and engineering challenges.

2.3.1 Nuclear Fission

In 1942, Enrico Fermi demonstrated the first nuclear chain reaction at the University of Chicago. After Fermi's experiment, the secret Manhattan project started that developed and built the atomic bomb. It was the atomic bomb that destroyed Hiroshima and Nagasaki in August 1945. Since its development, mankind has also focused on the peaceful uses of nuclear fission. In July 1957 an experimental reactor in California produced electricity for the first time using a nuclear reactor, and in 1958, the first large-scale nuclear power plant was commissioned in Pennsylvania, United States.

Example 2.1 *Nuclear fission*
A typical fission reaction, illustrated in Figure 2.7, is

$$^{235}_{92}U + ^{1}_{0}n \rightarrow ^{236}_{92}U \rightarrow ^{92}_{36}Kr + ^{141}_{56}Ba + 3^{1}_{0}n \tag{2.14}$$

A uranium nucleus is excited by the capture of a neutron, causing it to split into two nuclei (krypton and barium) and emit three neutrons. There is a difference

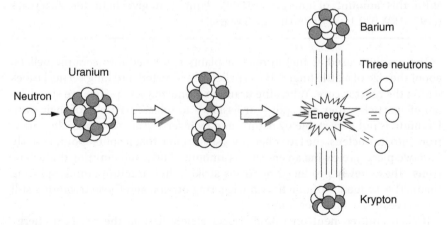

Figure 2.7 A typical fission reaction.

in the mass of the particles before and after the reaction:

$$235.0439231 - 91.9261528 - 140.9144064$$
$$- (2 \cdot 1.0086649) = 0.1860341 \, \text{amu} \tag{2.15}$$

The unit "amu" is the atomic mass unit and equals 1 amu $= 1.6605402 \times 10^{-27}$ kg. This loss of mass is released as energy conforming the famous equation

$$E_{\text{nucleus}} = mc^2 = 0.1860341 \cdot 1.6605402 \times 10^{-27} \cdot (3 \times 10^8)^2$$
$$= 2.7802539 \times 10^{-11} \, \text{J} \tag{2.16}$$

We can convert the unit joule to kWh by means of the following equality:

$$1 \, \text{J} = 1 \, \text{Ws} = \frac{1 \, \text{h}}{3600 \, \text{s}} \cdot \frac{1 \, \text{kW}}{1000 \, \text{J/s}} = 2.7777778 \times 10^{-7} \, \text{kWh} \tag{2.17}$$

Therefore, per fission, an energy is released of

$$E_{\text{nucleus}} = 2.7802539 \times 10^{-11} \cdot 2.7777778 \times 10^{-7} = 7.7229276 \times 10^{-18} \, \text{kWh} \tag{2.18}$$

This number looks far from "spectacular," but let us see how much energy is released by the fission of one gram of $^{235}_{92}$U. First, we use Avogadro's number ($N_A = 6.02 \times 10^{23}$ mol^{-1}) to compute the number of uranium nuclei in one gram of $^{235}_{92}$U:

$$N = \frac{6.02 \times 10^{23} \ \text{nuclei/mol}}{235 \ \text{g/mol}} \cdot 1 \ \text{g} = 2.56 \times 10^{21} \ \text{nuclei} \tag{2.19}$$

Therefore, the energy released by the fission of one gram of $^{235}_{92}$U amounts to

$$E = N E_{\text{nucleus}} = 2.56 \times 10^{21} \cdot 7.7229276 \times 10^{-18} = 19770 \, \text{kWh} \tag{2.20}$$

With this amount of energy, a 100 W lamp can give light for 22.5 years ((19770 kWh/0.1 kW)/8760 h $\approx$ 22.5 years)!

The uranium used as fuel in nuclear plants is formed into ceramic pellets, about the size of a little finger. These pellets are inserted into long vertical tubes within the reactor core. When the uranium atoms in these pellets are struck by atomic particles, they can split – or fission – to release particles of their own. Uranium is not very stable by nature, and by inserting an extra particle, a neutron, into its nucleus, the latter becomes so unstable that it splits spontaneously into two pieces, releasing an enormous amount of heat and emitting three neutrons. These newly produced neutrons strike other uranium atoms, splitting them. This sequence of one fission triggering others, and those triggering still more, is called a chain reaction.

If each emitted neutron would trigger a new fission, the released energy would increase too rapidly. The nuclear reaction inside the reactor is controlled by rods that are inserted among the tubes that hold the uranium fuel. These

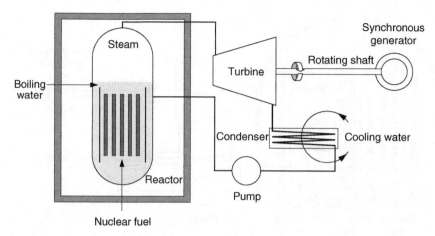

Figure 2.8 The boiling water reactor.

control rods are made up of material that absorbs neutrons and prevents them from hitting atoms that can fission. The control rods are made from hafnium, a very good absorber of neutrons. In this way, the nuclear reaction can be speeded up or slowed down by varying the number of control rods that are withdrawn/inserted and to what degree they are withdrawn/inserted.

There are two main types of commercial power plants: boiling water reactors and pressurized water reactors. In boiling water reactors, schematically illustrated in Figure 2.8, the water is heated by the nuclear fuel and boils to steam in the reactor vessel. It is then piped directly to the steam turbine. In pressurized water reactors, the water is heated by the nuclear fuel but kept under pressure to prevent it from boiling. Outside the reactor the heat of the water is transferred to a separate supply of water that boils and makes steam. The pressurized water reactor is schematically illustrated in Figure 2.9.

The solid uranium fuel contains two kinds, or isotopes, of uranium atoms: ^{235}U and ^{238}U. ^{235}U makes up less than 1% of natural uranium and fissions easily, but ^{238}U, which makes up most of natural uranium, is practically non-fissionable. Through a process known as "enrichment," the concentration of ^{235}U in the uranium is increased to about 3–5% before it is used as reactor fuel.

Most of the fragments of fission, the particles left over after the atom has split, are radioactive. During the life of the fuel, these radioactive fragments collect within the fuel pellets. The fuel remains in the reactor for 3–4 years; most of the ^{235}U is fissioned then and trapped fission fragments reduce the efficiency of the chain reaction. The fuel removed from the reactors, which we call "nuclear waste," is stored underwater in large concrete and stainless steel containers or above ground in steel and lead containers.

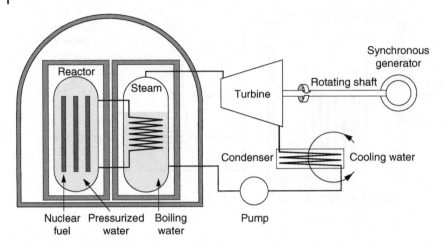

Figure 2.9 The pressurized water reactor.

In 2016, 450 nuclear reactors in 30 countries around the world were used for electricity generation. These nuclear power plants provided in 2012 some 11% of the world's electricity production. France, Ukraine, Slovakia and Hungary produce more than 50% of its electricity with nuclear power plants (see also Table 2.1).

2.3.2 Nuclear Fusion

The enormous potential of nuclear fusion is hidden in Einstein's relation $E = mc^2$: a small amount of matter can result in an enormous amount of energy. In 1920 the astronomer Arthur Eddington reasoned that the nuclear fusion from hydrogen to helium was the source of energy that powers the sun. Two atoms of hydrogen combine together, or fuse, to form an atom of helium. In the process of fusion of the hydrogen, matter is converted into energy. The easiest fusion reaction is combining deuterium (or heavy hydrogen) with tritium to make helium and a neutron, as visualized in Figure 2.10. Deuterium is plentifully available in ordinary water, and tritium can be produced by combining the fusion neutron with the light metal lithium. So the fuel for the fusion process is not a problem, and that makes it so very attractive. The bad news is that to make fusion happen, the atoms of hydrogen must be heated to some 100 million degrees so they are ionized and form a plasma. At this high temperature, the ionized atoms have sufficient energy to fuse, but the plasma has to be held together for the fusion to occur. The sun and stars do this by gravity. To achieve this on Earth, a strong magnetic field is used to hold the ionized atoms together while they are heated by, for instance, microwaves. The technical concept to perform this was developed by the Russians in 1959.

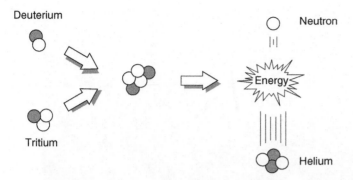

Figure 2.10 A typical fusion reaction.

They called their design "tokamak," which means toroidal magnetic chamber in the Russian language. The tokamak forms the basic element in the majority of the fusion reactors.

In 1997, 16 MW of power was released during one second in the Joint European Torus (JET) in Culham (near Oxford) in the United Kingdom. In 2006, the decision was taken to build a 500 MW experimental reactor in Cadarache in the south of France. In this International Thermonuclear Experimental Reactor (ITER) project, the countries of the European Union, Japan, China, the former Soviet Union, and the United States are working together toward the next step on the long road to realizing the dream of nuclear fusion as the solution to the world's energy problems.

2.4 Renewable Energy

2.4.1 Wind Energy and Wind Turbine Concepts

Wind power has now overcome initial teething problems. Wind power systems converting the kinetic energy of the wind into electricity with turbines of 3–5 MW nominal power have matured in technology and lowered the cost of wind energy significantly. In Denmark 50% of the electric power consumption is delivered by wind. The wind turbines are also going offshore, and Denmark, the Netherlands, the United Kingdom and Sweden have gathered experience with near-shore wind farms, using existing technologies from the wind turbine industry and civil engineering contractors. A photo of a wind park is shown in Figure 2.11.

A wind turbine consists of a rotor that extracts kinetic energy from the wind and converts this into a rotating movement, which is then converted into electrical energy by a generator. A wind turbine is a rather complex system combining high-level knowledge of aerodynamics and mechanical and electrical

Figure 2.11 A wind park. Reproduced with permission of TenneT TSO B.V.

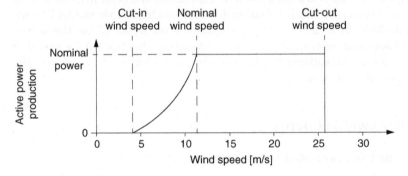

Figure 2.12 Typical power curve of a wind turbine.

engineering. The amount of energy in airflow depends on the wind speed. The wind speed at which the turbine begins to deliver its nominal rated power is called the nominal wind speed. A typical power curve of a wind turbine is illustrated in Figure 2.12. When the wind speed increases above the nominal wind speed, the pitch angle of the blades (pitch control) should be adjusted to avoid mechanical or electrical overload of the turbine components. Another way to reduce the overall efficiency of the turbine is called stall control and makes use of the aerodynamic phenomenon stall. Stall automatically increases the drag and thus reduces the mechanical torque on the blades at increasing wind speed; when designed properly, the blades come into stall at high wind speeds without

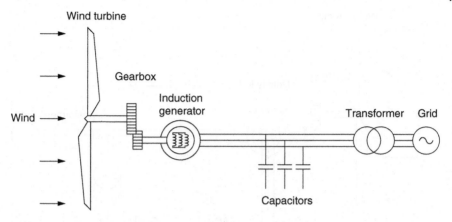

Figure 2.13 The Danish wind turbine concept.

the necessity of pitch control. When we look from the generator side at a wind turbine, three different concepts can be distinguished [4, 5].

The Danish concept, so-called because this type of wind turbine has been widely adopted by the Danish manufacturers, uses a squirrel cage induction generator for the conversion of mechanical energy into electrical energy (see Figure 2.13). This concept uses a gearbox to adapt the operating speed of the turbine rotor and the squirrel cage generator. A squirrel cage generator, being an induction generator, has a slip that slightly varies with the amount of generated power, but in spite of that slip this type of wind turbine is called a constant speed turbine. A constant speed wind turbine is relatively simple in its design, robust in its construction, and therefore reliable in its operation. The disadvantages are the lack of control of active and reactive power and the occurrence of large fluctuations in output power because there is no energy buffer in the form of a rotational mass present when the rotational speed varies. Furthermore, the gearbox is a weak point in the design, because fluctuations in the wind power are transformed in torque pulsations on the shaft.

Another widely used concept is the doubly fed induction generator, again in combination with a gearbox (see Figure 2.14). The rotor winding is fed from the grid by a power-electronic converter. The power-electronic converter, a back-to-back voltage source converter with current control loops, compensates for the difference between the mechanical rotor speed and the power frequency by injecting a rotor current with a variable frequency. The mechanical rotor speed and the power frequency are in fact decoupled, and this enables – within certain limits – variable speed operation in order to minimize the noise or to maximize the electrical power output. In practice, the reactive power of the generator is controlled by the power-electronic converter, and the wind power input is controlled by adjusting the pitch of the rotor blades.

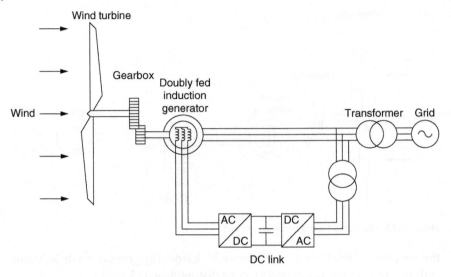

Figure 2.14 A wind turbine with a doubly fed induction generator.

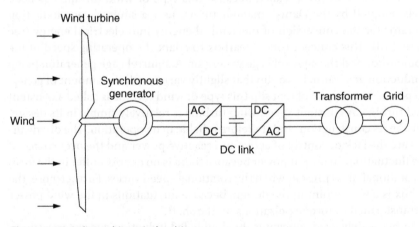

Figure 2.15 A direct drive wind turbine.

The latest wind turbine concept is the direct drive wind turbine, which does not need a gearbox (see Figure 2.15). A low-speed multipole synchronous generator, having the same mechanical speed as the turbine rotor, converts mechanical energy into electrical energy. The rotor of the synchronous generator can be either a wound rotor or a rotor with permanent magnets. The stator windings are not directly coupled to the grid but through a power-electronic converter. This can be a back-to-back voltage source converter or a diode rectifier with a single voltage source converter. This concept enables variable speed operation and the wind power is limited by pitch control.

Power electronics become cheaper and more reliable, and it can therefore be expected that variable speed wind turbines have the future. The energy output of variable speed wind turbines is higher than the energy output of constant speed wind turbines, because for each wind speed the optimal rotor speed can be achieved and this outweighs the losses of the power-electronic converter. The advantage of the variable speed wind turbine with the doubly fed induction generator over the direct drive wind turbine is that the power-electronic converter has a rating of one-third of the nominal power of the wind turbine, but a gearbox is still necessary, decreasing the reliability and increasing the cost. In the direct drive wind turbine concept, there is no gearbox, but the generator is larger and therefore more expensive, and the power-electronic converter necessarily has a higher rating.

2.4.2 Hydropower and Pumped Storage

Hydro is nowadays a mature technology and has been developed all around the world. In countries blessed with large rivers, hydro plays a major role in the generation of electrical energy. The Itaipú hydroelectric power plant in Brazil has a generating capacity of 12600 MW. Manitoba Hydro in Canada has a total generating capacity of 5200 MW. The abundant hydropower resources in southwest China will be exploited on a large scale during the next three decades. A number of large hydropower projects will be completed during this period. China ranks first in the world of hydro resources with a potential of 680 GW. An estimated 370 GW can be developed to provide electric energy production of 1900 TWh/year. Exploitable hydropower resources in southwestern China account for 53% of China's total [6].

The running costs of hydropower plants are very low as energy is free, but the civil engineering component of the capital cost is very high. The type of water turbine used depends on the head of water available. The energy is extracted from the water falling through the available head. The lower the head, the larger the quantity of water necessary for a given turbine rating. High-head turbines operate with low water volumes and low-head turbines with high water volumes.

There are three basic types of water turbines. The Kaplan turbine has variable pitch blades that can be adjusted for optimum regulation (see Figure 2.16). This type of turbine has been built for heads up to 60 m and is applied in river and pondage stations. In the Francis turbine the guide and runner blades are designed for higher heads up to 500 m, and the regulation is done by adjustment of the guide vanes at the water inlet (see Figure 2.17). The third type is the Pelton wheel, a turbine suitable for high heads and small quantities of water. The water is injected through one or more nozzles on the buckets on the turbine wheel. The regulation is done by means of needle valves on the nozzles and jet deflectors at the water inlet (see Figure 2.18).

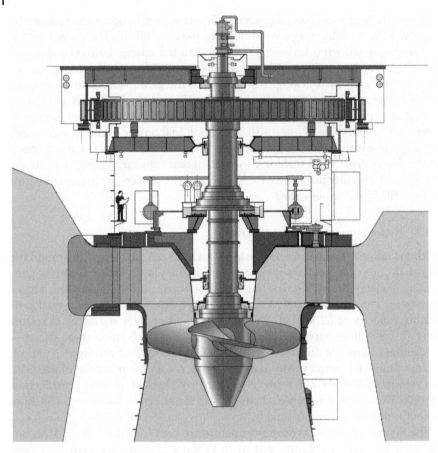

Figure 2.16 The Kaplan water turbine. Reproduced with permission of Voith Siemens Hydro Power Generation GmbH & Co. KG.

In comparison with a thermal power plant, a hydropower plant has the advantage of quick starting and effectively meeting an abrupt increase in load demand. When water supplies are not available, a pump storage scheme provides the advantages of a hydro plant. A pump storage scheme has two reservoirs at different heights, with the hydro plant situated at the level of the lower reservoir. During periods of low demand, usually at night, water is pumped from the lower to the higher reservoir using the cheap electricity from the thermal power plants. The generators act as motors and drive the turbines that act as pumps. During periods of peak demand, usually in daytime, the water turbine drives the generator in the normal manner. The overall efficiency of the operation is not very high, in the order of 65%. One of the larger pump storage plants in Europe is located in Vianden in Luxembourg.

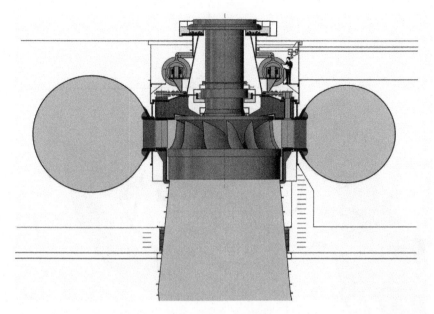

Figure 2.17 The Francis water turbine. Reproduced with permission of Voith Siemens Hydro Power Generation GmbH & Co. KG.

2.4.3 Solar Power

The sun is our source for light, heat, and food. Fossil fuels, wood, and crops are in fact an accumulation of solar energy. Also hydropower and wind power are in their origin forms of solar energy. But in the daily conversation, when we speak about solar energy, we mean the direct conversion from solar radiation into heat or electricity. In the case of heat production, we speak of thermal solar energy, and when the output is electricity, we call it photovoltaic (PV) energy. One of the most widespread uses of thermal solar energy is solar water heating. Water is pumped through a collector, being a black-painted body with tubes inside. The collector absorbs the solar radiation and transfers the heat to the circulating water. The water rises in temperature and the hot water is stored in a boiler. The boiler acts as a heat exchanger to heat up tap water. This works well for domestic and small industrial applications, and the utility bills are reduced.

The so-called concentrating solar power systems use the sun's heat to generate electricity. In California nine commercial concentrating solar power systems with a total generating capacity of 354 MW have been in operation since the mid-1980s. These systems consist of rows of highly reflective parabolic troughs, and each parabolic trough focuses and concentrates sunlight on a central tube with heat-absorbing fluid, which is used in a heat exchanger to produce steam. The steam expands in a steam turbine, which is the prime mover for a synchronous generator. Of similar design is the power

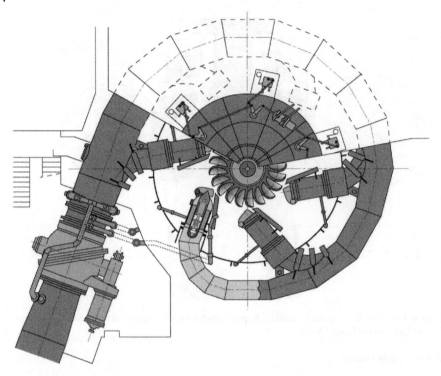

Figure 2.18 The Pelton wheel turbine. Reproduced with permission of Voith Siemens Hydro Power Generation GmbH & Co. KG.

tower, which can generate large amounts of electricity using a tall fluid-filled tower located at the focal point of a large field of mirrors (see Figure 2.19). Dish Stirling units use a small array of mirrors to focus sunlight on a Stirling engine, which produces from 5 to 50 kW of electricity for grid consumption or stand-alone applications [7].

PV technology transforms the energy of solar photons into direct current using semiconductor materials. The smallest unit is called a solar cell or a PV cell. When photons enter the solar cell, electrons in the semiconductor material are freed, generating a direct current. The most common semiconductor materials used in PV cell manufacturing are single-crystal silicon, amorphous silicon, polycrystalline silicon, cadmium telluride, copper indium diselenide, and gallium arsenide. The most important PV cell technologies are crystalline silicon and thin films, including amorphous silicon.

PV cells connected together and sealed form a PV module or panel (see Figure 2.20). Their size ranges from a few watts to around 100 W. Power-electronic converters convert the DC output of a PV panel to AC. Only a portion of the sunlight received by a PV cell is converted into electricity.

Figure 2.19 Power tower. Reproduced with permission of DOE/NREL.

PV technology has a wide range of applications:

Stand-alone off-grid systems are applied when it is too expensive to extend power lines to a remote area. They power, for instance, remote telecommunication systems, where reliability and low maintenance are the principal requirements. PV systems are also widely applied in developing countries to serve the substantial rural populations who do not have otherwise access to basic energy services.

Grid-connected systems in buildings are used when a public grid is available and a PV system can be connected to it. When more electricity is required than the PV system is generating, the need is automatically met by power from the grid. The owner of a grid-connected PV system may sell excess electricity production to the utility.

Large-scale PV power plants consisting of many PV arrays installed together can provide bulk electricity. Utilities can build PV plants faster than conventional power plants and can expand the size of the plant as the demand increases.

2.4.4 Geothermal Power

Geothermal technology uses the heat of the Earth to produce steam to drive a turbine. It is almost 6500 km from the surface to the center of the Earth, and the deeper you go, the hotter it gets. The outer layer of the Earth, the crust, is 5–55 km thick and insulates us from the hot interior. From the surface down through the crust, the normal temperature gradient (the increase

Figure 2.20 Solar panels at Aspen Mountain. Reproduced with permission of DOE/NREL.

of temperature with the increase of depth) in the Earth's crust is about 17–30 °C/km of depth. Below the crust is the mantle, consisting of highly viscous, partially molten rock with temperatures between 650 and 1250 °C. At the Earth's core, which consists of a liquid outer core and a solid inner core, temperatures may reach 4000–7000 °C. Since heat always moves from hotter regions to colder regions, the Earth's heat flows from its interior toward the surface. In some regions with high temperature gradients, there are deep subterranean faults and cracks that allow rainwater and snowmelt to seep underground, sometimes for several kilometers. There the water is heated by the hot rock and circulates back up to the surface, to appear as hot springs, mud pots, or geysers. If the ascending hot water meets an impermeable rock layer, however, the water is trapped underground where it fills the pores and cracks, forming a geothermal reservoir. Much hotter than surface hot springs, geothermal reservoirs can reach temperatures of more than 350 °C and are powerful sources of energy. If geothermal reservoirs are close enough to the surface, we can reach them by drilling wells, sometimes over 3 km deep.

Geothermal power plants consist of one or more thermal wells. A single thermal well usually ranges from 4 to 10 MW. To avoid interference, a spacing of 200–300 m between wells is kept. Since it is not practical to transport high-temperature steam over long distances by pipelines due to heat losses, most geothermal plants are built close to the resources. Three power plant technologies are being used to convert hydrothermal fluids to electricity. The type of conversion depends on the state of the fluid (steam or water) and on its temperature.

Dry steam power plants, schematically illustrated in Figure 2.21, use hydrothermal fluids primarily in the form of steam, which goes directly to a turbine, which drives a generator that produces electricity. This is the oldest type of geothermal power plant and was originally used at Larderello in 1904. It is a very effective technology and is used today at The Geysers in Northern California, the world's largest geothermal field.

Flash steam power plants use hydrothermal fluids above 175 °C. The fluid is sprayed into a tank (called the separator), which is held at low pressure. A part of the fluid vaporizes very rapidly (flashes) to steam, and the steam then drives the turbine. A schematic drawing of a flash steam power plant is shown in Figure 2.22.

Binary-cycle power plants use hot hydrothermal fluid with a temperature below 175 °C and a secondary fluid (hence the name binary cycle) with a much lower boiling point than water. The geothermal fluid delivers its heat to the secondary fluid in a heat exchanger. The secondary fluid evaporates very rapidly and drives the turbine. This principle is illustrated in Figure 2.23.

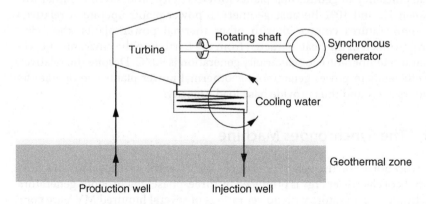

Figure 2.21 A dry steam power plant.

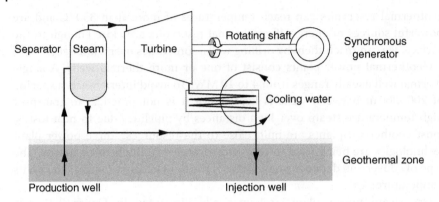

Figure 2.22 A flash steam power plant.

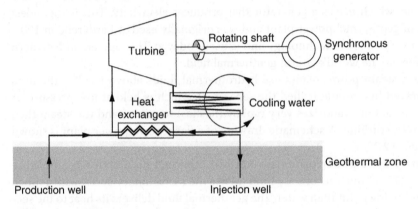

Figure 2.23 A binary-cycle power plant.

The efficiency of geothermal plants for electricity production is rather low, between 7% and 10%. Because geothermal power plants operate at relatively low temperatures compared with other thermal power plants, they eject nearly as much of the heat extracted from the Earth into the environment. The minimum temperature for electricity generation is 90 °C. Despite the relatively low efficiency in power generation, geothermal power plants can operate for 24 hours a day and thus provide base load capacity.

2.5   The Synchronous Machine

The workhorse for the generation of electricity is the synchronous machine. The bulk of electric energy is produced by three-phase synchronous generators. Synchronous generators with power ratings of several hundred MVA are common; the biggest machines have a rating up to 1500 MVA. Under steady-state conditions, they operate at a speed fixed by the power system frequency, and therefore they are called synchronous machines. As generators, synchronous machines operate in parallel in the larger power stations. A rating of 600 MVA is then quite common.

In a power plant the shaft of the steam turbine is mounted to the shaft of the synchronous generator. It is in the generator that the conversion from mechanical energy into electrical energy takes place. The two basic parts of the synchronous machine are the rotor and the armature or stator. The iron rotor is equipped with a DC-excited winding, which acts as an electromagnet. When the rotor rotates and the rotor winding is excited, a rotating magnetic field is present in the air gap between the rotor and the armature. The armature has a three-phase winding in which a time-varying EMF is generated by the rotating magnetic field.

Synchronous machines are built with two types of rotors: cylindrical rotors that are driven by steam turbines at 3000–3600 RPM (see Figure 2.24 (a)) and salient-pole rotors that are usually driven by low-speed hydro turbines (see Figure 2.24 (b)). In the cylindrical rotor the field winding is placed in slots, cut axially along the rotor length. The diameter of the rotor is usually between 1 and 1.5 m, and this makes the machine suitable for operation at 3000 or 3600 RPM. These generators are named turbo generators. A turbo generator rotor has one pair of poles. Salient-pole machines have usually more pair of poles, and to produce a 50 Hz or 60 Hz frequency, they can operate at a lower rotational speed.

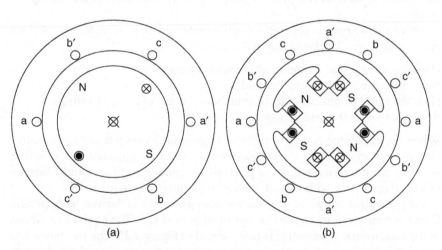

(a) (b)

Figure 2.24 Cross sections of an elementary three-phase generator with (a) a two-pole cylindrical rotor and (b) a four-pole salient-pole rotor. The black dot indicates that the positive current is directed out of plane of the paper. The cross indicates that the positive current is directed into plane of the paper.

The frequency of the EMF generated in the stator windings and the rotor speed are related by

$$f = \frac{np}{60} \tag{2.21}$$

f the electrical frequency [Hz]
n the speed [RPM]
p the number of pairs of poles (i.e., the number of poles divided by two)

Hydraulic turbines in hydro plants rotate at a few hundred RPM, depending on the type, and therefore need many pole pairs to generate 50 or 60 Hz.

The efficiency of the generators is very important. Synchronous generators in power plants have an efficiency of 99%. This means that for a 600 MW generator 6 MW heat is produced, and therefore the machine has to be cooled. Large turbo generators are cooled with hydrogen or water. Hydrogen has 7 times the heat capacity of air and water 12 times. The hydrogen and/or water flows through the hollow stator windings. Cooling equalizes the temperature distribution in the generator, because temperature hot spots affect the life cycle of the electrical insulation. When the temperature gradient is small, the average temperature of the machine can be higher, and this means that the generator can be designed for a higher output. The evolution of big turbo generators has been determined by better materials, novel windings, and sophisticated cooling techniques. Low-speed generators in hydro plants are always bigger than high-speed machines of equal power in thermal power plants, and a good air-cooling system with a heat exchanger usually does the job for low-speed machines.

Before a synchronous generator can be connected to the grid, four conditions must be satisfied. The generator voltage must:

Have the same phase sequence as the grid voltages
Have the same frequency as the grid
Have the same amplitude at its terminals as the one of the grid voltage
Be in phase with the grid voltage

When the generator is connected to a large grid, its output voltage and frequency are locked to the system values and cannot be changed by any action on the generator. We say that the generator is connected to an infinite bus: an ideal voltage source with a fixed voltage amplitude and frequency.

The equivalent circuit of the synchronous generator is derived in Appendix C and is connected to an infinite bus in Appendix C.5. The equivalent circuit of the synchronized generator is represented in Figure 2.25. The reactance X is called the synchronous reactance and is constant during normal steady-state conditions. The resistance of the stator coil is neglected in the equivalent circuit. Immediately after synchronization and coupling, the generator is neither feeding power to nor absorbing power from the grid. The steam going into the

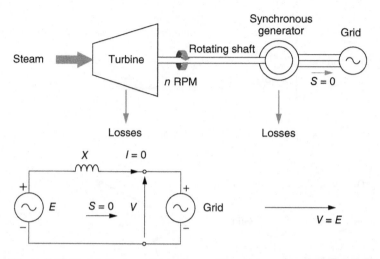

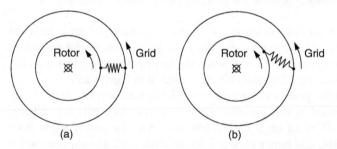

Figure 2.25 The equivalent circuit of a synchronous generator connected to an infinite bus with the corresponding phasor diagram; the generator is coupled to the grid, but there is no power exchange.

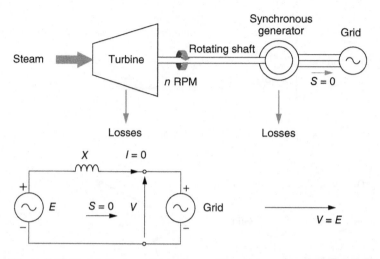

(a)   (b)

Figure 2.26 Generator and grid represented as two objects tied together by an elastic spring; (a) no power exchange between the generator and the grid (the situation of Figure 2.25); (b) the generator injects power into the grid (the situation of Figure 2.27).

turbine is the same as before the coupling action and is just enough to drive the rotor and compensate for the losses of the turbine and the generator.

If more steam is fed into the turbine, one might expect the generator to speed up, but this is not possible because the generator is connected to an infinite bus. It is much like having two objects tied together via an elastic spring. The infinite bus is one end of the spring and moves with a constant speed, and the synchronous generator is connected to the other end of the spring as shown in Figure 2.26 (a) for the situation where there is no power exchange between the generator and the grid. What happens if more steam is fed into the turbine is shown in Figure 2.26 (b). The torque supplied to the generator axis by the prime mover tries to speed up the generator but can only extend the spring a bit as it

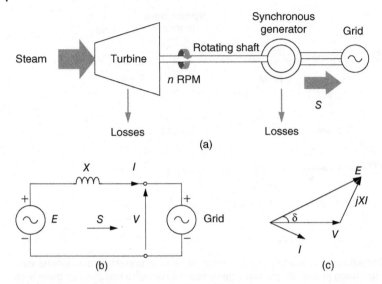

Figure 2.27 The equivalent circuit of a synchronous generator connected to an infinite bus with the corresponding phasor diagram; the generator injects power into the grid.

is tied to the infinite bus, and all the extra energy is transformed into electrical energy: the machine injects power into the grid. In relation to the extension of the spring, the internal EMF E now leads the terminal voltage; in a way, the internal EMF "drags along" the grid voltage as shown in Figure 2.27.

The amplitude of the generator internal EMF E is a function of the field current I_f and is controlled by the operator of the generator. In Figure 2.27 E leads the system voltage V by an angle δ, which is known as the power angle. As a consequence, current, and hence power, is fed into the grid. The expression for the current is

$$I = \frac{E - V}{jX} \tag{2.22}$$

I the current phasor: $I = |I|\angle{-\varphi}$ [A]
E the machine internal EMF phasor: $E = |E|\angle\delta$ [V]
V the system voltage phasor: $V = |V|\angle0$ [V]
X the machine reactance [Ω]

The three-phase complex power supplied to the power system equals

$$S = 3VI^* = P + jQ \tag{2.23}$$

Expressions for the active and reactive power can be found by either eliminating V or eliminating I from Equation 2.23. If we eliminate V, the three-phase complex power supplied to the power system can be written as

$$S = 3VI^* = 3(E - jXI)I^* = 3EI^* - 3jXII^* = 3|E||I|\angle(\delta + \varphi) - 3jX|I|^2$$
$$= 3|E||I|(\cos(\delta + \varphi) + j\sin(\delta + \varphi)) - 3jX|I|^2 = P + jQ \quad (2.24)$$

The real part of the complex power S is the active power:

$$P = \text{Re}(S) = 3|E||I|\cos(\delta + \varphi) \quad (2.25)$$

and the imaginary part is the reactive power:

$$Q = \text{Im}(S) = 3|E||I|\sin(\delta + \varphi) - 3X|I|^2 \quad (2.26)$$

If we eliminate I, the three-phase complex power supplied to the power system can be written as

$$S = 3VI^* = 3V\left(\frac{E - V}{jX}\right)^* = \frac{3VE^* - 3VV^*}{-jX}$$
$$= 3j\frac{|V||E|\angle(-\delta)}{X} - 3j\frac{|V|^2}{X} \quad (2.27)$$
$$= 3j\frac{|V||E|}{X}(\cos(\delta) - j\sin(\delta)) - 3j\frac{|V|^2}{X} = P + jQ$$

The real part of the complex power S is the active power:

$$P = \text{Re}(S) = 3\frac{|V||E|}{X}\sin(\delta) \quad (2.28)$$

and the imaginary part is the reactive power:

$$Q = \text{Im}(S) = 3\frac{|V||E|}{X}\cos(\delta) - 3\frac{|V|^2}{X} = 3\frac{|V|}{X}(|E|\cos(\delta) - |V|) \quad (2.29)$$

A closer look at the active power equation (Equation 2.28) shows us that the sign of the active power is determined only by the power angle δ:

$\delta > 0 \rightarrow P > 0$, the machine supplies active power to the grid and is a generator.
$\delta = 0 \rightarrow P = 0$, the machine has no active power exchange with the grid.
$\delta < 0 \rightarrow P < 0$, the machine absorbs active power from the grid and is a motor.

Given those relations, it is evident that we can project an "active power axis" in the phasor diagram as shown in Figure 2.28. We can follow a similar approach by investigating the reactive power equation (Equation 2.29). This equation shows us that the sign of the reactive power is determined by the following relations:

$|E|\cos(\delta) > |V| \rightarrow Q > 0$, the machine supplies reactive power to the grid and is overexcited.
$|E|\cos(\delta) = |V| \rightarrow Q = 0$, the machine has no reactive power exchange with the grid.
$|E|\cos(\delta) < |V| \rightarrow Q < 0$, the machine absorbs reactive power from the grid and is underexcited.

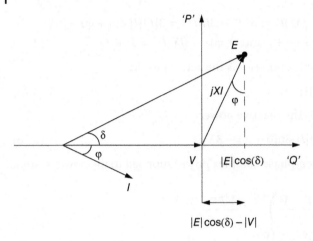

Figure 2.28 "Active and reactive axis" projected in the phasor diagram.

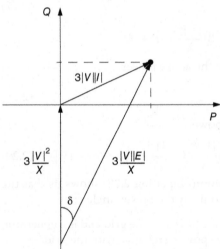

Figure 2.29 "Phasor diagram" projected in the three-phase active and reactive power coordinate system.

A "reactive power axis" can now be projected in the phasor diagram as well as that shown in Figure 2.28. The working point of the synchronous machine, of which the phasor diagram is shown in Figure 2.28, is indicated by the black dot. This machine injects both active and reactive power into the grid and is an overexcited generator.

When we want to project the voltage phasor diagram in a three-phase active and reactive power coordinate system, as shown in Figure 2.29, the phasors need to be multiplied by $3|V|/X$ as we can see from the active and reactive power equations (Equations 2.28 and 2.29). The working point of the synchronous machine is indicated by the black dot, and the amount of active and reactive power that the machine injects into the grid can be read from the diagram easily. The generator is of course limited in its output, and it is convenient to indicate the region for safe operation of the machine in the diagram, given a constant terminal voltage V – the so-called loading capability diagram of the generator.

The region of operation of the synchronous generator is restricted by the maximum permissible heating ($|I|^2R$ losses) in the armature and in the field windings. These two limits can be constructed for the "phasor diagram" that is shown in Figure 2.30. Let us assume that the lengths of the two phasors that indicate the working point (the black dot) correspond to the maximum allowed stator current ($|I|$) and the maximum allowed field current ($I_f \rightarrow |E|$) (note that both $|V|$ and X are constant). Now we can find the heating limits by rotating the two phasors around their point of origin so that circles are described in the diagram. This is illustrated in Figure 2.30. In the area confined by the circle pieces, the maximum permissible heating in the armature and in the field windings will not be exceeded.

The region of operation of the synchronous generator is restricted not only by the maximum permissible heating in the armature and in the field windings but also by some other factors, such as the steady-state stability limit. The active power varies sinusoidally with the angle δ (see Equation 2.28) as shown in Figure 2.31. The generator can be loaded to the limit value $P_{\max}$, known

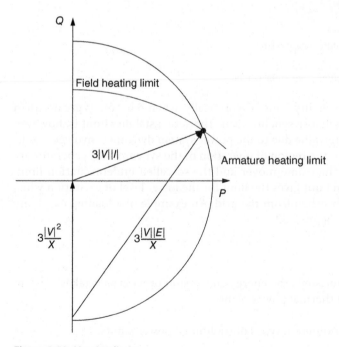

Figure 2.30 Heating limits.

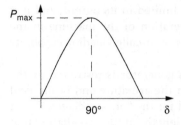

Figure 2.31 Active power of a synchronous generator as a function of the power angle δ.

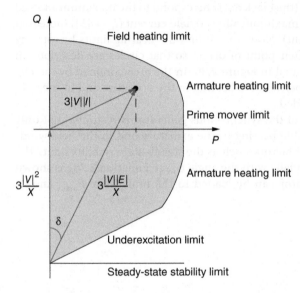

Figure 2.32 Example of a loading capability curve of a synchronous generator.

as the steady-state stability limit. Theoretically, when δ becomes greater than 90°, the generator will lose synchronism. The actual stability limit is, however, more difficult to determine due to the power system dynamics involved. Additional constraints on the region of operation of the synchronous generator are the power limit of the prime mover and the so-called underexcitation limit. The underexcitation limit gives the limit for the lower level of excitation when reactive power is absorbed from the grid. An example of a loading capability diagram is given in Figure 2.32.

Problems

2.1 Explain in four steps the energy conversion process as it takes place in conventional thermal power plants.

2.2 Mention four different types of traditional power plants.

2.3 Mention two different technologies for the generation of renewable energy and describe the energy conversion that takes place.

2.4 The following questions deal with geothermal power plants.
   a. What are common power ranges for a single thermal well?
   b. Why are geothermal power plants built close to the resources?
   c. What is the efficiency of a geothermal power plant for the generation of electricity?
   d. Are geothermal power plants used to supply base load capacity or are they used to match peak load?

2.5 Different power plant technologies are used to convert hydrothermal fluids to electricity. Describe the conversion, depending on the state of the fluid, for a:
   a. Dry steam power plant
   b. Flash steam power plant
   c. Binary-cycle power plant

2.6 What is the meaning of the expression "connected to an infinite bus?"

2.7 Draw for a synchronous generator, connected to an infinite bus, that injects power into the grid:
   a. A diagram including the turbine, the rotating shaft, and the synchronous generator
   b. An equivalent circuit
   c. A phasor diagram

2.8 What are the four conditions that must be satisfied before a synchronous generator can be connected to the grid?

2.9 A three-phase synchronous generator has a terminal voltage of 1 pu and is connected to an infinite bus by means of a reactance of 0.4 pu. The generator supplies an active power of 0.8 pu. The excitation voltage is 1.3 pu, and the machine reactance is 0.8 pu. Calculate the power angle δ of the generator.

2.10 A three-phase synchronous machine, with a synchronous reactance of 1.2 pu, is connected to an infinite bus by a reactance of 0.3 pu. The machine operates at its maximum power of 1.1 pu. The voltage of the infinite bus is $V = 1\angle0°$ pu. Calculate the excitation voltage of the synchronous machine.

2.11 Describe the operation of a synchronous machine for:
a. A power angle $\delta > 0$
b. A power angle $\delta = 0$
c. A power angle $\delta < 0$

2.12 What is, in theory, the maximum power angle at which a synchronous machine can be operated without losing synchronism?

References

1 Meadows, Donella H., Meadows, Dennis L., Randers, Jorgen, and Behrens III, William W.: *The Limits to Growth*, Universe Books, New York, 1972, ISBN 0-87663-165-0.
2 U.S. Energy Information Administration: *How much U.S. electricity is generated from renewable energy?*, U.S. Energy Information Administration, U.S. Department of Energy, Washington, May 5, 2016. http://www.eia.gov/energy_in_brief/article/renewable_electricity.cfm (Accessed on 3 January 2017).
3 ENTSO-E: *Statistical Factsheet 2015 - Provisional values as of 4 May 2016*, ENTSO-E, 2016.
4 Akhmatov, Vladislav: Analysis of dynamic behaviour of electric power systems with large amount of wind power, Ph.D. thesis, Technical University Denmark, Denmark, 2006, ISBN 87-91184-18-5.
5 Slootweg, Johannes G.: Wind power: modelling and impact on power system dynamics, Ph.D. thesis, Delft University of Technology, The Netherlands, 2006, ISBN 90-9017239-4.
6 Yang, H., Yao, G.: 'Hydropower development in Southwestern China', *IEEE Power Engineering Review*, Vol. **22**, Issue 3, March 2002, pp. 16–8.
7 Bull, S.R.: Renewable Energy Today and Tomorrow, *Proceedings of the IEEE*, Vol. **89**, Issue 8, August 2001, pp. 1216–1226.

3

The Transmission of Electric Energy

3.1 Introduction

One could raise the question: why do we have transmission and distribution networks at all? Imagine a world without overhead lines, without underground cables, and no transformers, in which each residential home and business enterprise generates and consumes its own electricity. The consequence of such a system would be that each and every customer should own and operate enough generating capacity to be able to supply the maximum possible load, whereas this generating capacity would not be used when the load was less. This situation exists on board of ships, aeroplanes, and so on. One could argue that it could be an option to store energy to supply a part of the peak load by the stored energy. However, to store energy in large quantities is still, if technologically spoken possible, very expensive. A better alternative is to interconnect one's system with those of neighbors. This is a good solution as long as the peak demands of the various customers do not occur at the same time.

Example 3.1 *Interconnection*
Two neighbors decide to interconnect their electricity systems. They each have a 5 kW generator installed to be able to run either the washing machine or the washer dryer together with other household loads. When they interconnect their systems, and plan their washing and drying on different days, they can cope with one generator of 5 kW instead of two.

In practice, the various customers' peak demands for electricity do not occur simultaneously, and therefore it is far more economical to pool the generation. The variation in peak demand allows us to install generators in an interconnected power system with a lower peak capacity, and so the capital expenditure per customer can be considerably lower than it would be in the case of a great number of isolated, independent power producers.

Electrical Power System Essentials, Second Edition. Pieter Schavemaker and Lou van der Sluis.
© 2017 John Wiley & Sons Ltd. Published 2017 by John Wiley & Sons Ltd.

The advantages of an interconnected power system are as follows [1]:

- It leads to a better overall system efficiency, because the total installed power can be less than the sum of the loads.
- It brings an improved system reliability, because when one of the partners in the power pool has a problem, for example unforeseen loss of generation power, the other partners can supply the missing generation.
- It results in smaller frequency deviations. In a large interconnected system, a great number of synchronous generators run in parallel. When there is a mismatch in balancing the production with the consumption during a short time, only a small frequency deviation in the system will occur because there is enough rotating mass in the system (see also Section 5.1).
- It covers a large geographical area. The European grid and the North American power pool, for example, span an area with multiple time zones: the peak load in the morning will start at different times in the various time zones. Hydropower can be connected, and nuclear and thermal power plants can be built where cooling water is present or where the fuel is available. Windmills can be erected at a location with a steady wind throughout the year (onshore or offshore). A geographical spread of the wind production will reduce the chance that no wind power is produced at all: if there is no wind in Denmark, it is not likely that at the same time this is also the case in Spain, for instance.
- It facilitates the dealing and wheeling of power. It becomes possible to exchange the power from the connected power producers and to create in this way a market for electrical energy.

An interconnected system also has some disadvantages, because:

- There must be sufficient interconnection transport capacity between the different partners in the power pool.
- There are power losses because of the energy exchanges.
- In large interconnected systems parallel transports can occur that put higher constraints on the system.
- There are organizational matters in order to operate the interconnected system smoothly.

Some of these disadvantages can be resolved by using DC links as interconnection between large supply areas or power pools (see also Section 1.3.1) or by applying flexible AC transmission system (FACTS) devices to control the flow of power (Section 5.6).

3.2 Transmission and Distribution Network

Electricity is transported and distributed by overhead transmission lines and underground cables (see also Section 3.9). The total amount of power and the

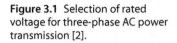

Figure 3.1 Selection of rated voltage for three-phase AC power transmission [2].

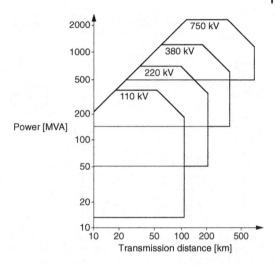

distance over which it has to be transported determine the basic design of the transmission and distribution system; the larger the distance and the more power to be transported, the higher the rated system voltage, as illustrated in Figure 3.1. The transmission network has the highest voltage rating and connects the largest power plants with the transmission substations (see Section 3.4), which in turn supply the distribution subsystems. The distribution network distributes the electrical energy between the connected loads. Sometimes we speak about sub-transmission systems (or regional transmission systems or primary distribution systems). It is not always clear where to draw the line.

The power system has developed over the years, voltage levels have increased, and the original function of certain system parts in the design has changed accordingly. In Figure 3.2, the voltage levels and the interconnecting transformers in the Netherlands are shown. The 380 and 220 kV networks form the backbone of the Dutch grid; they transport the bulk of the power. The 150, 110, and 50 kV networks also have a transport function and are therefore called sub-transmission networks. The 10 and 20 kV networks supply the low-voltage networks and distribute the electrical energy to the consumers. The transformation from 380 to 220 kV takes place in three substations with one or two transformer banks with a capacity of between 500 and 750 MVA. There are about 20 substations where the voltage is stepped down from 380 or 220 kV to 150 or 110 kV. Each substation has one to four power transformer banks with a capacity of between 200 and 500 MVA. Approximately 40 substations bring the voltage down to 50 kV with power transformers of 100 MVA, and roughly 180 substations directly step the voltage down to the 10 and 20 kV distribution level with power transformers ranging from 20 till 80 MVA. The northeastern part of the Netherlands, less densely populated than the rest of the country,

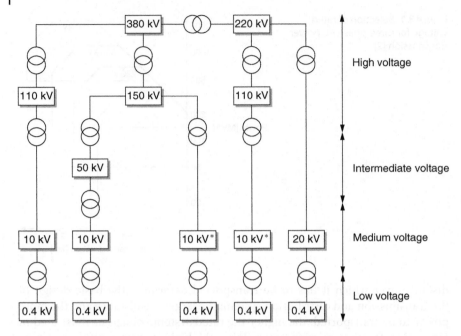

Figure 3.2 Voltage levels and transformation steps in the Dutch power system; *this voltage level can be 20 kV as well.

has locally concentrated loads and the voltage is transformed directly from the 220 kV transmission level to the 20 kV distribution level.

The average distance between the 380 kV substations is 50 km and between the 220 kV stations is 35 km. The substations at the intermediate voltage levels 50, 110, and 150 kV lie 10–15 km apart, while the average distance between the 10 kV supply stations is a kilometer or less, depending on the local situation. Each 10/0.4 kV supply station serves between 50 and 100 clients (houses connected to the low-voltage network). A single high-voltage substation supplies between 250 and 500 10/0.4 kV supply stations.

The Dutch high-voltage network is shown in Figure 3.3. The 380 and the 220 kV networks are loop systems, in which each and every substation is fed from two directions or from two sources. The 380 kV loop transports the electrical energy to the western, southern, and central part of the Netherlands, and the 220 kV loop serves the northeastern part of the country. The larger power plants feed directly into the 380 kV grid, while smaller power plants are connected at intermediate- and medium-voltage levels.

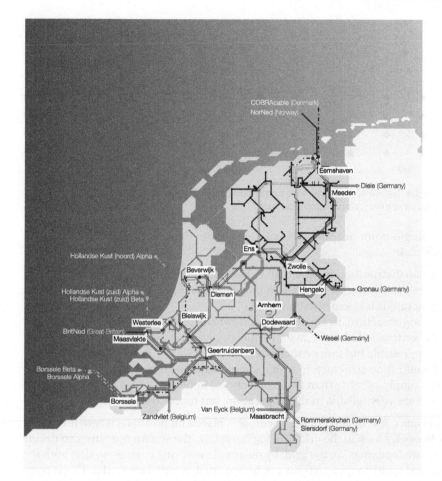

Figure 3.3 The Dutch high-voltage network. Reproduced with permission of TenneT TSO B.V.

3.3 Network Structures

The network structure is formed by the overhead lines, the underground cables, the transformers, and the buses between the points of power injection and power consumption. The number of voltage transformations from the highest voltage level to the lowest voltage level determines the principle network structure of a power system. Network structures can be distinguished in system parts with single-point feeding and with multiple-point feeding.

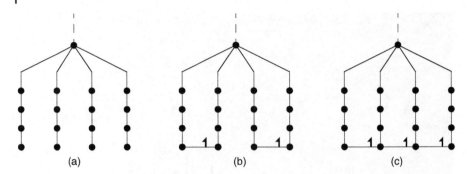

Figure 3.4 Network structures with single-point feeding; (a) radial structure; (b) loop structure; (c) multi-loop structure.

A single-point feeding network can have three layouts, as depicted in Figure 3.4:

- A radial structure, in which all substations (or consumers) are fed by lines or cables connected directly to one central supply; a network with a radial structure is less expensive to build.
- A loop structure, in which each and every substation (or consumer) within the system is fed from two directions; networks with a loop structure are more reliable but more expensive to build.
- A multi-loop structure, in which the substations (or consumers) are fed from the supply by more than two connections; networks with a multi-loop structure are very reliable in their operation, but more costly.

In Figure 3.4 (b) and (c), the **1** symbol is placed on positions where it is possible to open a loop in the grid. During operation, the system operator can decide to create "openings" in the grid, by means of switching devices, so that both the loop and multi-loop structures can be operated as radial networks. This is common practice in the Dutch distribution networks; most of these networks have a (multi-)loop structure but are operated as radial networks as this keeps the protection of the network simple. After a fault situation, for example a short circuit that has been cleared, a grid opening can be "relocated" in order to change the network configuration and restore the energy supply as shown in Figure 3.5.

A multiple-point feeding network nearly always has a multi-loop structure (see Figure 3.6). Transmission networks are in general operated in a multi-loop structure, as the multi-loop network gives a rather high reliability of the power supply. In the case that a fault occurs in a multi-loop system, the power supply can (usually) be continued. Let us imagine that the left feeder (the dashed line) in Figure 3.6 is short-circuited. The feeder will be isolated from the network (by the protective devices and the circuit breakers), which implies that the power supply from the left side is interrupted. However, we still have a power supply from the right side, which feeds all the substations in the multi-loop structure.

Figure 3.5 Restoration of energy supply in a faulted, radially operated system.

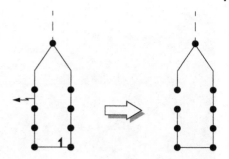

Figure 3.6 Network structure with multiple-point feeding.

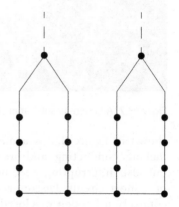

3.4 Substations

The simplest way to look at the power system is to consider it as a collection of nodes, which we call substations, and connecting power carriers, such as overhead lines and underground cables. By means of substations, the power of a generating plant can be supplied to the system, the power can be divided over the connected lines, and the power can be distributed to the consumers. Furthermore, transformers can be installed in the substations in order to interconnect different voltage levels. Substations play an important role in the protection of the power system. In the substations, the protection equipment (voltage transformers, current transformers, and protective relays; see Section 3.6) is installed together with the circuit breakers and disconnectors that perform the switching operations. Also the system grounding is established in the substations, and from the individual stations, measurement signals are guided to the control center (see also Section 6.1).

A substation basically consists of a number of incoming and outgoing power carriers that are connected to one (or more) common bus(es)/busbar(s) by circuit breakers, disconnectors, and instrument transformers: the feeders. In Figure 3.7, a photo of an open-air substation is depicted. A circuit breaker,

Figure 3.7 An open-air substation. Reproduced with permission of TenneT TSO B.V.

shown in Figure 3.8, is a mechanical switching device capable of not only making, conducting, and breaking currents under normal circuit conditions but also interrupting currents under abnormal conditions as in the case of a short circuit. Disconnectors are primarily used to visualize whether a connection is open or closed. An example of a pantograph disconnector is shown in Figure 3.9. Different from circuit breakers, disconnectors do not have current-interrupting capability. Therefore, a disconnector cannot be opened when it conducts a current and when a recovery voltage builds up across the contacts after opening. A disconnector can interrupt a small current when, after opening, a negligible voltage appears over the contacts. The instrument transformers in the substation, like voltage and current transformers, provide measured values of the actual voltage and current to the protective relay and the metering equipment. The protective relays have the task to detect and locate disturbances in the system, such as short circuits, and to switch off only the faulted part of the network by opening the appropriate circuit breaker(s).

When space is available, substations are erected in the open air. The ambient air serves as insulating medium and insulators support the live parts. These open-air substations do require quite some space but offer advantages: quick assembly and easy repair and expansion, and the possibility to install components of different manufacturers. When pollution is an issue, for example, the substation is planned close to an industrial area or in a coastal region, the substation can be placed indoors. If the available space is limited, the choice is made for an SF_6 gas-insulated station. In such a gas-insulated station, the live parts are located inside an earthed metal enclosure. Pressurized SF_6 gas serves as insulating medium in the enclosure. Pressurized SF_6 is a very good insulator

Figure 3.8 Three single-phase circuit breakers. Reproduced with permission of TenneT TSO B.V.

Figure 3.9 Pantograph disconnector in open position. Reproduced with permission of TenneT TSO B.V.

that can be used with electric field strengths that (at a pressure of 5 atm) are about 12 times higher than in atmospheric air. A gas-insulated substation that is filled with SF_6 gas requires only 20% of the space of a comparable open-air substation. In Figure 3.10, a feeder of an SF_6-insulated substation is depicted.

3.5 Substation Concepts

A substation basically consists of a number of incoming and outgoing feeders that are connected to one (or more) common bus(es)/busbar(s). The simplest

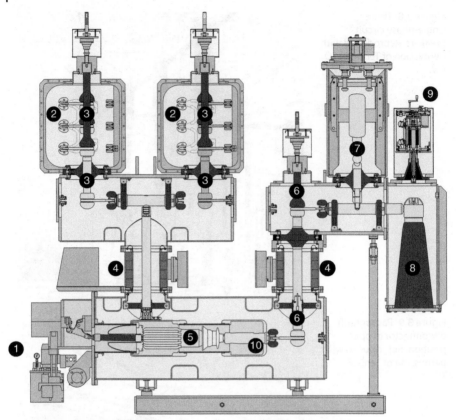

1. Interrupter driving mechanism
2. Busbars
3. Busbar disconnectors
4. Current transformer
5. Interrupter

6. Cable disconnector
7. Voltage transformer
8. Cable bushing
9. Earthing disconnector
10. Interrupter-pole enclosure

Figure 3.10 A feeder of an SF_6-insulated substation (E-SEP 245 kV). Reproduced with permission of Eaton Holec.

way to make the interconnection between the feeders is by connecting them to one single busbar. This layout is, however, rather vulnerable and can be improved in terms of security, maintenance, and flexibility. In the following sections, some of these improved substation concepts are presented.

3.5.1 Single Bus System

The single bus system with a bus disconnector has a simple layout (see Figure 3.11) and is easy to operate and rather simple to protect. This single bus system has, however, the disadvantage that:

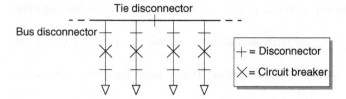

Figure 3.11 Single bus system.

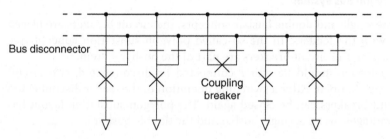

Figure 3.12 Double bus system.

- When maintenance has to be carried out at the tie disconnector, the complete substation has to be taken out of service.
- When the tie disconnector is closed, a bus fault causes disconnection of all feeders.
- When maintenance has to be done at a bus disconnector in one of the feeders, at least two feeders have to be taken out of service.

When the tie disconnector, which is shown in Figure 3.11, is replaced by a tie circuit breaker with disconnectors in series, the risk of disconnecting all feeders in the case of a bus fault is smaller, but the complexity, which was the original advantage of the single bus system, increases.

3.5.2 Double Bus System

The double busbar system is an often applied concept (see Figure 3.12). Each feeder has two bus disconnectors that can connect the feeder with either one of the two busbars. Both buses can be coupled by a coupling breaker.

In the case of a bus fault, the coupling breaker and the circuit breakers of the feeders that are connected to the faulted bus are tripped. Now the main advantage of the double bus configuration is that the supply can be restored by switching the disconnected feeders on the other (healthy) busbar. There are further advantages to the double bus system, namely:

- When maintenance has to be done at one of the two busbars, no feeders have to be taken out of service.

- There is a reasonably selective protection of the individual feeders and the busbars.
- The switchyard layout is fairly straightforward and its operation is not too complicated.
- Only a failing coupling circuit breaker can result in the disconnection of both busbars.

3.5.3 Polygon Bus System

In both the single and double busbar concepts, the circuit breakers are placed in series with the feeders. In the so-called polygon substation concept (see Figure 3.13), all the circuit breakers are part of the busbar system.

When a feeder should be either connected or disconnected, two circuit breakers have to open. After a switching operation of the feeder disconnector, both circuit breakers can be closed again. The polygon substation layout has three advantages over the single busbar and the double busbar concept:

- Maintenance on a circuit breaker can be carried out without disconnecting a feeder section and without an additional bypass.
- A bus fault results in disconnecting only a single feeder.
- A faulty circuit breaker puts only two feeders out of service, independent of the number of connected feeders.

3.5.4 One-and-a-Half Circuit Breaker Concept

The so-called one-and-a-half circuit breaker concept (see Figure 3.14) combines the reliability and the bypass possibility of the polygon substation concept while maintaining some of the flexibility of the double busbar substation. Compared with the double busbar system, each feeder has an extra, shared (half), circuit breaker available, but the coupling circuit breaker is missing.

Similarly to the polygon topology, two circuit breakers have to operate when a feeder has to be connected or disconnected. There are also more disconnectors (and earthing switches) needed. The result is a complicated switchyard layout and consequently less simple switching procedures. Advanced substation

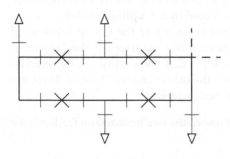

Figure 3.13 Substation layout according to the polygon concept.

Figure 3.14 The one-and-a-half circuit breaker concept.

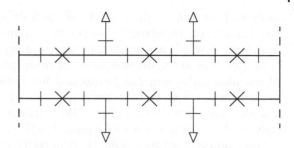

automation and protection equipment is therefore a requirement. The advantages of the one-and-a-half circuit breaker concept are the same as for the polygon concept:

- Circuit breaker maintenance with all feeders in service
- Only one feeder out of service in the case of a bus fault
- Only two feeders are lost when one of the station circuit breakers is out of service

3.6 Protection of Transmission and Distribution Networks

With the increasing dependency of our society on electricity supplies, the need to achieve an acceptable level of reliability, quality, and safety at an economic price becomes important to customers. The power system as such is well designed and also adequately maintained to minimize the number of faults that can occur. In normal operating conditions, a three-phase power system can be treated as a single-phase system when the loads, voltages, and currents are balanced (see also Section 1.3.3), and one can successfully use a single-line lumped-element representation of the three-phase power system for calculation. A fault brings the system to an abnormal condition. Short-circuit faults are especially of concern because they result in a switching action, which often results in transient overvoltages.

Line-to-ground faults are faults in which an overhead transmission line touches the ground because of wind, ice loading, or a falling tree limb. A majority of transmission line faults are single line-to-ground faults. Line-to-line faults are usually the result of galloping lines because of high winds (see also Section 3.9.1) or because of a line breaking and falling on a line below. Double line-to-ground faults result from causes similar to that of the single line-to-ground faults but are very rare. Three-phase faults, when all three lines touch each other or fall to the ground, occur in only a small percentage of the cases but are very severe faults for the system and its components. In the case of a symmetrical three-phase fault in a symmetrical system, one can still use a single-phase representation for the short-circuit and transient analysis. However, for the majority of the fault situations, the power system has become unsymmetrical. Symmetrical components and, especially, the sequence networks are an elegant way to analyze faults in unsymmetrical three-phase power systems because in many cases the unbalanced portion of the physical system can be isolated for a study, the rest of the system being considered to be in balance. Unbalanced systems and the method of symmetrical components applied to fault analysis are outside the scope of this book, and the interested reader is guided to [3].

Protection systems are installed to clear faults, such as short circuits, because short-circuit currents can damage the cables, lines, busbars, and transformers. The voltage and current transformers provide measured values of the actual voltage and current to the protective relay. The relay processes the data and determines, based on its settings, whether or not it needs to operate a circuit breaker in order to isolate faulted sections or components. The classic protective relay is the electromagnetic relay, which is constructed with electrical, magnetic, and mechanical components. Nowadays, computerized relays are taking over as they have many advantages: computerized relays can perform a self-diagnosis, they can record events and disturbances in a database, and they can be integrated in the communication, measurement, and control environment of the modern substations.

A reliable protection is indispensable for a power system. When a fault or an abnormal system condition occurs (such as over-/undervoltage, over-/underfrequency, overcurrent, and so on) the related protective relay has to react in order to isolate the affected section while leaving the rest of the power system in service. The protection must be sensitive enough to operate when a fault occurs, but the protection should be stable enough not to operate when the system is operating at its maximum rated current. There are also faults of a transient nature, a lightning stroke on or in the vicinity of a transmission line, for instance, and it is undesirable that these faults would lead to a loss of supply. Therefore, the protective relays are usually equipped with auto-reclosure functionality. Auto-reclosure implies that the protective relay, directly after having detected an abnormal situation leading to the opening of the contacts of the circuit breaker, commands the contacts of the circuit breaker to close again in order to check whether the abnormal situation is still there. In case of a fault of a transient nature, the normal situation is likely to be restored again so that there is and was no loss of supply. When the abnormal situation is still there, the protective relay commands the circuit breaker to open its contacts again so that either the fault is cleared or consecutive autoreclosure sequences can follow. In most cases, the so-called backup protection is installed in order to improve the reliability of the protection system.

When protective relays and circuit breakers are not economically justifiable in certain parts of the grid, fuses can be applied. A fuse combines the "basic functionality" of the current transformer, relay, and circuit breaker in one very simple overcurrent protection device. The fuse element is directly heated by the current passing through and is destroyed when the current exceeds a certain value, thus leading to an isolation of the faulted sections or components. After the fault is repaired/removed, the fuse needs to be replaced so that the isolated grid section can be energized again.

3.6.1 Protective Relay Operating Principles

The power system comprises many expensive components and so the complete power system represents a large capital investment. The system must therefore be utilized as much as possible in a secure and reliable way, but safety is at all times number one. No matter how well designed, faults will always occur, and they are a risk to life and property. The protection has to detect a fault or an abnormal operation condition of the system. The most common parameters, which give information about the presence of a fault, are the voltages and currents at the terminals of the protected component or the protected zone. Also information about the state of circuit breakers and switches, whether they are open or closed, can be used.

A protection system must use the appropriate input signals, process these input signals, come to the conclusion that a problem exists (and where it exists), and then initiate an action. It is clear that the protection should be able to distinguish between normal and abnormal system conditions. The destructive power of a fault arc, carrying a high current, is very large. The arc can burn copper conductors or weld together the laminations in a transformer core in a time span of tens or hundreds of milliseconds. Apart from the fault arc itself, heavy fault currents cause strong forces between line and bus conductors and at the terminals of the transformers and generators, if they continue for more than a few seconds. The provision of an adequate protection system to detect faults and to disconnect components from the system is therefore an integral part of power system design.

The protection system is the complete arrangement of protection equipment and other devices that is required to activate a specific function based on a protection principle as is specified in the standards. A collection of protection devices, such as relays and fuses, is called protection equipment. A set of protection equipment that provides a defined function and that includes relays, current transformers, circuit breakers, and so on is called protection scheme.

There are different types of protective relays that respond to various functions of the power system quantities. For example, only observing the current

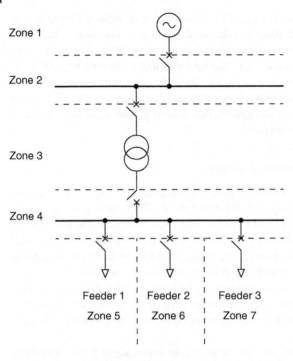

Figure 3.15 Division of power system into protection zones.

Zone 1

Zone 2

Zone 3

Zone 4

Feeder 1   Feeder 2   Feeder 3

Zone 5   Zone 6   Zone 7

amplitude might be sufficient in some cases, but in other cases measuring the transmitted power or the impedance could be necessary.

To limit the extent of the power system that is disconnected when a fault occurs, the protection is arranged in zones (see Figure 3.15). In the ideal case the zones should overlap (see Figure 3.16). This means that the circuit breaker is included in both zones as is depicted in Figure 3.17 (a): one current transformer is connected with the feeder protection, and the other with the busbar protection. In practice current transformers are installed on one side of a circuit breaker. This situation is shown in Figure 3.17 (b). The current transformers are now both at the feeder side of the breaker. A fault at F would cause the busbar protection to operate and to open the breaker, but the fault may continue to be fed through the feeder if the feeder protection is of the type that responds only to faults in its own zone, and it will not operate because the fault is outside its zone.

In general, currents increase in amplitude and voltages go down during a short circuit. Besides these changes in amplitudes, not only phase angles of current and voltage phasors can change but also harmonic components, active and reactive power, and the system frequency. Relay operating principles are based on detecting these changes and on identifying in which zone the fault occurs.

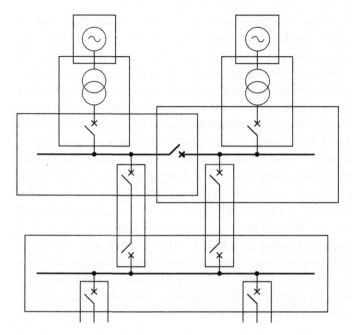

Figure 3.16 Overlapping zones of protection systems.

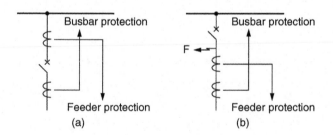

Figure 3.17 Locations of the current transformers.

   Level detection, or discrimination by current magnitude, is the simplest relay operating principle. The amplitude of the fault current is almost always greater than the amplitude of the load current during normal system operation. Faults in different parts of the power system will cause fault currents of different magnitudes on account of the different impedances between the source of supply and the point of fault. When the controlling relays of the various circuit breakers are set to trip at suitable set current values, the breakers near the fault will trip and will leave other breakers, traversed by the same fault current, undisturbed. The supply will be maintained to those parts of the system that are healthy. Such a relay is called an overcurrent relay. The level above which the overcurrent relay operates is named the pickup setting of the relay. For currents above the pickup value, the relay operates, and the relay takes no action for currents smaller than the pickup value. It is also possible to arrange the relay to operate for values smaller than the pickup value and to take no action for values above the pickup level. An example of such a relay is the undervoltage relay.

Instead of discrimination by current magnitude, an alternative is discrimination by time. In this the basic idea is to add time lag features to the controlling relays of a number of circuit breakers in the power system, so that the breaker(s) nearest to a fault on the system always trips first. Let us have a look at the simple radial line in Figure 3.18. The line passes through transformer stations A, B, C, and D, at each of which a breaker is installed at the outgoing line. The breakers are identical, and the protection setting is such that they will trip when the current through the line exceeds 2000 A. If there is a fault on line section CD, and the fault current exceeds 2000 A, the breakers at transformer stations A, B, and C will open, and all supply beyond A will be lost.

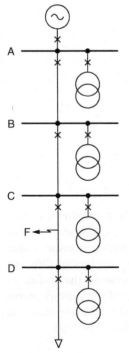

Figure 3.18 A simple radial line.

Suppose, instead, that we are able to add time lag features to the control of the breakers. After that the protection recognizes that a fault occurs the tripping is delayed:

D – no added delay
C – 0.4 seconds added delay
B – 0.8 seconds added delay
A – 1.2 seconds added delay

In this case, after the occurrence of a fault in line section CD, the breaker at C will open after 0.4 seconds and will disconnect the fault before the breakers at A and B can trip. The power supply to transformer stations A, B, and C is maintained. A time step interval in the order of 0.4 seconds is necessary to give the circuit breaker and its protective relays time to operate fully before the next breaker, with the longer time, can receive an impulse to trip.

The methods of discrimination by current magnitude and by discrimination in time have limitations, in that they are applicable only to simple power systems, and in the case of time discrimination, faults near the source, which are more severe, are held on the system for relatively long periods.

These limitations can be reduced if the relay is able to measure the distance from the breaker to the fault location. If this distance is less than that to the next circuit breaker out from the source, the fault is within the section controlled by the breaker concerned, and this breaker will trip. If the distance is greater than that to the next circuit breaker out from the source, the fault is beyond the section controlled by the breaker concerned, and it will not trip. For the simple radial line in Figure 3.18, the breaker at transformer station A will trip only for a fault whose measured distance is less than A to B, the breaker at transformer station B will trip only for a fault whose distance is less than B to C, and so on.

Power systems are complex, and this complexity means that the discriminative methods so far outlined may not be adequate. The current arising from a fault traverses non-faulted circuits, and this current tends to cause the protection of these circuits to operate. It is virtually impossible to achieve a correct discrimination under all conditions, for example, when there are parallel paths for the faulted current. This explains the need for a protection that is limited in its scope to one system element and that will cause disconnection of this element if it becomes faulted. This type of protection is called unit protection. Such a unit protection can be achieved by making use of the principle that current values at two (or more) points in the system can be continuously balanced by a suitable relay. This concept is called differential comparison and can be best understood by referring to the generator winding shown in Figure 3.19.

The current I'_1, entering one end must equal the current I'_2 leaving the other end. When a fault occurs between the two ends, the two currents are no longer equal, and the secondary current from the current transformers will not balance. The difference between the two secondary currents $(I_1 - I_2)$ will flow in

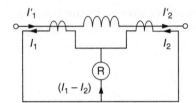

Figure 3.19 Differential comparison protection applied to a generator winding.

the relay. The drawback of the differential relay is that it requires currents from the extremities of a zone of protection, which restricts its application to components such as transformers, generators, motors, buses, capacitors, reactors, and so on.

3.6.2 Fuses

The fuse is the oldest and simplest of all protective devices. The early pioneers, such as Michael Faraday, had observed that a wire could be fused by an electric current, and so the fuse was born. Primitive fuses consisted of an open wire between two terminals (e.g., Thomas Alva Edison used a lead wire), but soon improvement was sought and the wire was replaced by a strip, made up of different materials, usually of lower melting point than copper, such as zinc. But as the available power increased, the behavior of the fuse became increasingly violent until attempts were made to screen the fuse element by a tube, and fuse designers concentrated on how to limit the emission of flame from the ends of the tube. A useful degree of breaking capacity was achieved, but the introduction of the filled cartridge fuse marked the greatest advantage.

A fuse is a weak link in a circuit and as such has one important advantage over circuit breakers. Because the element in the fuse has a much smaller cross section than the cable it protects, the fuse element will reach its melting before the cable. The larger the current, the quicker the fuse element melts. The fuse interrupts a very large current in a much shorter time than a circuit breaker does – so short in fact that the current will be cut off before it reaches its peak value, which in a 50 Hz system implies operation in less than 5 ms, and serious overheating and electromechanical forces in the system are avoided. This current-limiting action is an important characteristic that has application in many industrial low-voltage installations. The single-shot feature of a fuse requires that a blown fuse has to be replaced before service can be restored. This means a delay, the need to have a spare fuse and qualified maintenance personnel who must go and replace the fuse in the field. In a three-phase circuit, a single-phase-to-ground fault will cause one phase to blow and the other two phases stay connected.

Fuses for high-voltage applications require a high breaking capacity. The cartridge is made up of tough material, usually ceramic, and the cartridge contains, apart from the fuse element, a filler, such as powdered quartz, as is shown in

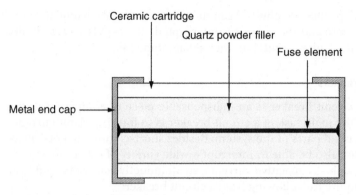

Figure 3.20 Current-limiting fuse link.

Figure 3.20. The purpose of the quartz filler is to condense as quickly as possible the metal vapor that is produced when a large overcurrent blows the fuse element. The filler prevents a dangerous pressure rise in the hermetically sealed enclosure. The filler should be neither too fine nor too coarse: an intermediate grain size provides the optimum cooling. The main classification of fuses is into current-limiting and non-current-limiting types.

Current-limiting fuses describe a class of fuses defined by the behavior that occurs when the current is so high that the fuse element melts before the peak of the fault current. The element is heated so rapidly that there is no time for heat loss to the surroundings; there is then a uniform temperature along the element, and all parts reach their melting temperature simultaneously. The wire thus becomes a liquid cylinder that becomes unstable and breaks up into a series of droplets. The fuse element has been replaced by a line of globules. The current is constrained to pass through them, because the surrounding filler has a nearly infinite resistance. In consequence, the voltage drop at the beginning of arcing is considerable and causes rapid suppression of the current. The number of globules per centimeter is about 10–12, and as the voltage drop in a short arc is about 20 V, the voltage drop across the element is approximately 200–250 V/cm. The maximum voltage rise across a fuse depends on the length and the design of the fuse element. Upon melting, this type of fuse introduces resistance in the circuit so rapidly that the current stops rising and instead is forced quickly to zero, before a natural current zero would occur. The fuse limits the current in magnitude as well as in duration, hence the name current limiting. The current-limiting fuse introduces an overvoltage, called the fuse switching voltage, into the system during the current-limiting action.

Non-current-limiting fuses or expulsion fuses melt under the same circumstances but add only a small resistance into the circuit, so that the current continues to about the same peak as would occur if the fuse had not melted. An expulsion action (i.e., where gas is generated by the arc and expelled along with ionized material) produces a physical gap such that, at natural current zero, the arc does not reignite and the current is interrupted. The expulsion fuse limits the duration of the fault current, but not its magnitude [4].

3.6.3 Circuit Breakers

A high-voltage circuit breaker is an indispensable piece of equipment in the power system. The main task of a circuit breaker is to interrupt fault currents and to isolate faulted parts of the system. Besides short-circuit currents, a circuit breaker must also be able to interrupt a wide variety of other currents at system voltage such as capacitive currents, small inductive currents, and load currents. We require the following from a circuit breaker:

- In closed position it is a good conductor.
- In open position it behaves as a good isolator between system parts.
- It changes in a very short period of time from close to open.
- It does not cause overvoltages during switching.
- It is reliable in its operation.

The electric arc is, except from power semiconductors, the only known element that is able to change from a conducting to a nonconducting state in a short period of time. In high-voltage circuit breakers, the electric arc is a high-pressure arc burning in oil, air, or sulfur hexafluoride (SF_6). In medium-voltage breakers, more often, the low-pressure arc burning in vacuum is applied to interrupt the current. The current interruption is performed by cooling the arc plasma so that the electric arc, which is formed between the breaker contacts after contact separation, disappears. This cooling process or arc extinguishing can be done in different ways. Power circuit breakers are categorized according to the extinguishing medium in the interrupting chamber in which the arc is formed. That is the reason why we speak of oil, air-blast, SF_6, and vacuum circuit breakers.

In 1907, the first oil circuit breaker was patented by J. N. Kelman in the United States. The equipment was hardly more than a pair of contacts submersed in a tank filled with oil. It was the time of discovery by experiments, and most of the breaker design was done by trial and error in the power system itself. In 1956, the basic patent on circuit breakers employing SF_6 was issued to T. E. Browne, F. J. Lingal, and A. P. Strom. Presently the majority of the high-voltage circuit breakers use SF_6 as extinguishing medium. J. Slepian has done much to clarify the nature of the circuit breaker problem, because the electric arc proved to be a highly intractable and complex phenomenon. Each new refinement in experimental technique threw up more theoretical problems. The practical development of circuit breakers was, especially in the beginning, somewhat pragmatic, and design was rarely possible as deduction from scientific principles. A lot of development testing was necessary in the high-power laboratory. A great step forward in understanding arc–circuit interaction was made in 1939 when A. M. Cassie published the paper with his well-known equation for the dynamics of the arc, and then in 1943, O. Mayr followed with the supplement that takes care of the time interval around current zero. Much work was done afterward to refine the mathematics of those equations and to confirm their physical validity through practical measurements. It becomes clear that current interruption by an electrical arc is a complex physical process when we realize that the interruption process takes place in microseconds, the plasma temperature in the high-current region is more than 10000 K, and the temperature decay around current zero is about 2000 K/µs while the gas movements are supersonic. The understanding of the current interruption process has led to SF_6 circuit breakers capable of interrupting 63 kA at 550 kV with a single interrupting element.

3.6.4 The Switching Arc

The electric arc in a circuit breaker plays a key role in the interruption process and is therefore often addressed as switching arc. The electric arc is a plasma channel between the breaker contacts formed after a gas discharge in the extinguishing medium. When a current flows through a circuit breaker and the contacts of the breaker part, driven by the mechanism, the magnetic energy stored in the inductances of the power system forces the current to flow. Just before contact separation, the breaker contacts touch each other at a very small surface area, and the resulting high current density makes the contact material to melt. The melting contact material virtually explodes, and this leads to a gas discharge in the surrounding medium either air, oil, or SF_6. When the molecular kinetic energy exceeds the combination energy, matter changes from a solid state into a liquid state. When more energy is added by an increase in temperature and the van der Waals forces are overcome, matter changes from a liquid state into a gaseous state. A further increase in temperature gives the individual molecules so much energy that they dissociate into separate atoms, and if the energy level is increased even further, orbital electrons of the atoms dissociate into free moving electrons, leaving positive ions. This is called the plasma state.

Because of the free electrons and the heavier positive ions in the high-temperature plasma channel, the plasma channel is highly conducting, and the current continues to flow after contact separation. Nitrogen, the main component of air, dissociates into separate atoms ($N_2 \rightarrow 2N$) at approximately 5000 K and ionizes ($N \rightarrow N^+ + e$) above 8000 K. SF_6 dissociates into sulfur atoms and fluorine atoms at approximately 1800 K and ionizes at temperatures between 5000 and 6000 K. For higher temperatures, the conductivity increases rapidly. The thermal ionization, as a result of the high temperatures in the electric arc, is caused by collisions between the fast-moving electrons and photons and the slower-moving positively charged ions and the neutral atoms. At the same time, there is also a recombination process when electrons and positively charged ions recombine to a neutral atom. When there is a thermal equilibrium, the rate of ionization is in balance with the rate of recombination. The relation between the gas pressure p, the temperature T, and the fraction of the atoms that is ionized f is given by Saha's equation

$$\frac{f^2}{1-f^2}p = 3.16 \times 10^{-7} \cdot T^{5/2} \cdot e^{-eV_i/kT} \tag{3.1}$$

e the charge of an electron $(1.6 \times 10^{-19}\text{C})$
V_i the ionization potential of the gaseous medium [eV]
k the Boltzmann's constant $(1.38 \times 10^{-23}\,\text{J/K})$

Saha's relation is shown in graphical form for oxygen, hydrogen, and nitrogen and for the metal vapors of copper and mercury in Figure 3.21. The graph of Figure 3.21 shows that thermal ionization can be used to switch between a conducting state ($f = 1$) and an isolating state ($f = 0$). Because of the rather steep slope of the function between the temperature and the degree of ionization, reduction of the average kinetic energy level of the moving particles by cooling with cold gas is a very good way to bring the arc channel from a conducting to a nonconducting state. Equation 3.1 implies that because the temperature T cannot change instantaneously, it takes a certain amount of time before thermal equilibrium is reached after changing from the conducting to the nonconducting state. This conductivity time constant depends on both the ion–electron recombination speed and the particle velocity distribution. The time to reach a local molecule–atom velocity equilibrium is in the order of 10^{-8} s, and the

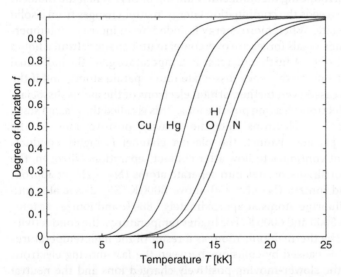

Figure 3.21 Degree of thermal ionization for some metal vapors and atomic gases.

time needed to reach a local electron velocity equilibrium is in the order of 10^{-10} s. The physical mechanisms that play a role in the electron–ion recombination process have time constants in the order of $10^{-7} - 10^{-8}$ s. This means that the time needed to reach ionization equilibrium is considerably shorter than the rate of change in the electrical phenomena from the power system during the current interruption period. For this reason, the circuit breaker arc can be assumed to be in a thermal ionization equilibrium for all electric transient phenomena in the power system.

3.6.5 Oil Circuit Breakers

Circuit breakers built in the beginning of the twentieth century were mainly oil circuit breakers. In those days, the breaking capacity of oil circuit breakers was sufficient to meet the required short-circuit level in the substations. Presently, oil and minimum-oil circuit breakers still do their job in various parts of the world, but they have left the scene of circuit breaker development. The first oil circuit breakers were of simple design – an air switch that was put in a tank filled with mineral oil. These oil circuit breakers were of the plain-break type, which means that they were not equipped with any sort of arc quenching device. In 1901, J. N. Kelman of the United States built an oil–water circuit breaker in this way, which is capable of interrupting 200–300 A at 40 kV. Kelman's breaker consisted of two open wooden barrels, each containing a plain-break switch. The two switches were connected in series and operated by one common handle. The wooden barrels contained a mixture of water and oil as extinguishing medium [5].

In the 1930s, the arcing chamber appeared on stage. The breaker, a metal explosion pot of some form, was fitted with an insulating arcing chamber through which the breaker contacts moved. The arcing chamber, filled with oil, fixes the arc, and the increase in pressure inside the arcing chamber improved the cooling effects on the arc considerably. Later, the design of the arcing chamber was further improved by pumping mechanisms, creating a cross flow of oil, giving extra cooling to the arc (Figure 3.22).

A next step in the development of oil circuit breakers was the minimum-oil circuit breaker. The contacts and arcing chamber were placed into a porcelain insulator instead of in a bulky metal tank. Bulk-oil circuit breakers with their huge metal tank containing hundreds of liters of mineral oil have been popular in the United States. Minimum-oil circuit breakers conquered the market in Europe.

3.6.6 Air-Blast Circuit Breakers

Air is used as insulator in outdoor-type substations and for high-voltage transmission lines. Air can also be used as extinguishing medium for current interruption. At atmospheric pressure, the interrupting capability, however,

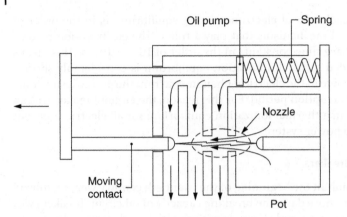

Figure 3.22 Cross section of an oil circuit breaker.

is limited to low-voltage and medium-voltage only. For medium-voltage applications up to 50 kV, the breakers are mainly of the magnetic air-blast type in which the arc is blown into a segmented compartment by the magnetic field generated by the fault current. In this way, the arc length, the arc voltage, and the surface of the arc column are increased. The arc voltage decreases the fault current, and the larger arc column surface improves the cooling of the arc channel. At higher pressure, air has much more cooling power, and air-blast breakers operating with compressed air can interrupt higher currents at considerable higher voltage levels. Air-blast breakers using compressed air can be of the axial-blast or the cross-blast type. The cross-blast type air-blast breaker operates similar to the magnetic-type breaker: compressed air blows the arc into a segmented arc-chute compartment. Because the arc voltage increases with the arc length, this is also called high-resistance interruption; it has the disadvantage that the energy dissipated during the interruption process is rather high. In the axial-blast design, the arc is cooled in axial direction by the airflow. The current is interrupted when the ionization level is brought down around current zero. Because the arc voltage hardly increases, this is called low-resistance interruption. When operating, air-blast breakers make a lot of noise, especially when the arc is cooled in the free air, as is the case with AEG's free-jet breaker (Freistrahlschalter) design.

3.6.7 SF$_6$ Circuit Breakers

The superior dielectric properties of SF$_6$ were discovered as early as 1920. It lasted until the 1940s before the first development of SF$_6$ circuit breakers began, but it took till 1959 before the first SF$_6$ circuit breaker came to the market. These early designs were descendants of the axial-blast and air-blast circuit breakers, the contacts were mounted inside a tank filled with SF$_6$ gas, and during the current interruption process, the arc was cooled by compressed SF_6 gas from a separate reservoir. The liquefying temperature of SF_6 gas depends on the pressure but lies in the range of the ambient temperature of the breaker. This means that the SF_6 reservoir should be equipped with a heating element that introduces an extra failure possibility for the circuit breaker; when the heating element does not work, the breaker cannot operate. Therefore, the puffer circuit breaker was developed, and the so-called double pressure breaker disappeared from the market. In the puffer circuit breaker (Figure 3.23), the opening stroke made by the moving contacts moves a piston, compressing the gas in the puffer chamber and thus causing an axial gas flow along the arc channel. The nozzle must be able to withstand the high temperatures without deterioration and is made from Teflon. Presently, the SF_6 puffer circuit breaker is the breaker type used for the interruption of the highest short-circuit powers, up to 550 kV–63 kA per interrupter.

Puffer circuit breakers require a rather strong operating mechanism because the SF_6 gas has to be compressed. When interrupting large currents, for instance in the case of a three-phase fault, the opening speed of the circuit breaker has a tendency to slow down by the thermally generated pressure, and the mechanism (often hydraulic or spring mechanisms) should have enough energy to keep the contacts moving apart. Strong and reliable operating mechanisms are costly and form a substantial part of the price of a breaker. For the lower voltage range, self-blast circuit breakers are now in the market. Self-blast breakers use the thermal energy released by the arc to heat the gas and to increase its pressure. After the moving contacts are out of the arcing chamber, the heated gas is released along the arc to cool it down. The interruption of small currents can be critical because the developed arcing energy is in that case modest, and sometimes a small puffer is added to assist in the interrupting process.

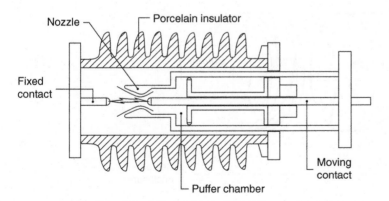

Figure 3.23 Operating principle of an SF_6 puffer circuit breaker.

In other designs, a coil carrying the current to be interrupted creates a magnetic field, which provides a driving force that rotates the arc around the contacts and thus provides additional cooling. This design is called the rotating arc circuit breaker. Both self-blast breakers and rotating arc breakers can be designed with less powerful (and therefore cheaper) mechanisms and are of a more compact design than puffer breakers.

3.6.8 Vacuum Circuit Breakers

Between the contacts of a vacuum circuit breaker, a vacuum arc takes care of the interruption process. The vacuum arc differs from the high-pressure arc because the arc burns in vacuum in the absence of an extinguishing medium. The behavior of the physical processes in the arc column of a vacuum arc is to be understood as a metal surface phenomenon rather than a phenomenon in an insulating medium. The first experiments with vacuum interrupters took place already in 1926, but it lasted until the 1960s when metallurgical developments made it possible to manufacture gas-free electrodes and when the first practical interrupters were built.

There are no mechanical ways to cool the vacuum arc, and the only possibility to influence the arc channel is by means of interaction with a magnetic field. The vacuum arc is the result of a metal-vapor/ion/electron emission phenomenon. To avoid uneven erosion of the surface of the arcing contacts (especially the surface of the cathode), the arc should be kept diffused or in a spiraling motion. The latter can be achieved by making slits in the arcing contacts (Figure 3.24) or by applying horseshoe magnets as used in the vacuum interrupters by Eaton

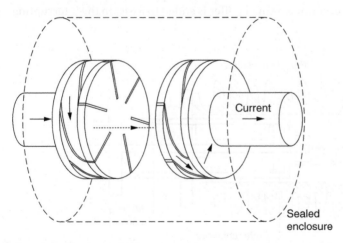

Figure 3.24 Vacuum interrupter with slits in the contacts to bring the arc in a spiraling motion.

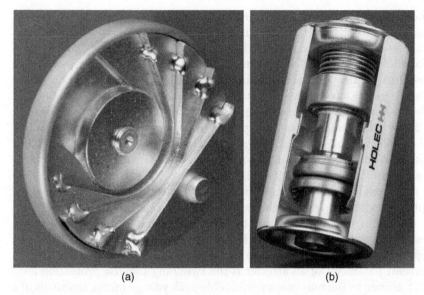

(a)   (b)

Figure 3.25 The use of horseshoe magnets as is done in the Eaton Holec interrupters. Reproduced with permission of Eaton Holec.

Holec (Figure 3.25). There is generally less energy required to separate the contacts of a vacuum circuit breaker, and the design of the operating mechanism usually results in reliable and maintenance-free breakers. Vacuum breakers are produced for system voltages up to 72.5 kV, and the short-circuit current rating goes up to 31.5 kA. When the arc current goes to zero, it does so in discrete steps of a few amperes to 10 A, depending on the contact material. For the last current step to zero, this can cause a noticeable chopping of the current. This current chopping in its turn can cause high overvoltages, in particular when the vacuum breaker interrupts a small inductive current, for example, when switching unloaded transformers or stalled motors.

3.7 Surge Arresters

Overvoltages, stressing a power system, can generally be classified into two categories regarding their origin:

- External overvoltages, generated by lightning strokes, which are the most common and severe atmospheric disturbances
- Internal overvoltages, generated by changes in the operating conditions of the network, such as switching

Surge arresters are placed in substations, and their job is to limit overvoltages to a specified protection level, which is in principle below the withstand voltage of the equipment. The ideal surge arrester would be one that starts to conduct at a specified voltage level; at a certain margin above its rated voltage, holds that voltage level without variation for the duration of the overvoltage; and ceases to conduct as soon as the voltage across the surge arrester returns to a value below the specified voltage level. Therefore, such an arrester would only have to absorb the energy that is associated with the overvoltage.

The design and operation of surge arresters has radically changed over the last 30 years from valve or spark-gap-type silicon carbide (SiC) surge arresters to the gapless metal-oxide (MO) or zinc-oxide (ZnO) surge arresters. Major steps in the development of surge arresters have been made, and modern surge arresters fulfill the present day requirements.

A metal-oxide surge arrester is essentially a collection of billions of microscopic junctions of metal-oxide grains that turn on and off in microseconds to create a current path from the top terminal to the Earth terminal of the arrester. It can be regarded as a very fast-acting electronic switch, which is open to operating voltages and closed to switching and lightning overvoltages. An important parameter of an arrester is the switching-impulse protection level (SIPL) defined as the maximum permissible peak voltage on the terminals of a surge arrester subjected to switching impulses under specific conditions.

In order to keep the power supplied to a metal-oxide arrester at the system operating voltage small, the continuous operating voltage of the arrester has to be chosen such that the peak value of the resistive-current component is well below 1 mA and the capacitive-current component is dominant. This means that the voltage distribution at operating voltage is capacitive and is thus influenced by stray capacitance. The voltage–current characteristic of the metal-oxide material offers the nonlinearity necessary to fulfill the mutually contradicting requirements of an adequate protection level at overvoltages and low current, that is, low energy dissipation at the system operating voltage. Metal-oxide surge arresters are suitable for protection against switching overvoltages at all operating voltages.

Traditionally, porcelain-housed metal-oxide surge arresters are used (see Figure 3.26). For satisfactory performance, it is important that the units are hermetically sealed for the lifetime of the arrester disks. The sealing arrangement at each end of the arrester consists of a stainless steel plate with a rubber gasket. This plate exerts a continuous pressure on the gasket against the surface of the insulator. It also serves to fix the column of the metal-oxide disks in place by springs. The sealing plate is designed to act as an overpressure relief system. Should the arrester be stressed in excess of its design capability, an internal arc is established. The ionized gases cause a rapid increase of the internal pressure, which, in turn, causes the sealing plate to open and the gases to flow out through the venting ducts. Since the ducts are directed toward each other, it results in an external arc, thus relieving the internal pressure and preventing a violent shattering of the insulator.

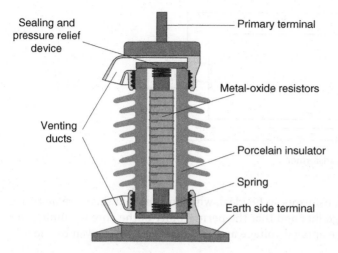

Figure 3.26 Porcelain-housed metal oxide surge arrester.

However, porcelain-housed distribution arresters have tended to fail due to problems with sealing. The benefits of a leak tight design, using polymers, have been generally accepted, leading to the changeover from porcelain to polymers. Polymer-housed arresters have a very reliable bond of the silicone rubber with the active parts. Hence, gaskets or sealing rings are not required. Should the arrester be electrically stressed in excess of its design capability, an internal arc is established, leading to rupture of the enclosure, instead of explosion. The arc will easily burn through the soft silicone material, permitting the resultant gases to escape quickly and directly. Hence, special pressure relief vents, with the aim of avoiding explosion of porcelain housing, are not required for this design. Moreover, polymer-housed distribution arresters are cheaper than porcelain-enclosed ones.

3.8 Transformers

Transformers are essential components in the AC power system as they make it possible to convert electrical energy to different voltage levels with an efficiency of more than 99%. That enables us to generate power at a relatively low voltage level (10–25 kV, limited by the insulation of the generator), to transport it at high voltage levels (110–420 kV and higher) to reduce the losses during transportation, whereas domestic consumption can take place at a low and (more or less) safe voltage level (400 V and below). Transformers consist essentially of two coils on a common iron core.

The iron core serves as the magnetic coupling between the two coils, such that nearly all the magnetic flux from one coil links with the other coil (Figure 3.27).

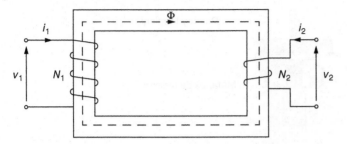

Figure 3.27 An ideal transformer.

If we assume the transformer to be ideal, which means that there are no resistance and no leakage flux and that the permeability of the core is infinite, the relation between the applied voltage and the resulting flux is given by the following expression:

$$v_1(t) = \sqrt{2}|V_1| \cos(\omega t) = N_1 \frac{d\Phi}{dt}$$

$$\Phi(t) = C + \frac{\sqrt{2}|V_1|}{\omega N_1} \sin(\omega t) = \sqrt{2}|\Phi| \sin(\omega t) \tag{3.2}$$

$$|\Phi| = |B|A$$

N_1 the number of turns of the primary transformer winding
Φ the magnetic flux $[\text{Wb} = \text{Vs}]$
C the integration constant $[\text{Wb}]$; zero in steady-state conditions
B the magnetic flux density $[\text{T} = \text{Wb/m}^2]$
A the cross-sectional area of the iron transformer core $[\text{m}^2]$

The induced voltage at the terminals of the secondary winding amounts to

$$v_2(t) = N_2 \frac{d\Phi}{dt} = \frac{N_2}{N_1} \cdot \sqrt{2}|V_1| \cos(\omega t) = \frac{N_2}{N_1} \cdot v_1(t) \tag{3.3}$$

N_2 the number of turns of the secondary transformer winding

As a result, the relation between the voltage at the primary side and the induced voltage at the secondary side can be written as (see also Appendix B.2)

$$\frac{v_1}{v_2} = \frac{V_1}{V_2} = \frac{N_1}{N_2} = n \tag{3.4}$$

n the turns ratio

How to select the number of turns for a given turns ratio? Let us examine two extreme cases in selecting the number of turns of the transformer windings.

Example 3.2 *Number of turns of the transformer windings*

In his book *Electric Energy: Its Generation, Transmission, and Use*, Laithwaite gives the following example [6] to illustrate the number of turns of transformer windings. Let us assume that we want a transformer that steps down from a primary voltage of 1000 V to a secondary voltage of 500 V, with a secondary current of 2 A. It is obvious that the turns ratio of the primary to secondary winding should be 2:1. But what number of turns to choose? Do we choose two primary turns and one secondary or two million primary turns versus one million secondary or something in between? In Equation 3.2, $|V| = 1000$ V and we take the maximum allowable value of the magnetic flux density to be 2.0 T: $|B| = 2.0/(\sqrt{2})$ T. If we look first at the design with two primary turns (i.e., $N_1 = 2$), we can calculate for the cross-sectional area of the iron transformer core:

$$|B|A = |\Phi| = \frac{|V|}{\omega N_1} = \frac{1000}{200\pi} = \frac{5}{\pi}\text{Wb} \rightarrow A = \frac{5/\pi}{\sqrt{2}} \approx 1.13\,\text{m}^2 \qquad (3.5)$$

The copper conductor required for the 2 A current at the secondary side must have a diameter of about 4 mm. So, our transformer, handling only 1 kVA, contains over 20 tons of steel (the iron core) and only 10 grams of copper (the windings)! If we consider the design with the two million primary turns (i.e., $N_1 = 2 \times 10^6$), the cross-sectional area of the iron transformer core becomes

$$A = \frac{5 \times 10^{-6}/\pi}{\sqrt{2}} \approx 1.13 \times 10^{-6}\,\text{m}^2 \qquad (3.6)$$

The steel core weighs only about 20 grams, but the primary and secondary copper windings have a mass of about 6 tons each! These two unpractical transformer "designs" are illustrated in Figure 3.28. It is evident that the market prices of steel and copper have their impact on the choice of the number of turns of the transformer windings and the transformer design.

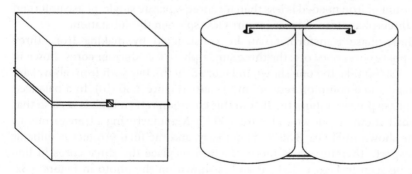

Figure 3.28 Two extreme transformer "designs."

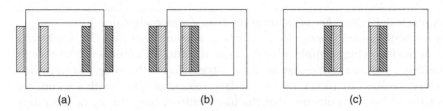

Figure 3.29 Single-phase transformer arrangements.

In Section 5.4.2, transformers are introduced that can alter the turns ratio so that they are able to adjust the voltage level and can be used for voltage control. A single-phase transformer can be built in a number of ways. Three common designs are shown in Figure 3.29. The arrangement with the separate windings, that is, the primary and secondary coils on a separate leg of the core (Figure 3.29 (a)), results in a rather high leakage flux, which can be reduced by putting the windings on top of each other (Figure 3.29 (b)). The magnetic circuit of this configuration is asymmetric: most of the leakage flux occurs on the open side. This can be circumvented by choosing the "shell"-type transformer as shown in Figure 3.29 (c). The flux divides evenly between the left and right limb, and therefore the cross-sectional area of those two outer limbs can be half that of the central limb.

Because the power system is a three-phase system, three-phase transformers are needed to realize the various voltage levels in the system. Such a three-phase transformer can be made by using three (identical) single-phase transformers. The advantage of this is that one creates redundancy in the system at a relatively low cost. When there are four identical single-phase transformers in a substation, one being a spare unit, this spare can be put into service quickly when one of the other transformers has a defect. A more common solution is the three-phase transformer, that is, one single three-phase unit, where the three phases share a common iron core. The advantage of such a single unit is that the amount of iron needed is less than for three separate single-phase units and that a three-phase transformer requires less space in the substation.

A three-phase transformer can be constructed by making from three single-phase cores one core: the three single-phase transformer cores shown in Figure 3.30 (a) (like the one shown in Figure 3.29 (b), but seen from above) are combined and a common return limb results (Figure 3.30 (b)). In a balanced three-phase power system, the flux in this common return limb is zero so that it can be left out, as shown in Figure 3.30 (c). Manufacturing a transformer as the one shown in Figure 3.30 (c) is not easy, and the final product is difficult to transport. Therefore, it is more practical to place the three cores in line as can be seen in Figure 3.31 (a) and is shown on the photo in Figure 3.32. Magnetically, this transformer arrangement is not symmetrical, but this can

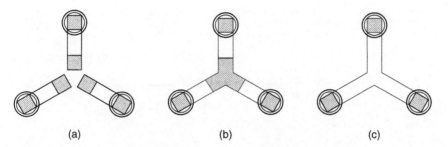

Figure 3.30 Construction of a three-phase transformer from three single-phase transformers (view from above).

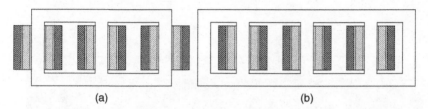

Figure 3.31 Three-phase transformer arrangements.

be overcome by adding two extra limbs at each side of the transformer core. This is illustrated in Figure 3.31 (b).

Many transformers that are in service have three windings per phase (instead of two); this is the so-called three-winding transformer, of which the third winding is named the tertiary winding. A single-phase three-winding transformer is shown schematically in Figure 3.33. The tertiary winding can be used as a point for reactive power injection (see also Section 5.4.3) or can supply a nearby distribution system. The tertiary winding can also be applied for harmonic suppression (more information on this topic is available in Section 3.8.2).

3.8.1 Phase Shifts in Three-Phase Transformers

The three coils at the primary or the secondary side of the three-phase transformer can be connected in wye (Y) or delta (D). An example of a Yy transformer (both the primary and the secondary sides are connected in wye) is shown in Figure 3.34. An example of a Yd transformer (primary side connected in wye and the secondary side in delta) is shown in Figure 3.35.

The advantage of wye-connected coils is that the star point, that is, the point where the three coils are connected to each other, can be earthed, so that the line-to-neutral voltages cannot become unacceptably high in case of system disturbances. Furthermore, in the case of wye-connected coils, the voltages across the coils equal the line-to-neutral voltages, whereas they equal the line-to-line voltages in the case of delta-connected coils; as a

Figure 3.32 A three-phase transformer under construction. Reproduced with permission of TenneT TSO B.V.

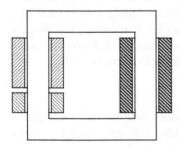

Figure 3.33 A single-phase three-winding transformer.

result, the voltage across a wye-connected coil is only $1/\sqrt{3}$ of the voltage across a delta-connected coil, so that the number of windings per phase can be smaller and the amount of insulation material needed can be less. Because of this, wye-connected coils are applied for the higher voltage levels. Delta-connected coils are advantageous when a Dy transformer serves as a distribution transformer: the single-phase loads, which are connected to the secondary wye-connected coils, "spread out" over two coils at the primary

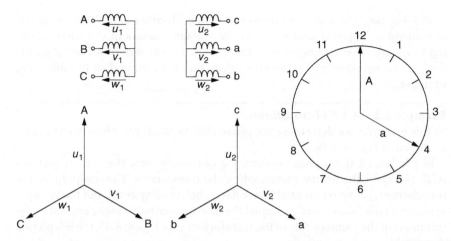

Figure 3.34 A Yy-4 transformer (the terminals of the secondary side of the transformer are labeled as c/a/b). When the terminals of the secondary side of the transformer are labeled as a/b/c, a Yy-0 transformer results. When the terminals of the secondary side of the transformer are labeled as b/c/a, a Yy-8 transformer results.

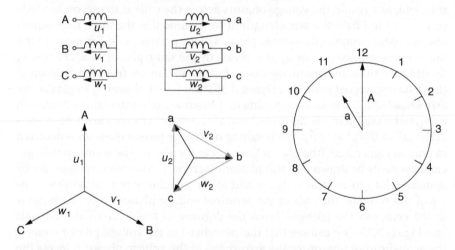

Figure 3.35 A Yd-11 transformer (the terminals of the secondary side of the transformer are labeled as a/b/c). When the terminals of the secondary side of the transformer are labeled as c/a/b, a Yd-3 transformer results. When the terminals of the secondary side of the transformer are labeled as b/c/a, a Yd-7 transformer results.

side, so that possible unbalances between the phases at the secondary side are smoothed at the primary side of the transformer. Another advantage is that the third harmonic current, as a result of the magnetization of the transformer core, remains trapped inside a delta-connected winding (more about this topic is explained in Section 3.8.2).

The way how the coils are connected with each other and to the terminals can introduce a phase shift between the voltage phasors of the corresponding terminals, that is, labeled with the same letter (or sign or something similar), at the primary and secondary sides. This is illustrated in the following example.

Example 3.3   *A Yd-11 transformer*
In this example, we determine the phase shift of the three-phase transformer as shown in Figure 3.35.

In a balanced three-phase system, we can easily draw the voltage phasors A, B, and C for the primary terminals of the transformer. The star point of the transformer is at zero potential (as we have a balanced system), and the voltages across the coils (u_1, v_1, and w_1) equal the line-to-neutral voltages applied at the terminals of the primary side of the transformer (see Figure 3.35, the left phasor diagram).

We have to find the terminal voltages at the secondary side of the transformer in order to determine the phase shift. We know that the coils at the secondary side of the transformer are magnetically coupled with the coils at the primary side, and, as a result, the voltage phasors across the coils at the secondary side (u_2, v_2, and w_2) have the same length (if we assume that the turns ratio equals one) and orientation as the voltage phasors of the primary side. Let us start with drawing the voltage phasor u_2: this is exactly the same phasor as u_1. Phasor v_2 should have the same orientation as v_1, but, as we can see from the diagram of the delta-connected coils (see Figure 3.35), the tail of phasor v_2 begins at the arrowhead of u_2. The same reasoning for phasor w_2: it has the same orientation as w_1, but its tail starts at the arrowhead of v_2, whereas its own arrowhead ends at the tail of the phasor u_2. The resulting triangular phasor diagram is depicted in gray in Figure 3.35 (the phasor diagram at the right). The terminal voltages can now easily be drawn into this phasor diagram. The terminal voltages are by definition line-to-neutral voltages, and as the imaginary neutral is in the center of the triangle, the tails of the terminal voltage phasors meet each other in the center of the triangle. From the drawing of the delta-connected coils (see Figure 3.35), we can see that the arrowhead of the voltage phasor a meets the arrowhead of phasor u_2, the arrowhead of the voltage phasor b meets the arrowhead of phasor v_2, and the arrowhead of the voltage phasor c meets the arrowhead of phasor w_2. The terminal voltage phasors are drawn in black in Figure 3.35.

We can see that the voltage phasors at the secondary side lead the corresponding phasors at the primary side by 30°. We also notice that the voltages at the secondary side are reduced in amplitude by a factor of $\sqrt{3}$.

It becomes evident from Figures 3.34 and 3.35 that a phase shift can occur when we apply a three-phase transformer to the power system, due to either the connection of the coils or the labeling of terminals. Because a coil at the secondary side of the transformer is magnetically coupled with one of the coils at the primary side, this phase difference between the primary and secondary voltages equals a multiplicity of 30°. That is why we can draw the voltage phasors in a clocklike diagram (the hours on the number plate are on a circle at an angle of 30°), and we can express the phase difference between the primary and secondary voltage phasors as a clock reference. The high-voltage line-to-neutral phasor A serves as the reference and points to 12 o'clock, whereas the low-voltage line-to-neutral phasor a points to the hour on the clock that indicates the phase shift.

The transformer configuration is expressed by an uppercase letter, a lowercase letter, and a number. The uppercase letter gives information about the connection of the coils at the high-voltage side: Y for a wye connection and D for a delta connection. The lowercase letter tells the connection of the coils at the low-voltage side: y for a wye connection and d for a delta connection. The number indicates the phase shift expressed as an hour on the clock. The phase shift in the case of the Yy-4 transformer in Figure 3.34, for example, equals $4 \times 30 = 120°$; the voltage phasors at the secondary side lag the corresponding phasors at the primary side by 120°.

In Section 5.5.1, a special type of transformer is introduced that is used to control active power flows in the system by creating a phase shift between the voltage at the primary and the voltage at the secondary side. This type of transformer is called a phase shifter.

3.8.2 The Magnetizing Current

The magnetic properties of the transformer core material are not ideal. The permeability (μ) is not infinite, and that means that not all of the flux is inside the core material. Also, the permeability is not a constant parameter, and therefore the relation between the magnetic flux density and the magnetic field intensity ($B = \mu H$) is not linear (see Figure 3.36 (a)). We can see from the B–H characteristic that at higher values for H, the core material saturates: an increase of H does not result in a proportional increase of B. In practice, the magnetic flux density of iron will not exceed a value of 1.3 T. One could put the question why a transformer is not operated in the linear part of the curve only, that is, around the origin of the B–H characteristic. The answer is that in that case the amount of core material would increase considerably as does the price of the transformer (we learn from Equation 3.2 that decreasing $|B|$ with a factor of 3 requires an increase of the cross-sectional area of the transformer core (A) with a factor 3 in order to achieve that $|\Phi|$ remains constant).

The B–H characteristic is not only nonlinear, but it also has hysteresis as can be seen in Figure 3.36 (b). Hysteresis comes from the Greek word *hysteros*,

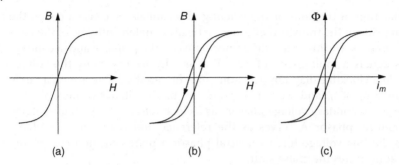

Figure 3.36 Transformer core characteristics: (a) nonlinear B–H characteristic, (b) nonlinear B–H characteristic with hysteresis, and (c) nonlinear Φ–i_m characteristic with hysteresis.

which means later or behind: the state of magnetization reflects the previous state of magnetization. This means that there is no such thing as a unique relation between an H-value and a B-value.

When a sinusoidal voltage is applied at the primary side of an unloaded transformer, a current will flow through the primary winding that causes the magnetization of the transformer core; we call this the magnetization current (i_m). The relationship between the magnetization current and the resulting flux in the transformer core, the Φ–i_m characteristic (Figure 3.36 (c)), has the same shape as the B–H characteristic. After all, the following equations apply: $\Phi = B/A$ and $i_m = i_1 = Hl/N_1$ (i.e., the application of Ampère's law in Figure 3.27, with the dashed line chosen as the contour of integration with length l). The applied sinusoidal voltage (v_1) drives a magnetic flux in the transformer core (Φ) that has also a sinusoidal shape (see Figure 3.37).

By making use of the Φ–i_m characteristic, we can draw the magnetization current as shown in Figure 3.37. It is obvious that the magnetization current is a periodic function but not a sinusoidal one. This current can be decomposed in a fundamental component (50 Hz) and higher-order components (such as 150, 250, 350 Hz, …), called the higher harmonics. The fundamental (50 Hz) component, being the first harmonic, and the third harmonic (150 Hz) form the dominant part of the current as depicted in the inset in Figure 3.37 (the first and third harmonic components are the dashed lines, and when they are added, they form the solid line).

We notice from this figure that the fundamental component of the current leads the flux a little. This means that the magnetization current contains a small component that is in phase with the driving voltage; in other words, the magnetization of the core results in ohmic losses. These losses are the so-called hysteresis losses and are dissipated as heat in the transformer core during the magnetizing and demagnetizing process; the dissipated energy is equal to the

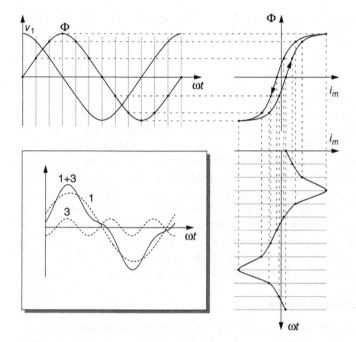

Figure 3.37 Construction of the magnetization current and a comparable wave shape built up by a first and third harmonic (inset).

area enclosed by the $\Phi-i_m$ characteristic. Because we complete the loop around the hysteresis characteristic during each cycle of the supply voltage, the hysteresis losses are proportional to the frequency of the supply voltage.

The third harmonic of the current needs extra attention. When we write down the equations of the first and the third harmonic currents that form the magnetization current, as displayed in the inset in Figure 3.37, we get for the three phases:

$$i_a = \sqrt{2}|I| \sin(\omega t - \varphi) - \sqrt{2}|I_3| \cos(3(\omega t))$$

$$i_b = \sqrt{2}|I| \sin\left(\omega t - \varphi - \frac{2\pi}{3}\right) - \sqrt{2}|I_3| \cos\left(3\left(\omega t - \frac{2\pi}{3}\right)\right)$$

$$= \sqrt{2}|I| \sin\left(\omega t - \varphi - \frac{2\pi}{3}\right) - \sqrt{2}|I_3| \cos(3(\omega t)) \qquad (3.7)$$

$$i_c = \sqrt{2}|I| \sin\left(\omega t - \varphi - \frac{4\pi}{3}\right) - \sqrt{2}|I_3| \cos\left(3\left(\omega t - \frac{4\pi}{3}\right)\right)$$

$$= \sqrt{2}|I| \sin\left(\omega t - \varphi - \frac{4\pi}{3}\right) - \sqrt{2}|I_3| \cos(3(\omega t))$$

$|I_3|$ the effective value of the third harmonic current [A]

We can see from Equation 3.7 that the third harmonic currents in the three phases are identical and that the sum of the currents i_a, i_b, and i_c is not equal to zero!

$$i_a + i_b + i_c = -3\sqrt{2}|I_3|\cos(3(\omega t)) \tag{3.8}$$

When the star point of the transformer is kept floating (i.e., not connected to the ground), the sum of the currents is forced to be zero, and the third harmonic in the current cannot flow. In this situation the magnetization current is forced to have no third harmonic, and the core's nonlinear behavior results in a deformed coil voltage. This problem does not exist if we have a transformer with at least one delta-connected winding. The third harmonic current cannot leave or enter the delta-connected winding, but it can circulate in this winding: the third harmonic current, induced by the magnetization of the transformer core, remains trapped within the loop of the delta-connected winding. This is one of the reasons to apply Yy transformers with a tertiary delta winding at the higher voltage levels; the tertiary delta winding creates a path for the third harmonic current, which is induced by the magnetization of the transformer core.

3.8.3 Transformer Inrush Current

When $i_m = 0$, the flux Φ has a certain value; this is called the remanent flux Φ_r (see Figure 3.38). Therefore, the transformer core of a power transformer that has been disconnected from the system contains a residual flux Φ_r. When, after some time, the power transformer is reconnected to the grid at an instant such that the system voltage initiates a flux in the same direction as the residual flux Φ_r, the total core flux becomes $\hat{\Phi} + \Phi_r$. This can be observed from the following equations, in which the integration constant equals the remanent flux ($C = \Phi_r$):

$$v_1(t) = \sqrt{2}|V_1|\cos(\omega t) = N_1 \frac{d\Phi}{dt}$$

$$\Phi(t) = C + \frac{\sqrt{2}|V_1|}{\omega N_1}\sin(\omega t) = \Phi_r + \hat{\Phi}\sin(\omega t) \tag{3.9}$$

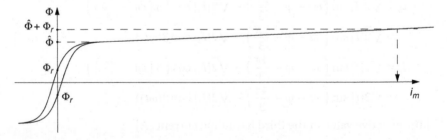

Figure 3.38 Remanent flux and transformer inrush current.

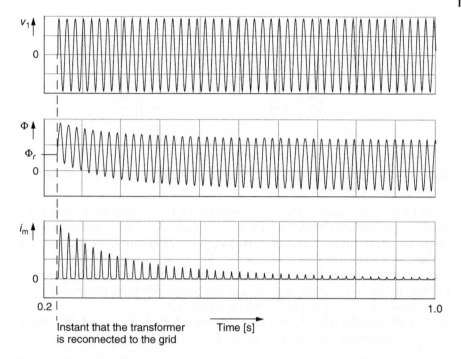

Figure 3.39 Voltage, flux, and current values after energization of a power transformer.

The core material goes into saturation, and, as a result, the transformer draws a large current from the supplying network, as illustrated in Figure 3.38: the transformer inrush current. It can take many cycles for the high initial flux level to decay, which is reflected in an asymmetrical transformer inrush current with a DC component that can take seconds before it disappears and a symmetrical magnetization current is left. An example where a power transformer is reconnected to the grid in an instant, such that the system voltage initiates a flux in the same direction as the residual flux, is illustrated in Figure 3.39. There is no saturation when the system voltage initiates a flux opposite to the direction of the residual flux.

Transformer inrush currents are inevitable, and the transformer protective devices are designed and tuned such that these temporarily high currents are recognized and ignored: the protective devices do not operate on the high inrush current, but do react in the case of a short-circuit current.

3.8.4 Open Circuit and Short Circuit Tests

The equivalent circuit of the transformer is derived in Appendix B. If the secondary circuit elements are referred to the primary side of the ideal transformer, we have the transformer equivalent circuit of Figure 3.40.

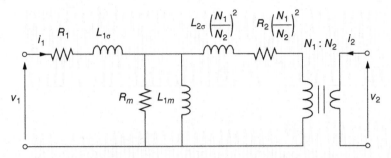

Figure 3.40 Transformer equivalent circuit with the secondary circuit elements referred to the primary side of the ideal transformer.

For most system studies the series resistance R can be neglected because its value is much smaller than the leakage inductance L_σ. During normal system operation, the leakage reactance causes a voltage drop across the transformer. When the transformer is short-circuited, the leakage reactance limits the current. For an ideal transformer, with no leakage flux, the current would become infinite, so the fact that the main flux Φ not completely links the primary and the secondary winding has definitely an advantage!

The leakage inductance L_σ, the magnetizing inductance L_{1m}, and the voltage ratio of a transformer can be determined by two rather simple tests:

- An open-circuit test
- A short-circuit test

In the open-circuit test, the secondary terminals of the transformer are left open, and the rated primary voltage is applied. From the voltages measured at the primary and the secondary terminals, the voltage ratio can be calculated. From the measured current that flows in the primary winding, the magnetizing inductance can be calculated.

In the short-circuit test, the secondary terminals are short-circuited, and a reduced voltage is applied until the rated primary current flows. From the simplified transformer equivalent circuit (Figure 3.40), we note that in the short-circuit test L_{1m} is effectively short-circuited, leaving only L_σ in the circuit; thus from the measured voltage and current, we can calculate the leakage inductance. The reduced voltage that has to be applied in order to supply the nominal rated primary current is a characteristic parameter for a transformer and is called the short-circuit voltage. The value of the short-circuit voltage $|V_k|$ is determined by the leakage flux of the transformer. To be able to make a comparison between transformers of different size and nominal power, the short-circuit voltage $|V_k|$ is expressed as a percentage of the nominal rated voltage $|V_n|$:

$$V_k(\%) = \frac{|V_k|}{|V_n|} \cdot 100 \tag{3.10}$$

3.9   Power Carriers

It is in principle an economical and an environmental issue whether to choose an overhead line (see Figure 3.41) or an underground cable for transmitting and distributing electrical power to densely populated areas. Underground systems are in general more reliable than overhead systems, because they are not exposed to wind, lightning, and vehicle damage. Underground systems do not disturb the environment, and they require less preventive maintenance. The main disadvantage, however, is its higher costs: for the same power rating, underground systems are in general 6–10 times more expensive than overhead systems. In the densely populated Netherlands, (almost) all conductors below the 50 kV level are underground cables, as shown in Table 3.1.

The conductor material can be either copper or aluminum, which depends on the type of power carrier. The resistivity of copper is only 60% of that of aluminum: $\rho_{Al} = 29 \times 10^{-9}\ \Omega\,m$ and $\rho_{Cu} = 18 \times 10^{-9}\ \Omega\,m$. However, the density of copper is much higher: $8900\ kg/m^3$ versus $2700\ kg/m^3$ for aluminum, and we can easily calculate that an aluminum conductor has only half the weight of a copper conductor with the same resistance ($29/18 \times 2700/8900 = 0.488$). So when we apply aluminum conductors as overhead transmission lines, our tower construction and insulator strings can be designed lighter and are therefore cheaper. Additionally, aluminum is a less expensive material than copper,

Figure 3.41 Overhead transmission lines. Reproduced with permission of TenneT TSO B.V.

Table 3.1 The power carriers in the Dutch power system [7].

Voltage Range [kV]	Aboveground [km]	Underground [km]
220/380	2719	16
50/110/150	5580	3614
3–25	–	105599
0.4	174	150623

and therefore overhead transmission lines are usually laid out with aluminum instead of copper conductors. In the case of underground cables, the low resistivity of copper is a major advantage because the heat, generated by the ohmic losses in the conductor, is less.

The rather low tensile strength of aluminum is a disadvantage to use it as conductor material for overhead transmission lines, and that is why steel is used as core material of the conductor (Figure 3.42). We see in Figure 3.42 that the conductor itself is not solid but is composed of strands (26 aluminum strands and 7 steel strands in this case), which gives the conductor the necessary flexibility. The strands are spiraled – each layer in opposite direction to avoid unwinding. Because of the skin effect, the steel core does not contribute to the conductor resistance. Skin effect occurs when an alternating current flows in a conductor: the current prefers to flow near the surface of the conductor, and the current density is not uniformly distributed over the conductor cross-sectional area. As a consequence of the skin effect, the AC resistance of a conductor is higher than its DC resistance, which is given by the following expression:

$$R_0 = \frac{\rho l}{A} \tag{3.11}$$

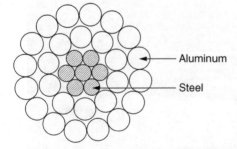

Figure 3.42 Cross section of an aluminum conductor with a steel core (ACSR: aluminum conductor steel reinforced).

Aluminum

Steel

R_0 the DC resistance of a conductor [Ω]
ρ the resistivity of the conductor material [Ωm]
l the length of the conductor [m]
A the cross-sectional area of the conductor [m^2]

The spiraling makes the strands longer than the conductor itself and increases the overall resistance by approximately 1% or 2%.

The modeling of overhead transmission lines and underground cables for steady-state power system analysis is covered in Appendix E, whereas more general characteristics of, and differences between, overhead transmission lines and underground cables are described in the two following sections.

3.9.1 Overhead Transmission Lines

High-voltage transmission lines are supported by transmission line towers, and the surrounding air serves as insulating medium (the breakdown field strength of air at atmospheric pressure is 3 kV/mm). An example of a 150 kV double-circuit transmission line tower is shown in Figure 3.43. The tower height is determined by the fact that the conductors should have a minimum clearance, which depends on the voltage level, from the ground. Therefore, the sag of the conductors in combination with the distance between the towers (the span) determines the height of the structure.

Insulators
It is obvious that the transmission lines cannot be suspended from a tower directly. Because the structure is at ground potential, there should be insulation between the tower and the conductor. Not every piece of insulating material, which is mechanically strong enough to carry the lines, is suited for this task. Imagine that we would use a cylindrical piece of insulating material to isolate the line from the tower; falling raindrops would immediately cause a short circuit, as the water film creates a conducting path between the tower and the line. Therefore, insulators are designed such that the creepage path is much longer.

The most common type of insulator is the cap and pin type made from either glass or from porcelain. This type of insulator is shown in Figures 3.44 and 3.45. The shape of the sheds under the insulator is determined by the environmental pollution conditions because, for instance close to the coast, salt deposit will affect the insulating properties. The number of disks applied for an insulator string is not determined by the nominal voltage alone. The maximum voltage that the line must be able to withstand without flashover while exposed to transient overvoltages plays a role, as well as the pollution conditions. Insulators are sensitive to high-current arcs that are initiated by a flashover, for instance

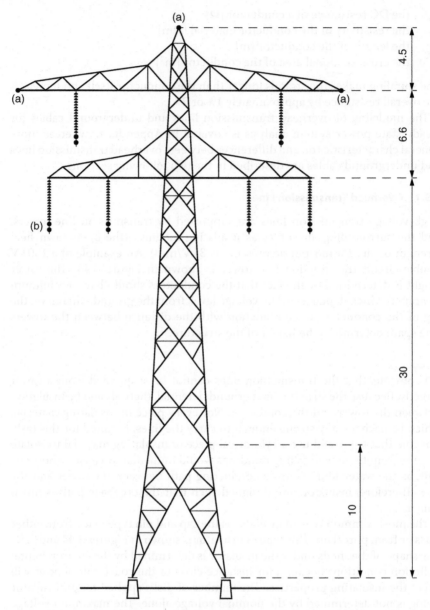

Figure 3.43 A 150 kV double-circuit transmission line tower (distances are given in meters). (a) Ground wires or shield wires. (b) A bundle of two conductors per phase.

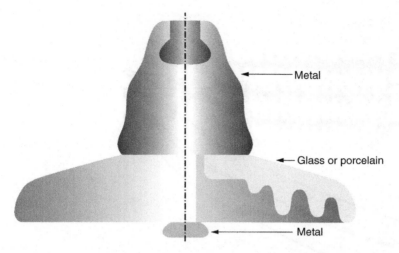

Figure 3.44 A single disk of an insulator string.

Figure 3.45 Three disks of an insulator string and a three-conductor bundle suspended from a tower by an insulator string. Reproduced with permission of TenneT TSO B.V.

when a line conductor is directly hit by lightning. High-current arcs can cause temperature cracks in the porcelain or the glass, and this reduces the insulating property of the insulator string. Arcing horns at the top and at the bottom of the insulator string, shown in Figure 3.46, serve as the foot points for the high-current arc and keep the arc channel away from the insulating body.

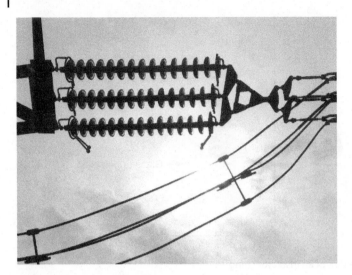

Figure 3.46 Arcing horns protect the insulator string from a high-current arc. Reproduced with permission of TenneT TSO B.V.

Bundled conductors

In particular under rainy or foggy weather conditions, one can hear a hissing sound nearby high-voltage transmission lines or other high-voltage components; this is called corona. The Latin word *corona* means crown because, apart from the noise, corona also produces light: a "crown of light."

Corona is caused by discharges in air that result from a locally strong electric field strength that ionizes the air around the conductor. A mechanism that can cause a locally strong electric field at the surface of a transmission line is illustrated in Figure 3.47. In Figure 3.47 (a), we see an irregularity on one of the plates of a parallel-plate capacitor and the corresponding equipotential lines. The electric field strength can be calculated from the equipotential lines by using the following equation:

$$E = \frac{\Delta V}{d} \tag{3.12}$$

E the electric field strength [V/m]
ΔV the potential difference between two equipotential lines [V]
d the distance between two equipotential lines [m]

Close to the sharp point, the distance d between the equipotential lines strongly reduces, and we can see from Equation 3.12 that locally we have a higher electric field strength. This is due to the fact that in the vicinity of the sharp point, the electric charges are close together so that locally we have

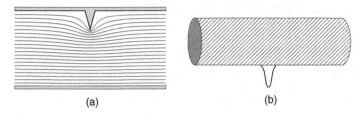

Figure 3.47 A locally strong electric field. (a) A parallel-plate capacitor with a sharp point on one of the plates. (b) A transmission line with a rain drop.

a high charge density and a high electric field strength. In Figure 3.47 (b), a transmission line with a raindrop on it is drawn. This situation resembles the irregularity on the plate and results in a locally strong electric field near the surface of the conductor. We describe hereunder how this locally strong electric field gives rise to the so-called negative corona, that is, corona where the driving voltage has a negative polarity. In the strong electric field, free electrons that are present in the air (freed by photons, for instance) are accelerated and will collide with the molecules of the insulating gas (for an overhead transmission line, the insulating gas is ambient air). The gas molecules fall apart in their constituent atoms (this is called dissociation), which, in turn, are hit by the accelerated electrons too. These collisions cause new electrons to be freed and positively charged ions to be created:

$$A + e \rightarrow A^+ + 2e \qquad (3.13)$$

A gas atom
e electron
A^+ positive ion

The released electrons are accelerated too, and they take part in the ionization process of the insulating gas around the sharp point. The positive ions that are left after the ionization are much heavier than the electrons, and a positive space charge is the result. Oxygen, which constitutes 20% of the ambient air, is slightly electronegative from itself and has the ability to attach free electrons to its molecules, resulting in the electrons being withdrawn from the ionization process and a cloud of negative space charge being formed that shields the point as shown in Figure 3.48. As a result the discharge stops, the space charge disappears, and a new discharge starts. Therefore, corona is a repetitive, high-frequent (0.1–5 MHz) phenomenon. Positive corona, that is, when the driving voltage has a positive polarity, follows another mechanism that we do not treat here (interested readers are referred to the high-voltage engineering literature; e.g., Ref. [8]). Corona is an unwanted effect because it

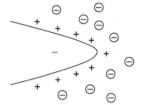

Figure 3.48 Negative corona [8].

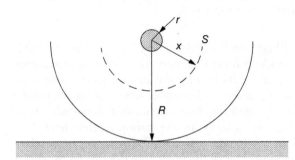

Figure 3.49 Front view of a transmission line conductor above the Earth's surface.

leads to considerable power losses in high-voltage transmission lines and it causes radio interference.

It is evident that in bad weather conditions it is hardly possible to avoid corona. But in order to minimize the effect of corona under normal operating conditions, the electric field strength at the surface of the conductor should be kept below the value of 1.5–2 kV/mm. In order to calculate the electric field strength at the surface of the conductor, we can model the conductor and the Earth's surface as two coaxial cylinders as illustrated in Figure 3.49. Gauss's law for the electric field states that the net flux of the electric flux density vector out of the closed surface S is equivalent to the net positive charge enclosed by the surface (see also Appendix A.1). In the two-dimensional situation that is drawn in Figure 3.49, we take a cylinder S with radius x and an axial length of l meter. The net positive charge enclosed by the surface of the cylinder is

$$q'l = \iint_S \boldsymbol{D} \boldsymbol{\cdot} \boldsymbol{n} \, dA = \varepsilon E_x 2\pi x l \rightarrow E_x = \frac{q'}{\varepsilon 2\pi x} \qquad (3.14)$$

q' the charge on the conductor per meter of length [C/m]

$\boldsymbol{D}$ the electric flux density vector [C/m^2]

ε the permittivity of the medium inside the cylinder, $\varepsilon = \varepsilon_0 \varepsilon_r$; the permittivity in vacuum is $\varepsilon_0 = 8.85 \times 10^{-12}$ F/m; for dry air, the relative permittivity is approximately $\varepsilon_r \approx 1$

E_x the electric field intensity at radius x [V/m]

The voltage between the inner cylinder (the conductor) and the outer cylinder (the Earth's surface) is

$$|V_{\text{LN}}| = \int_r^R E_x dx = \frac{q'}{\varepsilon 2\pi} \ln\left(\frac{R}{r}\right) \tag{3.15}$$

r   the radius of the inner cylinder [m]; in this case the radius of the conductor

R   the radius of the outer cylinder [m]; in this case the height of the conductor above the ground

When we combine Equations 3.14 and 3.15, we can write for the electric field intensity at radius x:

$$E_x = \frac{|V_{\text{LN}}|}{x \ln\left(\dfrac{R}{r}\right)} \tag{3.16}$$

The electric field strength is highest at the conductor surface:

$$E_r = \frac{|V_{\text{LN}}|}{r \ln\left(\dfrac{R}{r}\right)} \tag{3.17}$$

When we want a low electric field strength at the conductor surface, we have to choose for thick conductors. This is illustrated in Figure 3.50 (a); by increasing the diameter of the conductor, the distance between the equipotential lines increases and the electric field strength at the surface of the conductor decreases. However, thick conductors are heavy, and besides the fact that more material is needed to manufacture them, they require more rigid tower structures, which makes this solution an expensive one. A much better solution is to divide the conductor in bundles. Instead of one conductor per phase, multiple conductors are used per phase. In the photograph shown

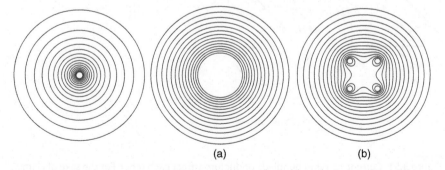

(a)   (b)

Figure 3.50 Reducing the electric field strength at the surface of the conductor (a) by increasing the diameter of the conductor or (b) by using a conductor bundle with 4 conductors.

in Figure 3.45, for instance, a three-conductor bundle is used. The electric field strength at the surface of the conductors of a bundle, shown in Figure 3.50 (b), is comparable with that of the thick conductor in Figure 3.50 (a).

Besides corona reduction, bundle conductors have more advantages compared with a single conductor with a larger diameter:

- Less line reactance
- Easier to transport and to assemble
- Better cooling of the conductors
- Increased power transfer capability

Bundling of conductors has one drawback: the current-carrying conductors attract each other. This is illustrated in Figure 3.51. That is why the individual conductors are separated by a so-called spacer. A spacer for a four-conductor bundle is shown in Figure 3.52.

Galloping lines

A dangerous situation can occur that is caused by a phenomenon known as "galloping lines." Galloping is a low-frequency (0.1–1 Hz), large-amplitude (between 0.1 and 1 times the sag of the span), wind-induced vibration of high-voltage overhead lines. A photograph of galloping lines is shown in Figure 3.53. Galloping is usually initiated by a moderately strong, steady crosswind blowing on an asymmetrically iced conductor surface. How an asymmetric ice shape is formed is shown in Figure 3.54. The surface of the conductor is not smooth, and ice particles will stick on it, so that gradually

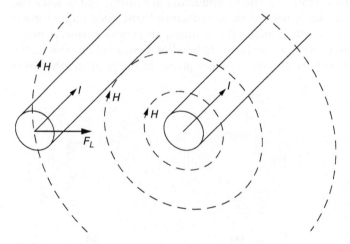

Figure 3.51 Current-carrying bundled conductors attract each other. For the sake of clarity, only the magnetic field surrounding the right conductor that causes an electromagnetic force on the left conductor is shown in the drawing; the interaction of the left conductor on the right one is not illustrated here. But if you hold your book upside down, you can see this effect.

Figure 3.52 A spacer for a four-conductor bundle. Reproduced with permission of Alcoa Conductor Accessories.

Figure 3.53 Galloping lines. The arrows indicate the minimum distance (when there is a risk of a short circuit) and the maximum distance between the conductors. Reproduced with permission of M. Tunstall.

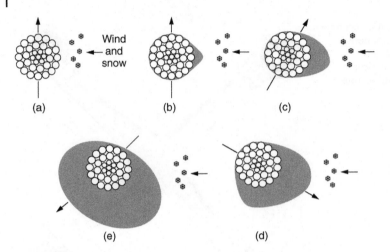

Figure 3.54 Snow and ice deposit on a conductor. The arrow indicates the spatial orientation of the line.

a deposit is formed on the conductor surface (Figure 3.54 (a) and (b)). When the conductor is able to rotate a little bit, the deposit grows on (Figure 3.54 (c)–(e)), and this can lead to a sleeve around the whole conductor surface. The sleeve can easily reach a mass of 5–10 kg/m, which raises the mass of the conductor by a factor of 2–20 times (!), and in severe cases, the conductors can break or the towers can even crack.

When the conductor is not able to rotate, which can be the case with bundle conductors because a torsional restraint is set by the spacers, a kind of "wing-shaped" ice deposit on the conductor may result, as shown in Figure 3.54 (b). It is this effect in combination with windy weather that can lead to galloping lines. Several devices have been developed to prevent galloping based on two principle types of countermeasures. The first type of devices tries to prevent the formation of a "wing-shaped" ice deposit on the conductor. This can be done by applying spacers that allow the conductors to rotate freely while maintaining the original geometry of the bundle system. The second type of devices accepts the shape of the ice deposit but modifies the conductor dynamics. A typical example is the conductor vibration damper as shown in Figure 3.55.

Galloping lines can lead to severe problems. The motion of the conductors can make the distance between two neighboring phases too small (see also the arrows in Figure 3.53) and a short circuit between the phases can be the result. When these short circuits repeat themselves, the system operators are forced to take the line out of service. Also the mechanical forces on the insulator strings (look at the insulator strings of the tower in Figure 3.53) and on the towers can get so strong that galloping can lead to breaking of an insulator or even cracking of the tower itself.

Figure 3.55 A conductor vibration damper. Reproduced with permission of Alcoa Conductor Accessories.

Ground wires or shield wires

The probability of a direct hit by lightning on an overhead transmission system is high compared with the vulnerability of other parts of the power system to this force of nature. For this reason, transmission line towers are interconnected by ground wires, also named shield wires (labeled "a" in Figure 3.43), that hang well above the phase conductors in the tower and are electrically connected with the tower frame and via the towers to the ground below. The ground wires shield the phase wires from a direct hit by lightning and reduce the tower ground resistance in dry or rocky soil.

Lightning is a part of the Earth–ionosphere electric system, depicted in Figure 3.56 [9]. The voltage between the ionosphere and the Earth's surface, in

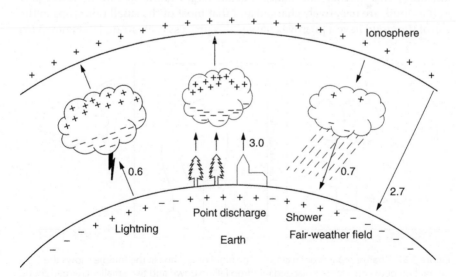

Figure 3.56 The Earth–ionosphere electric system [9]. Currents are in $\mu A/km^2$; average of the total Earth, over a long time.

fact an enormous capacitor, is roughly 300000 V. This "capacitor" has a small leakage current, the fair-weather current, which totals about 1400 A when we take the entire Earth's surface into account. Thunderstorms act as the "battery" in this electric system; the number of thunderstorms, any time of the day, all over the world, is approximately 1500. The power in the Earth–ionosphere electric system is, however, only 500 MW, equivalent to the power of a small generator station.

Lightning mostly occurs on summer days when the ambient temperature is high and the air is humid. Because of the temperature difference, and with that the difference in density, the humid air is lifted to higher altitudes with a considerable lower ambient temperature. Cold air can contain less water than warm air and raindrops are formed. The raindrops have a size of a few millimeters and are polarized by the electric field that is present between the lower part of the ionosphere and the Earth's surface (the strength of this atmospheric field is on summer days in the order of 60 V/m and can reach values of 500 V/m on a dry winter day). The charge separation in the clouds, shown in Figure 3.56, is a complex mechanism that is not yet fully understood. Here follows one "explanation." Larger raindrops fall down and meet smaller and lighter raindrops that are lifted up by the upward flow of air. These smaller raindrops can be positively or negatively charged. The negative drops are attracted by the positive charge in the lower part of the large raindrop, whereas the positive drops are pushed away. Even though the positive drops are attracted by the negative charge in the upper part of the large raindrop, the small positive raindrops cannot attach anymore. This mechanism makes that the larger raindrops, in the lower part of the cloud, are negatively charged and that most of the small raindrops in the top of the cloud have a positive charge. This process is illustrated in Figure 3.57.

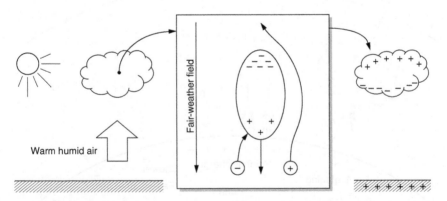

Figure 3.57 Charge separation in clouds. The highlighted box in the middle shows the interaction between a large polarized raindrop falling down and two smaller charged drops that are lifted up.

The clouds move at great heights and the average field strength is far below the average breakdown strength of air. Inside the thundercloud, the space charge formed by the accumulating negative raindrops creates a locally strong electric field in the order of 10000 V/m, and this electric field accelerates the quickly moving ions to considerable velocities. Collision between the accelerated negative ions and air molecules creates new negative ions, which in turn are accelerated, collide with air molecules, and free fresh negative ions. An avalanche takes place, and the space charge and the resulting electric field grow in a very short period of time. The strong electric field initiates discharges inside the cloud, and a negative stream of electrons emerges as a dim spark called a stepped leader or a dart leader that jumps in steps of approximately 30 m and reaches the Earth in about 10 ms. When the stepped leader approaches the Earth's surface, the electric field increases so strongly that a positive leader travels upward, preferably from a pointed object (which gives locally an even stronger electric field). After making contact, the main channel is formed.

Because of the stochastic behavior of the space charge accumulation, the stepped leader also creates branches to the main channel. The main channel carries initially a discharge current of a few hundred amperes, having a speed of approximately 150 km/s. This discharge current heats up the main channel, and the main discharge, the return stroke, is a positive discharge, and it travels at approximately half the speed of light, equalizing the charge difference between the thundercloud and the Earth. The main discharge current can be 100000 A or more, and the temperature of the plasma in the main channel can reach values as high as 30000 K. The pressure in the main channel is typically 20 bar. The creation of the return stroke takes place between 5 and 10 µs and is accompanied by a shock wave that we experience as thunder. A lightning stroke consists of several of these discharges, usually three or four, with an interval time of 10–100 µs. The human eye records this as flickering of the lightning, and our ear hears the rolling of the thunder. After each discharge, the plasma channel cools down to approximately 3000 K, leaving enough ionization to create a new conducting plasma channel for the following discharge.

The striking distance gives us insight into where the lightning will reach the Earth's surface. We can visualize the striking distance r_s as being a sphere with a radius r_s around the tip of the stepped leader: the first object at earth potential that is touched by the sphere will be struck by the lightning. The striking distance is related to the peak lightning current that will occur after contact with the object, often described by the following expression:

$$r_s = 10 I_{\text{pk}}^{0.65} \qquad (3.18)$$

r_s the striking distance [m]
I_{pk} the peak lightning current [kA]

Example 3.4 *Striking distance*
According to Equation 3.18, a peak lightning current of 10 kA corresponds to a striking distance of about 45 m, whereas a peak lightning current of 100 kA corresponds to a striking distance of about 200 m.

The ground wire is positioned such that the phase conductors are protected from too large lightning currents. In other words, the ground wires are positioned in such a way that, before the phase conductor is within the striking distance of a large lightning discharge, either the ground wire or the Earth's surface is touched by the sphere. This is illustrated in Figure 3.58. The sphere with radius r_{s1} corresponds to a large lightning current and touches either the ground or the shield wires; the phase conductor is protected from large lightning currents. Smaller currents, such as the sphere with radius r_{s2}, can hit either the ground or the phase conductor. Therefore, the ground or shield wires act as a kind of filter: the system is protected against too large lightning currents, but smaller currents can enter the system. In this way, the maximum amplitude of the current that is injected in the system is known, and the protection can anticipate it.

Transposition
In Section 1.3.3, we assumed that the three phases of the power system are balanced; this means, among other things, that the impedances in each of the phases are identical. When we look at the position of the conductors in the transmission line tower that is depicted in Figure 3.43, we see however that one of the conductors is mounted at a different height, and, consequently, it will have another value for the capacitance to the ground than the other two conductors. In general, the line geometries are diverse, and various positions for the three-phase conductors are possible: the three-phase conductors, for

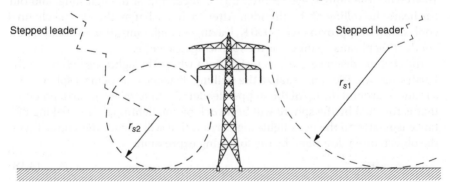

Figure 3.58 Lightning protection by means of shield wires; r_{s1} corresponds to a large lightning current that hits either the ground or the shield wires; r_{s2} corresponds to a smaller lightning current that hits either the ground or the phase conductor.

Figure 3.59 Transposition of overhead transmission lines.

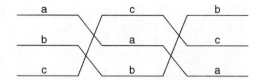

Figure 3.60 A twisting pylon.

example, can be mounted in one plane next to each other or above each other. This determines the line inductance and line capacitance per phase. When the values are different for each phase, the system is not balanced anymore. The balance can be restored by interchanging the conductors at regular intervals along the route so that the influence of the geometry on the line impedance is canceled out and equal parameters result for each phase. This interchanging of line conductors is called transposition and is illustrated in Figure 3.59. Transposition takes place in so-called twisting pylons, of which a photo is shown in Figure 3.60.

3.9.2 Underground Cables

When the power carrier is buried in the ground, the advantages of air for cooling and insulation of an overhead transmission line disappear. For a cable, the conductor must be insulated from the ground, and insulation is a very important topic here: the diameter of a cable must be limited in order to keep it

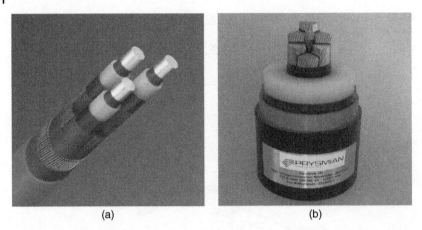

Figure 3.61 A three-core (a) and single-core (b) cable; (a) 6/10 kV with 3 × 240 mm² aluminum, circular solid, conductors and XLPE insulation; (b) 220/380 kV with 1 × 1600 mm² copper, circular stranded compacted, conductors and XLPE insulation. Reproduced with permission of Prysmian.

flexible enough to fit on a drum, but the electric field must be confined in between the limited space between the conductor and the sheath. As cables for the higher voltage levels require thicker conductors (to reduce the electric field strength at the conductor surface) and require a voluminous layer of insulating material, they are built up as single-core cables so that it is still possible to handle them and fit them on a drum (see Figure 3.61 (b)). For the lower voltage levels (about 60 kV and lower), three-core cables are manufactured, which can roughly be divided in two groups: the belted cable (three-phase conductors in a single sheath) and the Höchstädter or three-core cable where each of the three-phase conductor has its own sheath (i.e., electrically there are three single-core cables; see Figure 3.61 (a)). The two three-core cable designs with a "snapshot" of their equipotential lines, with one phase at maximum positive voltage and the other two at half the negative value, are shown in Figure 3.62. In the case of the belted cable, the electric field puts a higher demand on the (insulation) material inside the sheath: the equipotential lines are not completely concentric around the conductor, which results, in the case of paper insulation, in electric field lines that are tangential to the insulation surface, and that is the direction in which the insulation strength is the weakest. Also, the electric field is not confined to a relatively small area around the conductor only, but covers the whole area between the sheath and the conductors. Therefore, belted cables are manufactured only for the lower voltage ranges (10–20 kV and lower).

The conductors can either be stranded (similar as the overhead conductor) or solid (see Figure 3.63) and are made up of copper or aluminum. With a sector-shaped conductor, the area inside a three-core cable can be utilized more

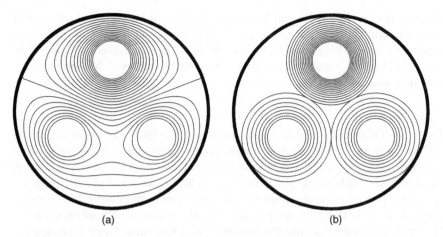

Figure 3.62 Belted cable (a) and a three-core cable where each of the three-phase conductors has its own sheath (b) and their equipotential lines.

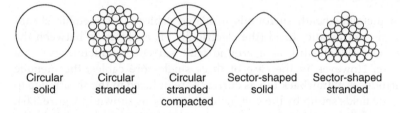

| Circular solid | Circular stranded | Circular stranded compacted | Sector-shaped solid | Sector-shaped stranded |

Figure 3.63 Conductor construction.

effectively, but at the surface of the sector-shaped conductor, a higher electric field strength is present than at the surface of a circular conductor. For the lower voltage levels, this is not a problem as the insulation must have a certain thickness to give the cable mechanical strength. At the higher voltage levels, the diameter of the sector-shaped conductor or the thickness of the insulation should be increased, but both result in a more voluminous cable, and the advantage of sector-shaped conductors is lost. Most cables are plastic insulated or have a paper–oil insulation.

Plastic insulation
As the insulating material, polyethylene (PE) and cross-linked polyethylene (XLPE) are often used. Plastic is a solid insulator that, during the manufacturing of the cable, is melted and pressed around the conductor (i.e., the plastic is extruded). The insulation must be free of cavities and inclusions (such as dust, fibers, and metal particles) in order to prevent partial discharges (such as corona; Section 3.9.1): inclusions can have a low dielectric strength (i.e., a weak

point in the insulation) or can be sharp, which leads locally to a high electric field strength. Polymer insulation is highly vulnerable for water and water vapor, as it lowers the dielectric withstand level. Therefore, the insulation must be sealed against water penetration. This can be achieved by applying, for example, a lead sheath.

Paper–oil insulation

Paper itself is an unsatisfactory insulator due to the spaces incorporated in the cellular structure of the cellulose fibers. In combination with oil, or some other impregnation compound that fills the spaces, an excellent insulator is obtained. In an oil-filled cable, or oil-pressurized cable, the center of the conductor is hollow to supply thin oil under moderate pressure that is maintained by reservoirs feeding the cable along the route. When the cable warms up, the oil expands and is driven from the cable into the reservoirs and vice versa. In this way, gaps in the insulating material are avoided so that no weak points are present. Oil-filled cables have proven to be the most reliable type of cable for the high voltage and extra high voltage levels.

Often a metallic sheath surrounds insulated cables. This metallic sheath serves as an electrostatic shield (the electric field is enclosed in between the conductor and the sheath), as a ground fault current conductor, and as a neutral wire. However, in the case of three single-core cables, the metallic sheath introduces a drawback too as currents are induced in the sheath due to the magnetic fields set up by the conductor currents, as shown in Figure 3.64. When the metallic sheaths of the cables are single-point bonded, that is, the sheaths of the three cables are connected and grounded at one point along their length as shown in Figure 3.65 (a), the voltage induced in the sheath is proportional to the cable length and can reach very high values. The accepted sheath voltage limits the length of single-bonded cables, and they can be used for limited route lengths only. Single-bonded cable sheaths do not form a closed loop and do not provide a path for the flow of circulating currents or external fault currents. This can be resolved by grounding the metallic cable sheaths at both ends (see Figure 3.65 (b)). In such a both-ends bonded system, a closed path exists for currents to flow through the sheath. These currents cause losses and heat, and they reduce the current-carrying capacity of the cable. When the metallic sheaths are grounded at both cable ends and sectionalized and cross-connected in between, a so-called cross-bonded cable is the result as illustrated in Figure 3.66. In this way, the total induced voltage in the consecutive sections is (approximately) neutralized. By cross-bonding the metallic sheaths, the current-carrying capacity is as high as with single-point bonding, but longer route lengths can be realized.

In Section 3.9.1, we derived the electric field distribution between two coaxial cylinders (see Figure 3.49). This is in fact what a single-core high-voltage cable

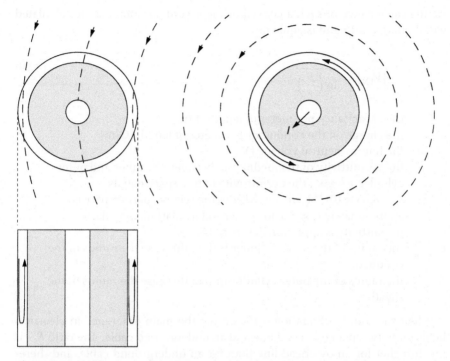

Figure 3.64 The flux lines of a current-carrying cable and the induced eddy currents in the sheath of a neighboring cable (without bonding) [10].

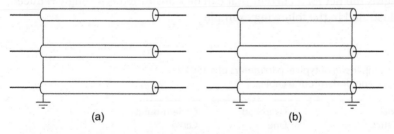

(a)   (b)

Figure 3.65 Single-point bonding (a) and both-ends bonding (b) of a metallic cable sheath.

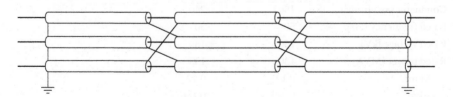

Figure 3.66 Cross-bonding of single-conductor cables.

is. The capacitance of such a coaxial cylinder configuration can be calculated with Equation 3.15 and is equal to

$$c = \frac{q'}{|V_{LN}|} = \frac{2\pi\varepsilon}{\ln\left(\frac{R}{r}\right)} \tag{3.19}$$

c	the capacitance per meter of length [F/m]
q'	the charge on the conductor per meter of length [C/m]
$|V_{LN}|$	the line-to-neutral voltage [V]
ε	the permittivity of the medium in between the inner and outer cylinder, $\varepsilon = \varepsilon_0\varepsilon_r$; the permittivity of free space equals $\varepsilon_0 = 8.85 \times 10^{-12}$ F/m; for (XL)PE, the relative permittivity is approximately $\varepsilon_r \approx 2.3$, for paper–oil insulation, the relative permittivity is approximately $\varepsilon_r \approx 3.5$
r	the radius of the inner cylinder [m]; in this case the radius of the conductor
R	the radius of the outer cylinder [m]; in this case the radius of the sheath

When we examine Equation 3.19, we see the main difference in electrical behavior between an overhead line and an underground cable. The ratio R/r is much higher for an overhead line than for an underground cable, and therefore, underground cables have a much higher capacitance (see Table 3.2). The current required to charge the cable capacitance can be that large that for long cable lengths the net load current that can be transmitted is strongly reduced. This is illustrated in the following example.

Table 3.2 Comparison of typical parameters of a 150 kV overhead line and underground cable.

Power carrier Characteristics	Overhead Line	Underground Cable
Conductor material	Copper	Copper
Conductor area [mm^2]	150	240
Rated frequency [Hz]	50	50
Rated voltage [kV]	150	150
Rated power [MVA]	130	135
r [Ω/km]	0.125	0.12
x [Ω/km]	0.425	0.166
c [nF/km]	11.7	210

Example 3.5 *Charging current of an underground cable*

Consider the parameters of the 150 kV underground cable that are specified in Table 3.2. From the rated power, we can compute the maximum current that is allowed to flow in this cable:

$$|I_{max}| = \frac{135 \times 10^6}{\sqrt{3} \cdot 150 \times 10^3} = 520\,A \tag{3.20}$$

The current required to charge the distributed capacitance of the cable amounts to

$$|I_c| = \omega c |V_{LN}| = 2\pi \cdot 50 \cdot 210 \times 10^{-9} \cdot \frac{150 \times 10^3}{\sqrt{3}} = 5.7\,A/km \tag{3.21}$$

If this cable is *unloaded* and 520/5.7 = 91 km long, the maximum allowed current has been reached at the sending end of the cable!

3.9.3 Gas-Insulated Transmission Lines

Besides the existing technologies of overhead transmission lines and solid insulated underground cables, gas-insulated transmission lines (GIL) offer an additional solution for high-power transmission. GIL can be applied for voltages from 100 up to 800 kV. Most applications of the GIL are at 420 and 550 kV. The upper ranges of 800 kV find only a few applications in China.

Gas-insulated lines were developed in the same period as gas-insulated switchgear and substations. The very good insulating capabilities of SF_6 reduces the dimensions of high-voltage substations, and the excellent arc-interrupting capability increased the current interruption capability of circuit breakers and the reliability of switchgear. The first 400 kV GIL project was the replacement of an oil cable in the pump storage power plant at Schluchseewerke in Germany. In generating mode the plant can generate up to 1000 MW by letting water from the reservoir run through the turbines, and in pumping mode it can store 700 MW of electricity by pumping water up the mountain into the reservoir. A failure in an oil cable destroyed the cables and damaged the tunnel. GIL, being a nonburnable solution, was chosen to replace the oil cables. This GIL was constructed in the years 1974 and 1975, can transmit 2000 A, and has a length of 700 m [11].

The GIL consist of a central aluminum conductor that rests on cast resin insulators, which center it within the outer enclosure (Figure 3.67). This enclosure is formed by a 400–600 mm diameter aluminum tube, which provides a solid mechanical and electrotechnical containment for the system. To meet up-to-date environmental and technical issues, GIL are filled with an insulating gas mixture of mainly nitrogen and a smaller percentage of SF_6.

Figure 3.67 Gas-insulated line. Reproduced with permission of Siemens.

3.10 High-Voltage Direct Current Transmission

As we have seen in Section 1.3.1, at an early stage in power system develop-ment, the choice for AC systems was made. The transmission of electricity over large distances requires high voltage levels. Because ohmic losses are propor-tional to the square of the current, every doubling of the voltage reduces losses to one-quarter. AC became the standard because transformers can quite easily transform the voltage from lower to higher voltage levels and vice versa. Nowa-days, power-electronic devices make it possible to convert AC to DC, DC to AC, and DC to DC with a high rate of efficiency, and the obstacle of altering the voltage level in DC systems has disappeared.

AC transmission systems have an important drawback: they require reactive power. We speak, in our daily practice, about voltages and currents when we analyze the power flow in the network. Current refers to the net flow of charge across any cross section of a conductor. The net movement of 1 C of charge through a cross section of a conductor in 1 s produces an electric current of 1 A. The potential difference or voltage between the terminals of a conductor cre-ates an electric field that forces the charges to move. The moving charges, which we call current, create a magnetic field around the conductor. The interaction between the electric and magnetic field is described by the Maxwell equations, and as we can see from Figure 1.2, the energy is stored in the electromagnetic field. For 50 or 60 Hz phenomena, the power system is, so to speak, of "electri-cally small dimensions" compared to the wavelength of the voltage, and Kirch-hoff's laws may fruitfully be used to compute the voltages and currents. But we have to realize that the electric energy transported by underground cables and overhead lines is stored in the electromagnetic field around the conductors and that that field has to be continuously charged to accommodate the oscillation of voltage and current. Reactive power is the flow of energy that continuously charges the electromagnetic field. This charge energy is not wasted, because the energy is recovered as the fields discharge, but to charge and discharge the field, a current has to flow through the conductor, and this current gives ohmic losses and causes a voltage drop across the line. As the electromagnetic field increases with the length of the conductor, the required reactive power also grows until a point is reached that for cables the reactive current level reaches the value of the nominal rated current and energy transmission is not possible anymore (see also Example 3.5).

Bringing DC transmission at higher voltages was done in Edison's days through the series connection of generators, but it was a cumbersome solution that could not compete with AC. Interest in DC transmission came back again when the mercury-arc valve came onto the scene. The mercury-arc valve is a sealed bulb filled with mercury vapor that uses a steel anode (later made up of carbon) and a mercury cathode (Figure 3.68) [12].

Once an arc is initiated between the anode and the cathode, the current flowing in the arc ionizes the mercury vapor by its heat. The bombardment of ions at the interface of the arc and the mercury causes ions to be released. The steel anode does not emit electrons at the operating temperature. There is a flow of electrons from the mercury cathode but not in the reverse direction. The mercury-arc valve acts as a diode and can be applied in a rectifier circuit: as soon the voltage across it becomes positive, the valve conducts the current, and it isolates when the current passes through zero (see also Section 4.2.4).

Mercury-arc valves can also be applied for DC to AC conversion: an auxiliary electrode inside the bulb receives a voltage pulse that initiates an arc between the anode and the cathode at an arbitrary point in the cycle. So mercury valves can perform both AC to DC and DC to AC conversion and made it possible to combine the transformation advantages of AC with the transmission advantages of DC.

The mercury-arc valve was first demonstrated by its inventor Peter Cooper Hewitt in 1902. In the 1960s a new type of valve became available: the thyristor. The thyristor displaced the mercury-arc technology by semiconductor technology. The advantages of the semiconductor technology are:

- Greater power density
- Higher switching speed
- Lower weight per volume
- Less losses
- Avoiding the toxic aspects of handling mercury

In the periodic table there are metals and nonmetals. In their pure form metals conduct electricity, whereas nonmetals mostly do not. In steel, for

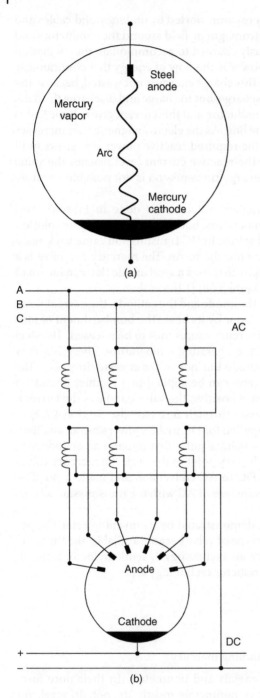

Figure 3.68 (a) Mercury-arc valve. (b) Mutator for rectification of three-phase AC.

Steel anode

Mercury vapor

Arc

Mercury cathode

(a)

A

B

C

AC

Anode

Cathode

DC

+

−

(b)

instance, many of the atoms lost their outermost electrons entirely, and those electrons float aimlessly inside the steel. Inside an insulator, glass, for instance, the outermost electrons are much more tightly bound to their atoms and have no interest to move under the influence of an electric field. There is also an interesting group of nonmetals that display intermediate levels of conductivity: germanium, silicon, gallium arsenide, and silicon carbide. It was Michael Faraday who, in 1833, recognized semiconduction in silver sulfide, but the phenomenon was not understood until in the 1930s the band theory of conduction was formulated.

The conductivity of semiconducting materials can be influenced by selectively inserting impurities into the crystal. These inserted atoms occupy positions in the crystal lattice that otherwise would be occupied by an atom of the substrate material. If the inserted atoms have more electrons in their outer band than the host material, we get n-type material. The additional electrons are free to move through the crystal and increase the conductivity. If the inserted atoms have fewer electrons in their outer band, we get p-type material, and this creates the so-called holes. Because the electrons from neighboring atoms are able to occupy these positions, the holes themselves have mobility and behave like positive charge carriers and also increase conductivity.

A diode is a simple example of a semiconductor device. A p-zone adjoins an n-zone on the same crystal, and if a voltage is applied, current can flow from the p-zone to the n-zone. In the p-zone, holes will flow toward the p–n junction, in the n-zone electrons will flow toward the p–n junction, and the holes and electrons recombine at the junction. If the polarity of the voltage reverses, charge carriers are depleted from the p–n junction and the conduction stops. To make a diode a switchable valve, the conductivity must be triggered externally. In 1947 the transistor was invented at Bell Laboratories by John Bardeen, William Shockley, and Walter Brattain. It uses an electric field to control the availability of charge carriers in a germanium crystal.

In 1950 William Shockley described the principle of the thyristor. A thyristor is an p–n diode with additional layers between the outer p- and n-zones. These layers prevent conduction, but when, at a third contact, the gate, a current is injected, this area floods with charge carriers and when a forward voltage is present between the anode and the cathode, a current can flow. Conduction does not cease until the current crosses zero. In the 1980s the gate turnoff thyristor (GTO) came in the market. A GTO is a thyristor that can be turned off by applying a current to the gate in the reverse direction to that required to turn it on. A further development in semiconductor technology was the introduction of the insulated gate bipolar transistor (IGBT). An IGBT is a power semiconductor controlled by voltage (instead of by current as the GTO) and capable of faster operation, permitting higher switching frequencies.

From AC to DC

The discovery of the phenomenon of electromagnetic induction by Michael Faraday in 1831 opened the door for large-scale generation of electricity. Ampère continued Faraday's and Oersted's experiments, and on the basis of his findings, Antoine-Hippolyte Pixii built a machine consisting of a fixed horseshoe armature and a rotating horseshoe magnet. The first machine produced AC: the rotating horseshoe magnet induced each half rotation a voltage of opposite polarity. The idea was, however, to generate DC. Pixii added a commutator to his original design, enabling the machine to supply electricity flowing in one direction. The first DC generators had a rather low efficiency, but they became an object of study and research. Many researchers contributed to the further development of DC generators in the 1860s and 1870s – Zenobe Theophile Gramme, Charles Wheatstone, and Werner Siemens were the most prominent ones (Figure 3.69).

The rotor of a DC generator produces an AC voltage as it spins. The commutator, basically a rotating switch, reverses the machine's output every half cycle. The commutator mechanically rectifies the current.

So far most of the electricity we use is generated in power plants, where three-phase synchronous generators convert the mechanical energy from the steam or gas turbines into electrical energy. Today's power system is an AC power system, as it has been for more than hundred years. In the last decades the challenges for the utilities and the transmission system operators have dramatically changed. The increasing costs of land for new transmission lines and substations, environmental concerns, and the price of the primary energy are making energy managers and power engineers to take another look at the transmission system. Innovations in DC technology have allowed us to consider for

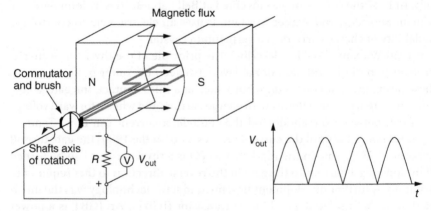

Figure 3.69 Principle diagram of a DC generator.

grid extensions a high-voltage DC (HVDC) connection as a serious alternative for AC transmission lines. To understand how HVDC works, it is necessary to understand the operation of the converter bridge. The simplest type of converter bridge is the 6-pulse bridge, also known as the Graetz bridge. Converter bridges are the building blocks of HVDC systems (Figure 3.70) [13].

The diode turns on when the anode voltage is more positive than the cathode and turns off when the cathode voltage is more positive than the anode. Each diode conducts for 120° in every 360° power frequency cycle. The amplitudes of the three phases change with time. When the voltage B–C becomes larger than the voltage A–C, valve 3 takes over the current that had been flowing in valve 1. This is called commutation. The successive conducting pairs of valves are valves 1 and 2, valves 2 and 3, valves 3 and 4, valves 5 and 6, and valves 6 and 1. The DC voltage resulting from the rectification with the 6-pulse diode converter bridge is shown in Figure 4.12.

The thyristor can conduct the current only in one way, but it can be controlled when it goes in conduction. The thyristor has an extra terminal, the gate, and in order to enter conduction, this gate terminal must also be energized. The thyristor is turned on by triggering the gate, but can only be turned off by the circuit connected between the anode and the cathode.

Figure 3.71 shows a thyristor circuit. When a pulse I_g is applied and a forward voltage V_{th} is present between the anode and the cathode, the thyristor will start to conduct current (I_a). The conduction continues as long as current flows in forward direction. When the current tries to reverse, the thyristor turns off. Hence a thyristor converter requires an alternating voltage supply in order to operate as an inverter, and that is why a thyristor-based converter is known as a line-commutated converter (LCC).

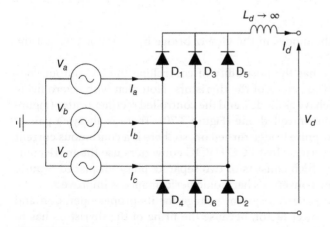

Figure 3.70 A 6-pulse converter.

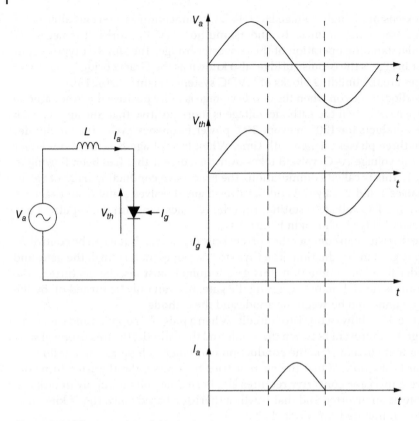

Figure 3.71 The operation of the thyristor.

When we replace the diodes in the Graetz bridge by thyristors, we get the scheme of Figure 3.72.

The thyristor bridge has the possibility of controlling the DC output voltage by changing the firing angle of the thyristors. Note that with a zero firing angle, the thyristors behave as diodes, and the controlled rectifier bridge (Figure 3.72) turns into the uncontrolled one (Figure 3.70). There is always a pair of thyristors within the 6-pulse bridge turned on, so there is a continuous current path through the converter. Most LCC HVDC converters use the more complex 12-pulse bridge, which consists of two separate phase-displaced 6-pulse bridges. In this way the converter's harmonic performance is improved.

An LCC depends on the AC system voltage for its proper operation and operates at a lagging power factor, because the firing of the thyristors has to be delayed relative to the voltage crossing to control the DC voltage. IGBTs and GTOs made it possible to build the voltage source converter (VSC). IGBTs

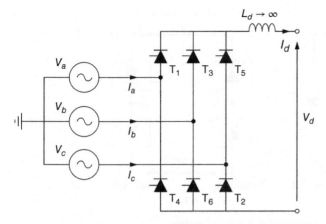

Figure 3.72 A 6-pulse thyristor bridge.

and GTOs are devices that can force commutation, and they enable the VSC to operate in all four quadrants of the P–Q plane. Commutation is independent of the AC system voltage.

Most HVDC links are point-to-point connections. The link can be monopolar or bipolar. The monopolar HVDC link consists of a return path either through the ground or through the sea. This method is mostly used for power transmission using cables, because it saves the cost of laying one cable. When the ground resistance is too high, a metallic return path is preferred instead of using the ground. The bipolar HVDC link consists of two poles: one with positive polarity and the other one with negative polarity and with their neutral points grounded. In steady-state operation the current flowing in each pole is the same, and hence no current flows in the grounded return. The poles may be operated separately. If one of the poles malfunctions, then the other pole can transmit power by itself with the ground return (Figure 3.73).

The back-to-back system is used for providing an asynchronous connection between two AC systems. The rectifier and the inverter are located in the same station.

HVDC technology has become a mature technology since the Gotland HVDC transmission link between the island of Gotland in the Baltic Sea and the Swedish mainland, with a rating of 20 MW, 200 A and 100 kV, went into operation in 1954. Until now HVDC links of 2000 MW and more have been designed and build for voltages in the ±500 to 600 kV range, but these voltage levels are not sufficient for distances around 2000 km from the giant power stations build in China and India. For these lines, 800 kV technology is applied. The power transmission link in China, between Xiangjiaba and Shanghai, operates at ±800 kV and is capable of delivering 6400 MW over a distance of 2071 km.

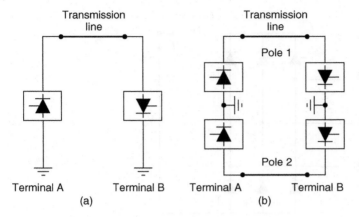

Figure 3.73 Monopolar and bipolar point-to point-system.

Problems

3.1 Explain why
   a. AC transmission and distribution systems are three-phase systems.
   b. transmission systems use mostly overhead lines.
   c. overhead lines use ACSR conductors.
   d. overhead line conductors are stranded.

3.2 Give reasons why
   a. transmission lines are three-phase three-wire circuits, while distribution lines are three-phase four-wire circuits.
   b. it is necessary to use high-voltage for transmission systems.
   c. at 110 kV and above, the transmission lines have bundled conductors.
   d. the tendency of corona formation is lesser in aluminum conductor lines than in copper conductor lines.

3.3 What are the advantages of an interconnected power system?

3.4 In a short-circuit test on a 100 kVA, 10 kV/380 V transformer, with a short-circuit voltage $V_k = 3.7\%$, the copper losses for this transformer are 1200 W. An open-circuit test gave the core losses: 200 W. Show that during the short-circuit test, the core losses can be neglected.

3.5 A three-phase transformer T has a short-circuit voltage $V_k = 12\%$ and is connected to an infinite bus. When there is no load connected at the secondary terminals of the transformer, the transformer draws 2 MVA from the grid at a power factor $\cos(\varphi) = 0.3$. When the secondary terminals are short-circuited, the infinite bus supplies 1200 MVA at a power factor $\cos(\varphi) = 0.07$.

a. What is the nominal power of the transformer?
b. Calculate the core losses of the transformer.

3.6 A 380/150 kV transformer T_1 feeds in on a 150 kV bus. The 150 kV bus supplies two 150/10 kV transformers T_2 and T_3. Transformer T_2 carries its nominal load at power factor $\cos(\varphi) = 1$, and transformer T_3 has a load of 50 MVA at power factor $\cos(\varphi) = 0.6$. The nameplate of the transformers reads:

T_1: 400 MVA, 380/150 kV, $V_k = 16\%$
T_2, T_3: 100 MVA, 150/10 kV, $V_k = 10\%$

a. What is the total apparent power transmitted through this configuration (neglect the short-circuit impedance of the transformers)?
b. What is the value of the current magnitudes through the 10 kV winding of transformers T_2 and T_3 and through the 150 kV winding of transformer T_1?

References

1 Working Group 37.12: The extension of synchronous electric systems: advantages and drawbacks, Paper 37–110, *Cigre 1994 Session*, Paris, France, August 28–September 3, 1994.
2 Siemens AG: *Siemens Power Engineering Guide, Transmission and Distribution*, Siemens Aktiengesellschaft, www.energy.siemens.com (Accessed on 13 October 2016).
3 Anderson, Paul M.: *Analysis of Faulted Power Systems*, IEEE Press, 1995, ISBN 0-7803-1145-0.
4 IEC-62655, Tutorial and application guide for high-voltage fuses, edition 1.0, 2013.
5 Wilkins, R. and Cretin, E.A.: *High-Voltage Oil Circuit Breakers*, McGraw-Hill, 1930.
6 Laithwaite, Eric R., and Freris, Leon L.: *Electric Energy: Its Generation, Transmission and Use*, McGraw-Hill, Maidenhead, 1980, ISBN 0-07-084109-8.
7 EnergieNed: *Energy in The Netherlands – Facts & Figures*, EnergieNed, Arnhem, 2007.
8 Kreuger, Fred H.: *Industrial High Voltage*, Vol. I, Delft University Press, Delft, 1991, ISBN 90-6275-561-5.
9 Wessels, H.R.A.: 'Luchtelektriciteit en Onweer', *Zenit*, Vol. **17**, 1990, pp. 258–64.

10 Weedy, Birron M.: *Electric Power Systems*, John Wiley & Sons, Ltd, Chichester, 1987, ISBN 0-471-91659-5.
11 Koch, Hermann: *Gas-Insulated Transmission Lines*, John Wiley & Sons, Ltd, Chichester, 2012, ISBN 978-0-470-66533-6.
12 ABB review 2|13: *Breakthrough technology* and ABB review 2|14: *Hundred years of ABB review.*
13 Alstom Grid: *HVDC: Connecting to the Future*, 2010, www.alstom.com (Accessed on 13 October 2016).

4

The Utilization of Electric Energy

4.1 Introduction

The line that has been chosen in this book, generation–transmission–distribution–utilization, stems from the pyramid structure of the power system: a relatively small number of large power stations supply the transmission network, which in turn supplies the distribution networks that fan out to provide the individual loads with the demanded energy. But the causality is in fact the other way around! The power system is and must be designed and organized in such a way that the demand can be fulfilled: the consumers are supplied with the requested amount of active and reactive power at constant frequency and with a constant voltage. Consumer demand is not constant but varies from hour to hour each day, from day to day within a week, and from season to season. An example of a daily load curve is shown in Figure 4.1. The minimum load of the day, which is called the valley load, is about 40–60% of the system peak load and usually occurs between 4 and 5 a.m. The generation should be able to fulfill the fluctuating demand, and the transmission and distribution systems have to be able to facilitate the flow of energy.

Most power systems are vertically operated, which means that the grid consists of large power plants feeding bulk power into the high-voltage transmission network that in turn supplies the distribution substations. A distribution substation serves several feeder circuits, and a feeder circuit supplies numerous loads of all types. A light to medium industrial customer can be supplied from the distribution feeder circuit primary busbar directly, while a large industrial load complex usually is served directly from the bulk transmission system. Residential and commercial customers are served from the distribution feeder circuit, which is connected to the secondary of the distribution transformers (see also Section 3.2).

Electrical Power System Essentials, Second Edition. Pieter Schavemaker and Lou van der Sluis.
© 2017 John Wiley & Sons Ltd. Published 2017 by John Wiley & Sons Ltd.

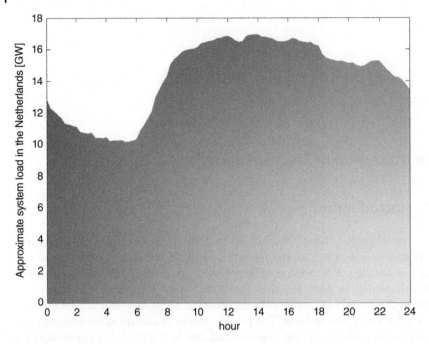

Figure 4.1 Example of a daily load curve in the Netherlands (Monday, May 15, 2006).

A load actually transforms the AC electrical energy into another form of energy. In this chapter we focus first on the various types of loads that transform the AC electrical energy into:

- Mechanical energy
- Light
- Heat
- DC electrical energy
- Chemical energy

After that, the individual loads in the system are clustered and classified as grid users according to the following three categories:

- Residential loads, comprising the domestic users
- Commercial and industrial loads, being the business users and factories
- Electric railways

4.2 Types of Load

Electricity is a very versatile energy carrier, and, since Thomas Edison designed the first commercial electric system in 1882, it has changed our society.

It is used to drive not only electric motors in factories but also our washing machines and refrigerators. Electricity lights our houses, is used for heating our houses, and powers our computers and other electronic equipment. When we take the train or subway, we enjoy the benefits of electric traction: all modern fast trains are driven by electric motors. The first trains had a 1500 V DC supply, but nowadays high-speed trains such as the TGV in France, the ICE in Germany, and the Shinkansen in Japan are supplied by a 25 kV AC supply.

4.2.1 Mechanical Energy

The transformation from electrical energy to mechanical energy is done by motors. The first motors to be developed were DC motors, and they are still applied in electric trains because of their excellent traction properties and as hard disk drives in computers. The majority of the motors, however, are AC machines: synchronous motors and induction motors. They are in principle three-phase machines, because a three-phase power supply will generate a rotating field, which in turn results in a torque on the rotor. In Section 1.3.3, we already discovered that a three-phase system is able to produce a rotating magnetic field as is illustrated here once more in Figure 4.2. When

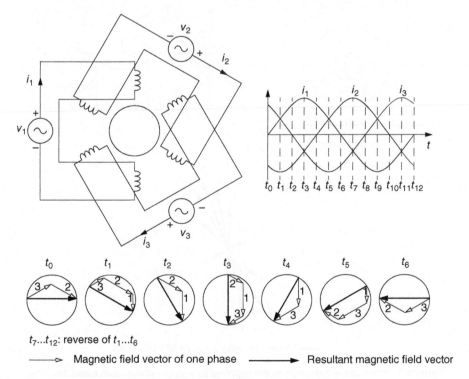

$t_7...t_{12}$: reverse of $t_1...t_6$

—————▷ Magnetic field vector of one phase   ————▶ Resultant magnetic field vector

Figure 4.2 Magnetic field generated by a three-phase coil system [1].

a compass needle is positioned in the middle of the three-phase coil system, the needle keeps pace with the rotating field, which is a crude equivalent of the synchronous motor. When a copper cylinder is placed in the center of the three-phase coil system, the rotating field drags the cylinder around with it, and we have a primitive equivalent of the induction motor.

In household appliances the squirrel-cage induction motor is the workhorse in our washing machines, dishwashers, and refrigerators. They are supplied by a single-phase circuit; how can we then create a rotating magnetic field such that the rotor rotates? The stator windings are supplied by a single-phase source through a Steinmetz connection as is shown in Figure 4.3. In between windings 2 and 3, a capacitor is added in order to establish a 90-degree phase shift between the currents i_2 and i_3. By doing this we are able to create a rotating magnetic field from a single-phase source that drags the rotor around with it.

Synchronous motors

In Appendix C the network model for a synchronous generator is derived from its principle operation and the electromagnetic field equations.

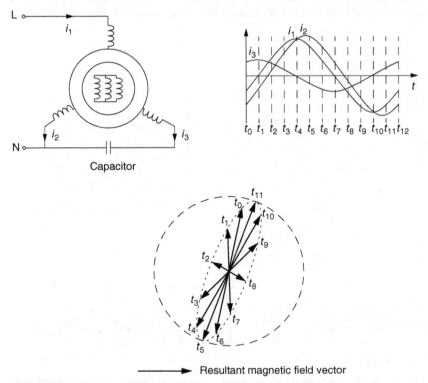

Figure 4.3 Three-phase induction motor supplied by a single-phase source and the resulting rotating magnetic field.

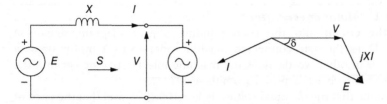

Figure 4.4 The equivalent circuit of a synchronous motor connected to an infinite bus with the corresponding phasor diagram; the resistance of the stator coil is neglected.

A synchronous machine can operate either as a generator or as a motor. The bulk of the electricity is produced by three-phase synchronous generators. Under steady-state conditions, they operate at constant speed and are therefore named synchronous generators. When operating as a motor, the synchronous machines are mainly used in constant speed drives but can be applied in variable speed drives as well, when the synchronous machine is supplied by a power-electronic converter that controls the frequency of the electrical power.

In Figure C.14, the equivalent circuit of a synchronous generator, connected to an infinite bus, with its corresponding phasor diagram, is given. In line with this, the equivalent circuit and the corresponding phasor diagram of a synchronous motor are shown in Figure 4.4.

The active and reactive power exchange between a synchronous machine and the infinite bus, given the direction of the complex power S and the current I as indicated in Figure 4.4, is described by the following two equations (Equations 2.28 and 2.29 in Section 2.5, respectively):

$$P = 3\frac{|V||E|}{X}\sin(\delta) \tag{4.1}$$

$$Q = 3\frac{|V||E|}{X}\cos(\delta) - 3\frac{|V|^2}{X} \tag{4.2}$$

The only parameter that can alter the sign of the active power P (Equation 4.1) is the angle δ, which is the phase difference between the terminal voltage V and the synchronous internal electromotive force (EMF) E; all the other variables in the equation are positive real numbers. When $\delta > 0$, the machine operates as a generator because $P > 0$ and the positive direction points out of the synchronous machine: it actually injects active power into the infinite bus. When $\delta < 0$, $P < 0$ and the machine operates as a motor: it absorbs active power from the infinite bus. There is no exchange of active power with the grid when the machine is operated such that the power angle is $\delta = 0$, and we say that the machine runs mechanically unloaded. In that case the expression for the reactive power is

$$Q = 3\frac{|V||E|}{X} - 3\frac{|V|^2}{X} \tag{4.3}$$

Example 4.1 *Motor or generator?*
Consider the circuit and the corresponding phasor diagram shown in Figure 4.4. Does the grid-connected machine operate as a motor or as a generator? Or in other words, does the machine inject active power into the grid or does it absorb active power from the grid?

Let us assume that the terminal voltage is $V = 100\angle 0°$ V and that the current that flows from the machine into the grid equals $I = 10\angle -150°$ A. The complex power that is injected into the grid equals $S = VI^* = 1000\angle 150°$ VA. Taking the real and imaginary part of this complex power gives us the active and reactive power that are injected into the grid: $P = -866$ W and $Q = 500$ var. Therefore, the machine consumes 866 W active power from the grid and runs as a motor. Furthermore, it injects 500 var reactive power into the grid and is overexcited.

Taking into account the sign convention for the complex power S and for the current I as shown in the circuit, we can obtain this information from the phasor diagram too. The internal EMF E lags the terminal voltage and $\delta < 0$. We can see from Equation 4.1 that in this case also $P < 0$. The direction of the complex power flow points from the machine toward the grid, so that $P < 0$ expresses that the machine consumes active power from the grid and operates as a motor.

Induction motors

Approximately 60% of the electric energy is consumed by electric motors: in the industry and in our houses, induction motors (in German-based literature, induction motors are referred to as asynchronous motors) are the workhorses in air conditioners, washing machines, dishwashers, and so forth. About 90% of the electric motors are induction motors. They are relatively inexpensive to make, easy to maintain (the squirrel-cage rotor has no brushes), and reliable and robust in their operation.

The induction motor does not operate at synchronous speed as the synchronous motor does and has no starting problem because it develops a torque at speeds other than the synchronous speed. Seen from the power system, the induction motor can cause some dynamic problems. These dynamic problems may occur when the machine starts and stops and also when the mechanical load on the motor shaft changes.

When the rotor of an induction motor is at stand still and a three-phase voltage is applied at the terminals of the stator windings, a rotating flux is produced in the air gap between the stator and the rotor as explained in Appendix D.2. The rotating air gap field induces a voltage in the rotor windings, which, since the rotor windings are shorted, causes a current flow in the rotor windings. This produces, in combination with the air gap field, the torque that spins the rotor. The initial starting current is high, easily as high as three to eight times the nominal rated current of the motor. The rotor current decays as the motor speeds up.

Figure 4.5 A model for an induction motor as derived in Figure D.9.

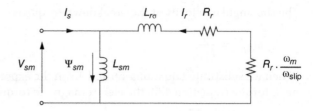

The power system experiences the starting current of large induction motors as an inrush current, similar to the inrush current of large power transformers (see also Section 3.8.3).

In the steady-state situation, the induction motor develops a torque equal in magnitude but opposite in direction to the mechanical load. The torque developed by the rotor needs a certain magnitude of the rotor current (see, e.g., Equation D.12). The magnitude of the rotor current needs a certain value of the induced voltage in the rotor windings, and this induced voltage requires in turn a certain relative speed (this is the difference between the rotor speed and the speed of the air gap flux). The steady-state speed of the rotor must always be sufficiently less than the synchronous speed to develop the torque needed to balance the load torque. This "lagging" of the rotor with respect to the rotating stator field is called the slip angular velocity (see Equation D.17): $\omega_{slip} = \omega_s - \omega_m$. The slip is larger for heavy-load torques and smaller for light-load torques. A practical circuit model for the induction motor is depicted in Figure 4.5.

From Figure 4.5, we can derive an expression for the current I_r:

$$I_r = \frac{-V_{sm}}{j\omega_s L_{r\sigma} + R_r + R_r \cdot \dfrac{\omega_m}{\omega_{slip}}} = \frac{-V_{sm}}{j\omega_s L_{r\sigma} + R_r \cdot \dfrac{\omega_s}{\omega_{slip}}} \tag{4.4}$$

When we substitute this equation in the expression for the electromagnetic torque of an induction motor (Equation D.62), we find

$$T_e = 3 \cdot \frac{R_r}{\omega_{slip}} \cdot |I_r|^2 = \frac{3R_r}{\omega_{slip}} \cdot \frac{V_{sm}^2}{(\omega_s L_{r\sigma})^2 + \left(R_r \cdot \dfrac{\omega_s}{\omega_{slip}}\right)^2}$$

$$= \frac{\dfrac{3}{L_{r\sigma}}\left(\dfrac{V_{sm}}{\omega_s}\right)^2}{\dfrac{\omega_{slip}L_{r\sigma}}{R_r} + \dfrac{R_r}{\omega_{slip}L_{r\sigma}}} \tag{4.5}$$

The maximum value of the torque is called the breakdown torque of the motor:

$$T_b = \frac{\dfrac{3}{2}}{L_{r\sigma}}\left(\frac{V_{sm}}{\omega_s}\right)^2 = \frac{3}{2} \cdot \frac{\Psi_{sm}^2}{L_{r\sigma}} \tag{4.6}$$

The slip angular velocity at the breakdown torque is

$$\omega_{slip,b} = \frac{R_r}{L_{r\sigma}} \tag{4.7}$$

When we substitute Equations 4.6 and 4.7 in the expression of the electromagnetic torque (Equation 4.5), the electromagnetic torque can be written as

$$T_e = \frac{2T_b}{\dfrac{\omega_{slip}}{\omega_{slip,b}} + \dfrac{\omega_{slip,b}}{\omega_{slip}}} \tag{4.8}$$

The electromagnetic torque can be plotted as a function of the angular rotor speed, as is shown in Figure 4.6.

Seen from the power system, the induction motor is a rather tricky load. When under heavy-load conditions the supply voltage decreases, also the electromagnetic torque initially reduces but has to maintain the balance with the torque required by the mechanical load so the slip increases. The increased slip enlarges the induced rotor current, and to maintain the balance with the mechanical load, the current in the stator windings must increase. The induction motor therefore has a negative load characteristic: when the supply voltage drops, the supply current grows larger and causes an extra voltage drop that

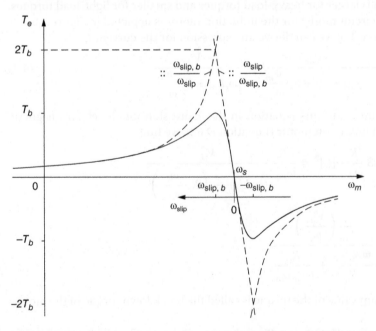

Figure 4.6 The electromagnetic torque of an induction motor as a function of the angular rotor speed.

lowers the supply voltage further. When the distribution feeders are heavily loaded with air conditioners (the compressor of an air conditioner is driven by a squirrel-cage motor), we have a situation that under circumstances can lead to a blackout. When, after disconnecting because of overload tripping, the utility wants to bring the feeder back to service again, the air conditioners are quite often still switched on, and the transient inrush currents can cause immediate tripping of the overcurrent relays that protect the feeder.

4.2.2 Light

One of the first applications of electricity was lighting. Voltage ratings of 110 and 220 V (see Section 1.3.4) and the standardization of the power frequency at 50 or 60 Hz (see Section 1.3.2) find their origin in electric lighting.

The oldest type of electric lamp that is still applied today is the incandescent lamp. The lamp consists of a thin filament that conducts the current. When switched on, the current heats the thin filament, which, in turn, radiates light. The incandescent light bulb produces a fair amount of heat and is not a very efficient source of lighting. To prevent the oxidation of the hot filament by the oxygen in the air, the filament is inside a glass bulb filled with an inert gas. An incandescent light bulb is a pure resistance, that is, the current through the lamp is in phase with the voltage applied (see Figure 4.7 (a)), and the power consumption is used to differentiate between the numerous light bulbs that are for sale. Dimmers can be used to control the brightness of the lamp by changing the RMS value of the voltage applied to the lamp. Devices such as adjustable transformers or resistors can be used for this purpose, but they are expensive, voluminous, and inefficient. Nowadays, (power-)electronic light dimmers are used. They chop in fact the applied AC voltage as is shown in Figure 4.7 (b). The current shape alters accordingly and the power supplied to the lamp reduces as does the light output.

We can see from Figure 4.7 (b) that the application of a dimmer results in non-sinusoidal currents in the grid. In Section 3.8.2, we already discovered that such a "distorted" periodic current consists of a fundamental (50 Hz)

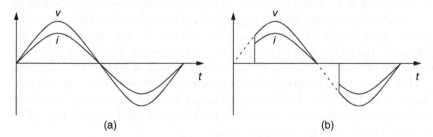

(a)   (b)

Figure 4.7 Voltage and current waveforms of an incandescent light bulb (a) without a dimmer and (b) with a dimmer (the grid voltage is shown as a dashed line).

sinusoidal current and higher harmonics (sinusoidal current components with frequencies that are a multiple of 50 Hz). Therefore, the dimmer causes harmonic currents in the grid. When there are a lot of loads bringing harmonic currents into the grid, for example, DC equipment (see Section 4.2.4), the voltage can become distorted as well. This means that the voltage supplied to the customers deviates from the ideal sinusoidal voltage shape. Simply said, the load voltage is the voltage that remains when we subtract the voltage drops across the series impedances in our grid (caused by the load currents) from the "ideal" voltage that is supplied by the generators. When the loads introduce harmonic currents in the system, these harmonic currents cause harmonic voltage drops across the series impedances in the supply, and this leads to "distorted" load voltages. When we realize that the (inductive) impedances in the system increase with the frequency, it becomes apparent that the voltage drop caused by higher-order harmonics can be considerable.

The halogen lamp is an "improved" incandescent lamp. In an incandescent lamp, the metal filament slowly evaporates and deposits on the glass bulb. Finally, the filament will break and the lamp has to be replaced. The housing of a halogen lamp is filled with a halogen gas that reacts with the evaporated metal and redeposits it on the filament again. Actually, a kind of self-repair or filament regeneration takes place, so that the filament temperature can be higher than for incandescent light bulbs. This results in higher luminosity.

Today the fluorescent lamp is the main source of lighting. In these lamps an electrical discharge takes place where the electrons collide with mercury ions, resulting in ultraviolet radiation. The fluorescent material that covers the inside of the lamp turns the ultraviolet radiation into visible light. Fluorescent lamps dissipate less heat and have a longer service life than incandescent lamps. Where an incandescent lamp can be connected to its supply source directly, a fluorescent lamp needs extra equipment. The fluorescent lamp is a gas discharge lamp, and an inductor (also called ballast) in series with the fluorescent lamp is required to limit the current flow. When switching a fluorescent lamp on, the gas in the tube has to be ignited first. This is taken care of by an ignition device called a "starter." The ballast and the starter are usually located in the armature of the lamp. Nowadays, electronic ballasts are available that control the power supply of the lamp and provide the "starter" function as well. Fluorescent lamps have an inductive nature, because of the current-limiting ballast, and the current lags the supply voltage by almost 90 electrical degrees. Therefore, fluorescent lamps consume a considerable amount of reactive power. In large office buildings the fluorescent lamps are therefore compensated by capacitors, so that the capacitors supply the reactive power and not the utility (see also Example 1.12 (p. 39) in Section 1.6.3).

Compact fluorescent lamps, usually called energy-saving lamps, have an integrated ballast in their housing and are of such a design that they take the same space as incandescent lamps; incandescent lamps can therefore easily be replaced by compact fluorescent lamps.

A light-emitting diode (LED) is a semiconductor that emits light when a current passes through. The main advantage of LEDs is their low energy consumption. In addition, LEDs have a long lifetime. These qualities contribute to the fact that the application of LEDs, or clusters of LEDs, steadily increases and that LEDs have the potential to gradually replace the previously mentioned lighting techniques in the future.

4.2.3 Heat

Electricity is also used for heating: the electric power is converted into heat by a resistor. We find this conversion very often in our houses: in heaters to warm parts of the house, in water cookers and in boilers for warm water, in the stove, and so on. The aluminum- and steelmaking industry applies electric heating on a large scale as well as for their electric arc furnaces. A midsized steelmaking furnace has an electrical rating of 60 MVA. It is in fact a transformer with a secondary voltage of about 800 V and, in the case of a 60 MVA rating, a secondary current of 44000 A. The melted metal short-circuits the secondary winding and the current heats up the furnace.

4.2.4 DC Electrical Energy

The conversion of AC electrical energy into DC electrical energy is a necessary step to drive DC appliances (almost all electronic equipment) from the AC grid. As such, it is not really a load type, but an interface between the DC appliance and the AC grid. Two basic layouts of rectifiers are shown in Figure 4.8. Those rectifier circuits are built up from diodes. A diode is a sort of switch: as soon as the voltage across it becomes positive, it conducts the current, and it isolates when the current through it passes zero. The circuit symbol for the diode illustrates this behavior: it conducts the current in the direction of the arrow, and the vertical line indicates that a current coming from the other direction will be blocked. A more detailed graph of the way the single-phase full-wave rectifier operates is shown in Figure 4.9.

We can see from Figure 4.8 that the output voltage is indeed a DC voltage (it does not change polarity), but it is not yet a constant DC voltage. This can be achieved by putting a smoothing filter across the DC output. The most simple filter is a capacitor at the DC output terminals, as illustrated in Figure 4.10. The output voltage is close to a constant DC voltage, and the capacitor serves in fact as the supply source for the resistance. The diodes only conduct when the instantaneous input voltage (V_{in}) is higher than the output voltage (V_{out}), which is only the case during a relatively small time interval (see Figure 4.10). This leads to a periodic, but non-sinusoidal, current (I_{in}) and introduces harmonic currents into the grid (see also Section 4.2.2).

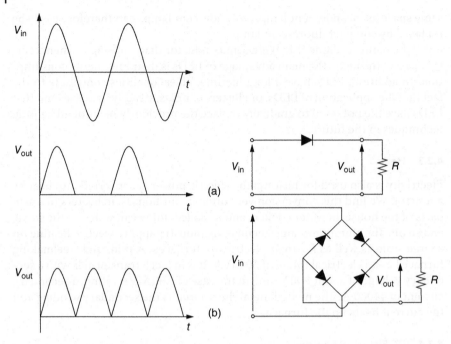

Figure 4.8 Single-phase half-wave (a) and full-wave (b) rectification.

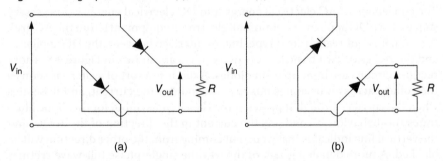

Figure 4.9 Single-phase full-wave rectifier during the positive (a) and the negative (b) half cycle of V_{in}.

The previously mentioned rectifiers are so-called uncontrolled rectifiers, as we have no means to control the diodes. When we replace the diodes by thyristors, being diodes that require (besides a positive voltage drop) a trigger signal before they become conductive, we have a controlled rectifier as shown in Figure 4.11. The circuit symbol for the thyristor (or, more generally, power-electronic switching device) is similar to that of the diode where the control possibility is indicated by means of the line with a right angle to the diode symbol. This gives us the possibility to control the DC power output in a similar way as shown in Figure 4.7 (b) for AC voltage and current.

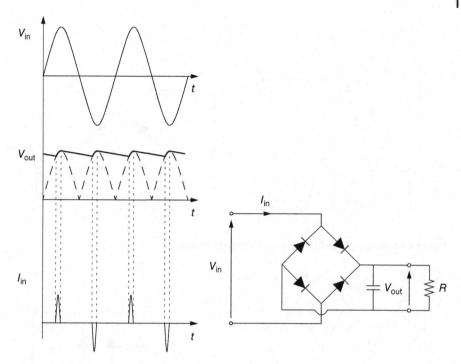

Figure 4.10 Full-wave rectification with a capacitive filter.

Up to now, we only discussed single-phase rectifiers. A three-phase full-wave (uncontrolled) rectifier is shown in Figure 4.12. We can see that the ripple in the DC output voltage (V_{out}) is smaller than in the case of single-phase full-wave rectification (Figure 4.8 (b)).

4.2.5 Chemical Energy

A battery is a device in which chemical energy is directly converted into electrical energy, and – in the case of a rechargeable battery – the other way around. When it is charging, the battery is a load for the power system. Rechargeable batteries are widespread and can be found in electric toothbrushes, laptop computers, mobile phones, cars, and so on.

A battery is built up from one or more voltaic cells. Each voltaic cell has a positive terminal and a negative terminal that are immersed in a solid or liquid electrolyte. The electrical potential across the terminals of the battery that is neither discharging nor being charged is called the open-circuit voltage and is equal to the EMF of the battery. The voltage produced by a voltaic cell depends on the chemicals used. Carbon–zinc and alkaline cells have EMFs of about 1.5 V, lead-acid cells of about 2 V, while lithium cells can deliver 3 V or more.

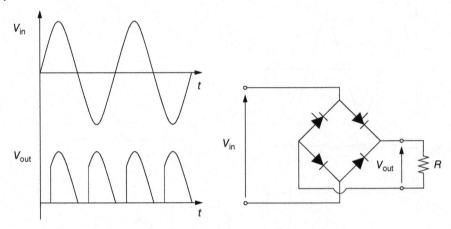

Figure 4.11 Single-phase full-wave controlled rectification.

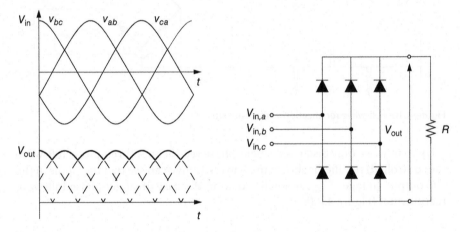

Figure 4.12 Three-phase full-wave uncontrolled rectification.

Because of the chemical reaction inside the cells, the capacity of a battery depends on the discharge conditions such as the duration of the current, the current magnitude, the ambient temperature, and so on. Battery manufacturers rate their batteries with a voltage and an ampere-hour rating: the higher the ampere-hour rating, the longer the battery will last for a certain loading.

Example 4.2 *Ampere-hour rating*
The well-known 1.5 V AA (or LR6, or penlite) battery has a typical capacity of 2000 mAh. When the battery is loaded by a device that draws 40 mA, the device will operate for approximately 50 hours.

4.3   Classification of Grid Users

In this section, the individual loads in the system are clustered and classified as grid users according to the following three categories:

- Residential loads, comprising the domestic users
- Commercial and industrial loads, being the business users and factories
- Electric railways

4.3.1   Residential Loads

The electricity generated in power plants travels to our homes and offices through the distribution grid. The voltage of the distribution grid is typically less than 20 kV. Voltage ratings of 12, 11, 10, and 7.2 kV are quite common. The distribution grid is fed by substations where step-down transformers make the connection with the transmission network. The power goes from the step-down transformer to the distribution bus, and smaller distribution transformers bring the power down to a lower voltage level. A typical distribution transformer has a primary line-to-line voltage of 10 kV and a secondary line-to-line voltage of 400 V. The primary windings are connected in delta, while the secondary low-voltage windings are connected wye. The low-voltage cable connected to the secondary winding has four wires: three wires carry 230 V (in a 400 V system) and the fourth wire is the neutral conductor, which is connected to the star point of the transformer. Residential loads are in general single-phase loads that are connected to the secondary of the distribution transformer in such a way that the total load seen by the distribution transformer can be regarded approximately as a balanced three-phase load. The usual way of doing this is to connect the residences alternately between one of the three phases and the neutral. Sometimes houses do have a three-phase low-voltage connection, when, for instance, cooking is done electrically. The wiring layout of four houses or buildings connected to a three-phase 400 V/230 V supply is shown in Figure 4.13. The current flows through the phase conductor to the electric load and back via the neutral conductor. The energy consumption is measured by the watt-hour meter.

The electric system in the building is protected by fuses, which are mounted on a distribution panel directly behind the watt-hour meter. Larger apparatus can be protected by fuses as well. In a fuse, a thin piece of foil or wire quickly vaporizes when a too high current runs through it. This interrupts the power to the conductor immediately and protects the equipment or the load from overheating. Fuses must be replaced each time they burn out. Quite often miniature circuit breakers are used instead of fuses. A miniature circuit breaker uses the heat from an overload to trip a switch, and circuit breakers are therefore resettable. The grounding (also called earthing) of the supply in a house or building serves as a protection for the users. It protects them from electrical shocks

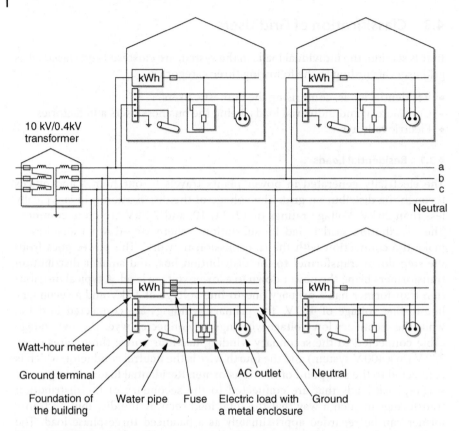

Figure 4.13 Wiring layout of four houses or buildings connected to a three-phase supply.

when a piece of electric equipment has an insulation failure to ground. When such an insulation failure occurs, a short-circuit current, which is many times higher than the normal operating current, flows through the safety ground wire and via the earth back to the star point of the distribution transformer. The fuse(s) of the electrical device will operate and interrupt the power immediately. When the device is not fuse-protected, the fuses, or miniature circuit breakers, mounted on the distribution panel after the watt-hour meter will operate. Besides fuses (or miniature circuit breakers), one or more residual current circuit breakers are installed on the distribution panel as well. Such a device measures the difference between the current in the phase conductor and the current in the neutral conductor and disconnects the circuit as soon as an imbalance occurs that exceeds a certain threshold value. The imbalance can be caused, for example, by a grounded person touching an energized part of the circuit so that the current returns through the ground conductor instead of the neutral conductor.

Safety for human beings and animals is very important, and therefore the safety earth and the neutral conductor are connected to the basement of the building and preferably also connected to the water supply, as long as the water board does not use PVC pipes for its supply. A single-phase current can flow through the building's basement only when it can find itself a return path to the grounding point of the delta/wye-connected distribution transformer. The conductivity of the earth plays an important role. The DC resistivity of wet soil is in the order of $10\,\Omega m$, but in a dry sand area it becomes much higher: $1\,k\Omega m$ up to $10\,k\Omega m$.

When it is not easy to dig a trench because of rocky ground, which makes it very unattractive to use an underground cable system, the electricity distribution is done at a higher voltage than 400 V. In North America, for instance, it is common practice that the distribution transformers step the voltage down to a standard line-to-line voltage of 7.2 kV. This is a four-wire system: three phases and a neutral. The 7.2 kV voltage has to be stepped down by smaller transformers somewhere down the line, either underground or in the air. These transformers regulate the voltage on the line to prevent under- and overvoltage conditions.

4.3.2 Commercial and Industrial Loads

Large factories and industries get their electricity supply at distribution level (10–72.5 kV) or are supplied at sub-transmission level (110–150 kV). The public utility brings underground cables or overhead transmission lines to the premises of the company. The company itself takes care of the supply to the different production facilities, office buildings, and so on, and it owns the medium-/low-voltage transformers.

The advantages of a medium-voltage distribution network over a low-voltage distribution network are the much lower transmission losses and the lower voltage drops. A simple calculation shows us what precisely influences the electrical losses. The simplified grid that is shown in the line-to-neutral diagram in Figure 4.14 supplies the following amount of three-phase active power to the load:

$$P = 3|V||I|\cos(\varphi) \tag{4.9}$$

The current through the grid equals

$$|I| = \frac{P}{3|V|\cos(\varphi)} \tag{4.10}$$

Figure 4.14 Line-to-neutral diagram of a load that is supplied by a simplified grid.

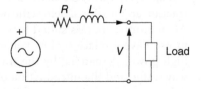

Therefore, the three-phase active power losses in the grid can be written as

$$P_{\text{loss}} = 3|I|^2 R = \frac{P^2 R}{3|V|^2(\cos(\varphi))^2} \tag{4.11}$$

This teaches us that the transmission losses can be reduced by choosing for a higher transmission voltage. Another option is to reduce the resistance of the conductors by increasing the conductor diameter (see also Equation 3.11 in Section 3.9), but this is in general a costly affair. A higher value of the power factor, that is, $\cos(\varphi)$ closer to one, gives us a reduction in the transmission and distribution losses as well. The loads at the end of the feeders are in majority inductive loads, in particular in factories because electrical machines convert electrical energy in a mechanical torque that does the work. Capacitor banks are applied on a large scale to provide the inductive loads with reactive power. In this way, the reactive power does not come from the distant supply but from a locally connected capacitor bank, which saves us the ohmic losses caused by the reactive current. This so-called power factor improvement was illustrated in Example 1.12 (p. 39) in Section 1.6.3.

Quite often, large industries do not only consume electricity but also supply electricity to the grid as an independent power producer (IPP). The chemical and food industry, paper mills, and oil refineries require in their production process not only electricity but also heat, often in the form of steam. When electricity is produced in a conventional thermal power plant, as we discussed already in Chapter 2, only 40–50% of the chemical energy of the fossil fuels is converted into electrical energy, and the rest becomes heat that is absorbed by the environment. A combined production of electricity and heat can therefore be very profitable for industries that also need heat. Most of the combined heat and power (CHP) installations are operated such that the steam generation is the dominant process and the electricity is the "by-product"; the required amount of steam that is required for the production process varies with time, and so will the amount of produced electricity.

4.3.3 Electric Railways

Basically, electrified rail (or road) systems are power systems in themselves that supply traction equipment such as trains, trolleybuses, trams, and metros. In general, those electrified rail systems do not have their own power generation and are interconnected with the main grid for their power supply. Therefore the electrified rail systems are grid users of the main grid, with rather specific needs.

It all started with DC because DC series motors have excellent properties for traction applications. Before the power-electronic converters became available, AC motors were less suited for traction purposes. The DC systems currently in use have a relatively low voltage and need thick conductors to supply the high power demanded by the traction equipment. The high intensity of railway traffic and the introduction of high-speed trains have caused AC systems to become more widespread. High-speed trains such as the TGV in France, the

Figure 4.15 High-speed train. Reproduced with permission of TenneT TSO B.V.

ICE in Germany, and the Shinkansen in Japan are supplied by a 25 kV AC supply. A photo of the ICE is shown in Figure 4.15; on the photo the train runs, non-high speed, on a Dutch 1500 V DC railway track. Typical ratings are 600 and 750 V for DC overhead line and conductor rail systems (trams and metros) and 1500 and 3000 V for DC overhead line systems (trains). For AC traction 15 kV, 16 2/3 Hz and 25 kV, 50 Hz overhead line systems are the standard. These AC systems are all single-phase systems.

An example of a DC railway system is shown in Figure 4.16. The AC grid voltage is transformed to a lower voltage level and then rectified. The DC power supply is connected to the overhead lines and the rails. The DC current flows from the overhead lines, to the train, via the rails back to the supply substation. The substations are equipped with three-phase rectifiers (see also Section 4.2.4) and comprise a balanced loading for the grid.

An example of an AC system is depicted in Figure 4.17. The AC grid voltage is transformed to a lower voltage level and connected to overhead lines and rails. The transformer can only be connected to two of the three phases of the grid,

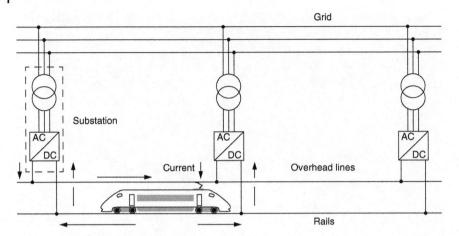

Figure 4.16 Supply principle of a DC railway system.

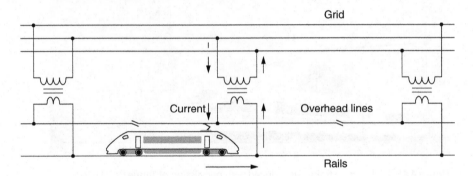

Figure 4.17 Supply principle of an AC railway system.

which leads to an asymmetrical load on the three-phase grid. This effect can be overcome by cyclically connecting the transformers between the different phases, as is illustrated in Figure 4.17. This, however, gives rise to a phase difference between neighboring supply areas. For this reason, the different supply areas cannot be connected with each other, and the train travels through small sections that are not supplied by electric power.

Problems

4.1 Explain the following terms and concepts:
  a. Peak load
  b. Valley load
  c. Distribution substation
  d. Vertically operated power systems

4.2 A squirrel-cage induction motor has a rotor resistance of $R_r = 0.1\,\Omega$ per phase and a rotor reactance of $X_r = 0.7\,\Omega$ per phase. The stator has eight poles and the motor operates at 230 V phase voltage and 50 Hz. Determine the stalled rotor current, being the current drawn from the grid when the rotor is at standstill, per phase.

4.3 Describe four different types of lamps used for lighting.

4.4 Give examples in which electricity is used for heating. Explain the mechanism.

4.5 Explain how a single-phase rectifier works. Draw the circuit diagram of a half-wave rectifier and a full-wave rectifier and draw the respective output voltages.

4.6 Batteries convert chemical energy into electricity.
a. Explain how a battery is built up.
b. How can the electromotive force (EMF) of a battery be measured? What are common EMF values for lead acid and for lithium cells?
c. Which conditions affect the capacity of a battery?
d. How do battery manufacturers rate their battery?

4.7 Give some common line-to-line voltage ratings for distribution grids at the primary and secondary windings of distribution transformers to which the residential loads are connected. Explain how the network voltages are brought down to low voltage levels.

4.8 Answer the following questions about traditional residential loads:
a. Give two characteristics of residential loads.
b. Explain how residential loads are connected in order to obtain a three-phase balanced network load.

4.9 Commercial and industrial loads are connected at distribution level (10–72.5 kV) or at sub-transmission level (110–150 kV).
a. What are the advantages of a medium-voltage distribution network over a low-voltage distribution network?
b. Are industrial loads at the end of the feeder commonly resistive, capacitive, or inductive? Explain why.
c. How is the power factor compensated?

4.10 An industrial three-phase motor, with a rated power of 10 kW and a power factor of 0.8, is Y connected to a 400 V power supply. Calculate the current in each phase.

4.11 Electrified rail systems supply traction equipment such as trains, trolleybuses, trams, and metros. Referring to their physical characteristics, typical ratings, and applications,

a. give two typical characteristics of DC-powered railways.
b. give two typical characteristics of AC-powered railways.

Reference

1 Laithwaite, Eric R., and Freris, Leon L.: *Electric Energy: Its Generation, Transmission and Use*, McGraw-Hill, Maidenhead, 1980, ISBN 0-07-084109-8.

5

Power System Control

5.1 Introduction

Electrical power systems can be regarded as one of the most complex systems designed, constructed, and operated by humans. The consumers are supplied with the requested amount of active and reactive power at constant frequency and with a constant voltage. Loads are switched on and off continuously, and because electricity cannot efficiently be stored in large quantities, the balance between the amount of generated and consumed electricity has to be maintained by control actions.

The active power balance is controlled by the generators. There is also another option, and that is to reduce the active power consumption by disconnecting parts of the load (this is called load shedding), but this is merely an emergency measure and not common practice. The reactive power balance can be controlled by rotating equipment (generators and motors) and by static components (capacitors and inductors). The synchronous generator is the most important component in the system for maintaining the active and the reactive power balance.

Let us look at what happens in a simple system, in which a generator is supplying a variable load, when suddenly more active or reactive power is consumed and no control actions are taken. When we neglect losses, we can write for the active power balance

$$P_m = P_e + P_a \tag{5.1}$$

P_m the mechanical power supplied to the generator axis by the prime mover [W]

P_e the electrical (active) power output of the generator [W]

P_a the power accelerating (or slowing down) the generator [W]

We start from the situation where the generated and consumed power are in balance (i.e., $P_m = P_e$ and $P_a = 0$) and a sudden increase of the active power consumption of the load occurs ($P_e \uparrow$). When no control actions are taken,

Electrical Power System Essentials, Second Edition. Pieter Schavemaker and Lou van der Sluis.
© 2017 John Wiley & Sons Ltd. Published 2017 by John Wiley & Sons Ltd.

the mechanical power supplied to the generator remains constant. In order to maintain the active power balance (Equation 5.1), a decelerating power arises ($P_a \downarrow$): the rotation of the generator slows down and the frequency of the electrical power output drops. Without control actions, in this case increasing the mechanical power input to the generator in order to restore the active power balance and thus the frequency, the deceleration continues. The following numerical example illustrates this.

Example 5.1   *More active power consumption from a non-controlled generator*

A generator is running on no load with a frequency of $f_1 = 50$ Hz (3000 RPM). Assume that the kinetic energy stored in the rotating parts of the generator and steam turbine equals $K_1 = 200$ MJ. Suddenly a 10 MW load is connected. Without control actions, the frequency drop after one second can be computed as follows.

In one second of time, the load consumes $10 \times 10^6 \cdot 1 = 10$ MJ of electric energy. As a result, the kinetic energy of the rotating part of the generator reduces to $K_2 = 190$ MJ. The kinetic energy of rotation equals $K = (1/2)J\omega^2 = 2 \cdot J \cdot \pi^2 \cdot f^2$, with J the moment of inertia relative to the axis of rotation. Therefore, the frequency after one second (f_2) has reduced to

$$\frac{K_1}{K_2} = \frac{200}{190} = \frac{f_1^2}{f_2^2} = \frac{50^2}{f_2^2} \rightarrow f_2 = 48.73\,\text{Hz} \tag{5.2}$$

The influence of an increasing reactive power consumption of the load is illustrated in the following example.

Example 5.2   *More reactive power consumption from a non-controlled generator*

Consider a generator that is connected to a variable load, as shown in Figure 5.1. In this 10 kV system, the load consumes a three-phase active power of 2 MW with a power factor of 0.9 lagging. The generator terminal voltage serves as a reference: $V_t = |V_t|\angle 0 = (10 \times 10^3/\sqrt{3})\angle 0$ V, and the current can be calculated from $P_{3\phi} = 3|V_t||I|\cos(\varphi) = 3 \cdot (10 \times 10^3/\sqrt{3}) \cdot |I| \cdot 0.9 = 2$ MW and equals $I = 128.3\angle{-25.8°}$ A. The reactance of the generator is $X = 3\,\Omega$ and the internal EMF of the generator can be calculated from $E = |E|\angle\delta = V_t + jXI = 5951\angle 3.3°$ V. The amount of reactive power consumed by the load follows from $\cos(\varphi)^2 = P^2/(P^2 + Q^2) = 0.81$ and is equal to $Q = 968.6$ kvar.

Starting from the aforementioned situation in which the power is in balance, the reactive power consumption of the load suddenly increases to a value of $Q = 1.5$ Mvar; the amount of consumed active power remains the same. When no control actions are taken, both the mechanical power supplied to the generator and the internal EMF remain constant. The voltage angle at the load serves

Figure 5.1 Generator connected to a variable load.

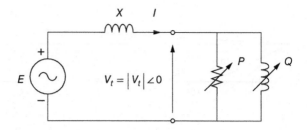

still as a reference: $V_t = |V_t|\angle 0$. The power factor has reduced to

$$\cos(\varphi) = \frac{P}{\sqrt{P^2 + Q^2}} = 0.8 \tag{5.3}$$

and the angle between the voltage phasor V_t and the current phasor I is $\varphi = 36.9°$. The current can be determined from the following two equations (Equations 2.25 and 2.26 in Section 2.5):

$$P_{3\phi} = 3|E||I|\cos(\varphi + \delta) = 3 \cdot 5951 \cdot |I| \cdot \cos(36.9 + \delta) = 2\,\text{MW} \tag{5.4}$$

$$Q_{3\phi} = 3|E||I|\sin(\varphi + \delta) - 3X|I|^2$$
$$= 3 \cdot 5951 \cdot |I| \cdot \sin(36.9 + \delta) - 9|I|^2 = 1.5\,\text{Mvar} \tag{5.5}$$

When these two equations are squared and added, the following equation results:

$$(3 \cdot 5951 \cdot |I|)^2 = (2 \times 10^6)^2 + (1.5 \times 10^6 + 9|I|^2)^2 \tag{5.6}$$

The current equals $I = |I|\angle{-\varphi} = 146.8\angle{-36.9°}$ A. The voltage at the terminals of the load can be calculated from

$$P_{3\phi} = 3|V_t||I|\cos(\varphi) = 3 \cdot |V_t| \cdot 146.8 \cdot 0.8 = 2\,\text{MW} \tag{5.7}$$

and equals $V_t = |V_t|\angle 0 = 5676.6\angle 0\,\text{V}$. As a result of the increased reactive power consumption, the terminal voltage dropped with almost 100 V!

The internal EMF is $E = |E|\angle\delta = V_t + jXI = 5951\angle 3.4°\,\text{V}$. The voltage and current phasors – before and after the increased reactive power consumption – are depicted in Figure 5.2.

It is clear from Examples 5.1 (p. 186) and 5.2 (p. 186) that without the appropriate control actions, both the power system frequency and the voltage would be far from constant.

5.2 Basics of Power System Control

In the previous section, we observed that an increased active power consumption reduces the frequency and that, in the example shown, an increased reactive power consumption reduces the voltage. In the steady-state situation,

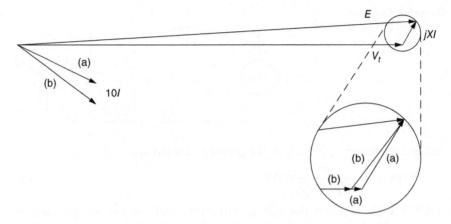

Figure 5.2 Phasor diagram: (a) reactive power consumption $Q \approx 1$ Mvar; (b) reactive power consumption $Q = 1.5$ Mvar.

the active power–frequency control and reactive power–voltage control are approximately independent from each other, as is shown in the following analysis. Consider the power transfer through an impedance as shown in Figure 5.3. The current flowing from node 1 to node 2 can be expressed as

$$I = \frac{|V_1|\angle\delta_1 - |V_2|\angle\delta_2}{|Z|\angle\rho} \quad \text{and} \quad I^* = \frac{|V_1|\angle(-\delta_1) - |V_2|\angle(-\delta_2)}{|Z|\angle(-\rho)} \quad (5.8)$$

The complex power at node 2 (which is transferred to a load or another part of the network) is

$$S = P + jQ = V_2 I^* = |V_2|\angle\delta_2 \cdot \frac{|V_1|\angle(-\delta_1) - |V_2|\angle(-\delta_2)}{|Z|\angle(-\rho)}$$

$$(5.9)$$

$$= \frac{|V_1||V_2|\angle(\delta_2 - \delta_1) - |V_2|^2\angle 0}{|Z|\angle(-\rho)} = \frac{|V_1||V_2|}{|Z|}\angle(\rho + \delta_2 - \delta_1) - \frac{|V_2|^2}{|Z|}\angle\rho$$

Accordingly, the active and reactive power components at node 2 are

$$P = \mathrm{Re}(S) = \frac{|V_1||V_2|}{|Z|}\cos(\rho + \delta_2 - \delta_1) - \frac{|V_2|^2}{|Z|}\cos(\rho)$$

$$(5.10)$$

$$Q = \mathrm{Im}(S) = \frac{|V_1||V_2|}{|Z|}\sin(\rho + \delta_2 - \delta_1) - \frac{|V_2|^2}{|Z|}\sin(\rho)$$

In transmission systems where most of the control actions take place, the following approximations can be made:

- Because the resistance of transmission links is much smaller than the reactance values $(R \ll X)$, the resistance of the transmission link can be neglected: $Z = |Z|\angle\rho = X\angle(\pi/2)$.

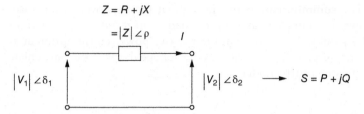

Figure 5.3 Power transport through an impedance.

- Because the difference between the voltage angles is rather small, we can replace $\sin(\delta_1 - \delta_2)$ by $\delta_1 - \delta_2$ and $\cos(\delta_1 - \delta_2)$ by the value 1.

When we substitute these approximations in the active and reactive power equations (Equation 5.10), the following expressions for the active and reactive power are obtained:

$$P = \mathrm{Re}(S) = \frac{|V_1||V_2|}{X}\sin(\delta_1 - \delta_2) = \frac{|V_1||V_2|}{X}(\delta_1 - \delta_2)$$

$$Q = \mathrm{Im}(S) = \frac{|V_1||V_2|}{X}\cos(\delta_1 - \delta_2) - \frac{|V_2|^2}{X} = \frac{|V_2|}{X}(|V_1| - |V_2|)$$

(5.11)

We can see that:

- The voltage depends on the reactive power (see also Example 5.2 (p. 186)).
- The angle of transmission depends on the active power.

This "decoupling" between the active power and voltage angles and the reactive power and voltage magnitudes is of great importance for power system computations as we see in Sections 6.2.3 and 6.2.4.

Example 5.3 *More reactive power consumption from a non-controlled generator (revisited)*
Consider again the generator that is connected to a variable load as in Example 5.2 (p. 186). The simplified reactive power equation (Equation 5.11) can be applied to determine the change in voltage when the reactive power consumption of the load increases to 1.5 Mvar:

$$Q = \frac{1.5 \times 10^6}{3} = \frac{|V_2|}{X}(|V_1| - |V_2|) = \frac{|V_t|}{3}(5951 - |V_t|)$$

(5.12)

The voltage drops to $|V_t| = 5687.3\,\mathrm{V}$. When we compare this voltage with the value that we calculated in Example 5.2 (p. 186): $|V_t| = 5676.6\,\mathrm{V}$, the difference is not more than 11 V. This shows that the approximating equations (Equation 5.11) can be fruitfully applied to get a quick insight in the voltage changes due to deviations in the power balance.

There is an important difference between the coupling of the active power and the frequency and the coupling of the reactive power and the voltage.

The frequency is a common parameter throughout the network: an increase of the active power consumption at a certain load point results in a reduction of the system frequency. An increase of the reactive power consumption at a certain load point is noticed only locally, as the voltage drops at the particular load point and at some nearby nodes.

5.3 Active Power and Frequency Control

5.3.1 Primary Control

A change in the balance between the generation of active power and the consumption of active power changes the kinetic energy of the rotating mass of the generators and alters the system frequency. Without control actions – such as increasing (decreasing) the input of mechanical power to the generator in order to restore the active power balance and thus the frequency, the deceleration (acceleration) continues as is already explained in the introduction of this chapter.

The active power balance is restored by a so-called speed governor, as is shown in Figure 5.4. The speed governor can be set with a characteristic (frequency–power relation) as is depicted in Figure 5.5 (a): the rotational speed (and thus the frequency) is independent of the generator loading. Such a control becomes problematic when generators are connected to run in parallel: when a frequency drop occurs, caused by an increased active power consumption in the system, each speed governor will try to restore the original frequency by increasing the power to the prime mover (in Figure 5.4 the steam valve will be opened further to supply more steam to the turbine). The generators equipped with fast-operating controllers will produce more active power than the generators with slow-operating controllers, and after a number of control actions, the active power to be produced is distributed more or less randomly over the generators. This is an unwanted situation, and this

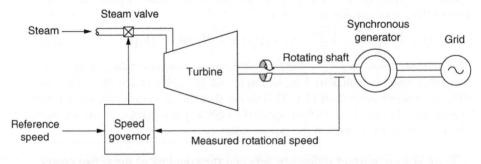

Figure 5.4 The basic principle of the speed governor control system of a generating unit.

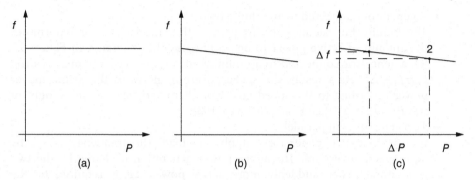

Figure 5.5 Speed governor characteristics.

problem can be resolved by changing the speed governor characteristic into a one-to-one relation between frequency and power as shown in Figure 5.5 (b). This characteristic has a negative slope and a higher (lower) active power output is established when the frequency drops (rises). The slope of the frequency–power characteristic is called the droop or regulation and can be expressed as

$$R_{gi} = -\frac{\Delta f / f_r}{\Delta P_{gi} / P_{gi,r}} \tag{5.13}$$

R_{gi} the droop or regulation of generator i [pu]
Δf the frequency change in the system [Hz]
f_r the nominal rated frequency [Hz]
ΔP_{gi} the change in active power of generator i [MW]
$P_{gi,r}$ the nominal rated power of generator i [MW]

In other words, the droop is the frequency drop, in per unit of the rated frequency, when the active power output of the generator rises from no load to full load (being the rated power).

For the speed governor, three different cases can be distinguished:

- A stand-alone generator
  The load dictates the amount of active power to be supplied by the generator; the speed governor determines the frequency. In Figure 5.5 (c), a stand-alone generator connected to a load experiences a sudden change of the active power balance: the load suddenly requires more active power $P_2 = P_1 + \Delta P$. To cope with this, the kinetic energy of the rotating mass of the generator, and thus the frequency, drops. To restore the active power balance, the speed governor increases the mechanical power of the prime mover with ΔP, in accordance with the speed governor characteristic. The new frequency, for which the active power balance is fulfilled, is lower than the original frequency: $f_2 = f_1 - \Delta f$. See Example 5.4 (p. 192) for the illustration of this behavior.

- A generator connected to an infinite bus
  The infinite bus dictates the frequency; the speed governor determines the amount of active power to be supplied by the generator. In Figure 5.5 (c), a generator connected to an infinite bus experiences a frequency drop $f_2 = f_1 - \Delta f$. As a result, the speed governor increases the prime mover power according to the speed governor characteristic with an amount of ΔP: $P_2 = P_1 + \Delta P$. See Example 5.5 (p. 193).
- Two generators in parallel
  In this case two generators supply the load; the frequency is set by both speed governors. The active power generation is shared by the two generators. When suddenly more active power ΔP is required by the load, the frequency drops with Δf (see Figure 5.6). To restore the active power balance, the speed governors increase the prime mover power according to the respective speed governor characteristics: $\Delta P = \Delta P_{g1} + \Delta P_{g2}$ (with $\Delta P_{g1} = P_{g1,2} - P_{g1,1}$ and $\Delta P_{g2} = P_{g2,2} - P_{g2,1}$). See Example 5.6 (p. 193).

Example 5.4 *A stand-alone generator*
A stand-alone generator feeding a single load supplies 50 MW at 50 Hz (indicated with "1" in the speed governor characteristic as shown in Figure 5.5 (c)). The generator has a nominal rated power of 200 MW, and its speed governor has a droop of $R = 0.02$ pu. The active power consumption of the load suddenly increases to 100 MW, and this event causes a frequency drop of

$$\Delta f = -R \cdot f_r \cdot (\Delta P / P_r) = -0.02 \cdot 50 \cdot (50/200) = -0.25\,\text{Hz} \tag{5.14}$$

The power is balanced again in the new working point "2" of the speed governor characteristic (Figure 5.5 (c)) with a frequency $f_2 = 50 - 0.25 = 49.75$ Hz and a generator output power of $P_2 = 100$ MW. Observe that the droop of a stand-alone generator is rather small in order to keep the deviation in the frequency small.

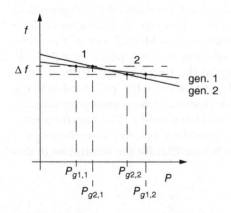

Figure 5.6 Speed governor characteristics of two generators in parallel.

Note that the generator, without a speed governor being present, would continuously decelerate as described in Section 5.1. The speed governor takes care that the active power balance is restored at a slightly lower frequency.

Example 5.5 *A generator connected to an infinite bus*
A generator connected to an infinite bus supplies 50 MW at the nominal rated frequency of 50 Hz (indicated with "1" in the speed governor characteristic as shown in Figure 5.5 (c)). The generator has a nominal rated power of 200 MW, and its speed governor has a droop of $R = 0.06$ pu. When the system frequency drops to a value of $f_2 = 49.5$ Hz, the active power supplied by the generator increases with

$$\Delta P = \frac{-\Delta f/f_r}{R} \cdot P_r = \frac{0.5/50}{0.06} \cdot 200 = 33.3 \text{ MW} \tag{5.15}$$

In the new working point, point "2" in the speed governor characteristic of Figure 5.5 (c), the generator supplies $P_2 = 50 + 33.3 = 83.3$ MW at a frequency of $f_2 = 49.5$ Hz. Observe that the change in output power of the generator is smaller when the droop of the generator is larger.

Example 5.6 *Two generators in parallel*
Consider a situation with two generators running in parallel. Generator 1 has a nominal rated power $P_{g1,r} = 200$ MW, and its speed governor has a droop $R_{g1} = 0.02$ pu. Generator 2 has a nominal rated power $P_{g2,r} = 200$ MW, and its speed governor has a droop $R_{g2} = 0.06$ pu. When the load suddenly consumes more active power ΔP, the frequency drops with an amount Δf. To restore the active power balance, the speed governors increase the prime mover power in accordance with the speed governor characteristics:

$$\left. \begin{aligned} \Delta P_{g1}/P_{g1,r} &= -\frac{\Delta f/f_r}{R_{g1}} \\ \Delta P_{g2}/P_{g2,r} &= -\frac{\Delta f/f_r}{R_{g2}} \end{aligned} \right\} \quad \frac{\Delta P_{g1}/P_{g1,r}}{\Delta P_{g2}/P_{g2,r}} = \frac{R_{g2}}{R_{g1}} \tag{5.16}$$

$$\frac{\Delta P_{g1}/200}{\Delta P_{g2}/200} = \frac{0.06}{0.02} = 3 \rightarrow \frac{\Delta P_{g1}}{\Delta P_{g2}} = 3 \tag{5.17}$$

In other words, the active power increase of the load will be distributed over the two generators ($\Delta P = \Delta P_{g1} + \Delta P_{g2}$) with a ratio 3:1, that is, generator 1 takes 75% of the load increase and generator 2 takes 25%. When both generators have an equal droop, the load increase will be distributed over the two generators according to the ratio of their nominal rated powers.

The case of two generators running in parallel, as described earlier, is in fact the most simple case of a multi-generator system. In a large-scale power system, with a large number of generators connected, it is the network power frequency characteristic that relates the difference between scheduled and actual system frequency to the amount of generation required to correct the power imbalance for that system:

$$\lambda = -\frac{\Delta P}{\Delta f} \tag{5.18}$$

λ the network power frequency characteristic [MW/Hz]
ΔP the amount of generation required to correct the power imbalance [MW]
Δf the difference between scheduled and actual system frequency [Hz]

The network power frequency characteristic λ is determined by the droop of the primary active power controls and the nominal rated power of all the generators in the system. A change in active power consumption is met by all the generators in the system (Equation 5.13 is substituted; note that the change in frequency and the nominal rated frequency are the same throughout the system):

$$\Delta P = \sum_i \Delta P_{gi} = \sum_i -\frac{1}{R_{gi}} \cdot \frac{P_{gi,r}}{f_r} \cdot \Delta f \tag{5.19}$$

R_{gi} the droop or regulation of generator i [pu]
Δf the frequency change in the system [Hz]
f_r the nominal rated frequency [Hz]
ΔP_{gi} the change in active power of generator i [MW]
$P_{gi,r}$ the nominal rated power of generator i [MW]

The network power frequency characteristic λ can be written as

$$\lambda = -\frac{\Delta P}{\Delta f} = \sum_i \frac{1}{R_{gi}} \cdot \frac{P_{gi,r}}{f_r} \tag{5.20}$$

Example 5.7   *The network power frequency characteristic of a system with two generators*

Consider again the two generators operating in parallel as in Example 5.6 (p. 193). Generator 1 has a nominal rated power $P_{g1,r} = 200$ MW, and its speed governor has a droop $R_{g1} = 0.02$ pu. Generator 2 has a nominal rated power $P_{g2,r} = 200$ MW, and its speed governor has a droop $R_{g2} = 0.06$ pu. The network power frequency characteristic of this system equals

$$\lambda = -\frac{\Delta P}{\Delta f} = \sum_i \frac{1}{R_{gi}} \cdot \frac{P_{gi,r}}{f_r} = \frac{1}{0.02} \cdot \frac{200}{50} + \frac{1}{0.06} \cdot \frac{200}{50} = 266.6\,\text{MW/Hz}$$

$$\tag{5.21}$$

This tells us that an increase of the active power consumption of 133.3 MW, which requires an amount of generation $\Delta P = 133.3$ MW to correct the power imbalance in the system, reduces the system frequency with half a hertz: $\Delta f = -0.50\,\text{Hz}$.

Motors cause the system load to be slightly frequency dependent in practice. That means that when the frequency drops, the load is reduced slightly too. This is referred to as self-regulation of the load, which influences the network power frequency characteristic as follows:

$$\lambda = -\frac{\Delta P}{\Delta f} = \sum_i \frac{1}{R_{gi}} \cdot \frac{P_{gi,r}}{f_r} + \frac{\mu}{100} \cdot P \qquad (5.22)$$

$$\mu = \frac{\Delta P'/P}{\Delta f} \cdot 100\% \qquad (5.23)$$

μ the self-regulating effect of the load [%/Hz]
$\Delta P'$ the change in the power consumption of the loads because of the change in frequency Δf [MW]
P the value of the original load of the system plus the change ΔP at the original system frequency (i.e., at the frequency before adding ΔP) [MW]

Example 5.8 *The network power frequency characteristic taking the self-regulating effect of the load into account*
Let us consider once more the system with the two generators from Example 5.7 (p. 194): in this example we calculated that an increase in active power consumption of 133.3 MW lowers the system frequency with half a hertz: $\Delta f = -0.50\,\text{Hz}$.

Now let us take into account the self-regulating effect of the loads in the system as well, for instance, $\mu = 2\%/\text{Hz}$. Assume that the load of the system rises from 100 MW to a value of $P = 100 + 133.3 = 233.3\,\text{MW}$. The network power frequency characteristic equals (Equation 5.22)

$$\lambda = 266.6 + 0.02 \cdot 233.3 = 271.3\,\text{MW}/\text{Hz} \qquad (5.24)$$

The increase of the active power consumption in the system with 133.3 MW reduces the frequency with $\Delta f = -133.3/271.3 = -0.49\,\text{Hz}$. The reduction of the system frequency is less than in our previous example ($-0.49\,\text{Hz}$ instead of $-0.50\,\text{Hz}$). This is because of the self-regulating effect of the loads ($\mu = 2\%/\text{Hz}$) that we have now taken into account. The frequency drop causes a slight reduction of the active power consumption of the loads (Equation 5.23):

$$\Delta P' = \frac{\mu}{100} \cdot \Delta f \cdot P = 0.02 \cdot -0.49 \cdot 233.3 = -2.3\,\text{MW} \qquad (5.25)$$

It is this reduced active power consumption that makes the reduction of the system frequency to be less ($-0.49\,\text{Hz}$ instead of $-0.50\,\text{Hz}$) than in the case where the self-regulating effect of the loads was not taken into account.

5.3.2 Secondary Control or Load Frequency Control (LFC)

When an imbalance between the active power generation and active power consumption occurs, the primary control (being the speed governor) detects a frequency deviation and increases or decreases the mechanical power supply to the generator in order to restore the active power balance and to prevent the frequency from deviating further. As the speed governor characteristics have a droop, the power balance is restored at a lower or higher frequency. The load frequency control (LFC) is used to modify the setting of the speed governor in such a way that the frequency is brought back to the original reference value. This is illustrated in Figure 5.7. Let us assume that a stand-alone generator supplies a load (point "1" on the I characteristic). When the load consumes more active power $P_2 = P_1 + \Delta P$, the power balance is distorted, and the kinetic energy of the rotating mass of the generator, and thus the frequency, will decrease. To restore the active power balance again, the speed governor increases the prime mover power, in accordance with the speed governor characteristic, with the value ΔP. The active power balance is restored (point "2" on the I characteristic) at a frequency that is lower than the original one: $f_2 = f_1 - \Delta f$. The LFC brings back the original frequency by increasing temporarily the prime mover power that raises the kinetic energy (and thus the frequency) of the generating unit. In the graph of Figure 5.7, this action is equivalent to shifting the I characteristic in upward direction to the level of the II characteristic (note that the power output of the generator remains constant during this action so that the horizontal position of working point "2" remains fixed).

An interconnected power system is divided into control areas. The transmission system operator (TSO) is responsible for maintaining the power balance in a specific control area. The balance in a control area is reached when the scheduled power exchange with the neighboring control areas equals the actual

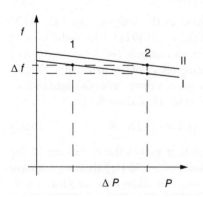

Figure 5.7 LFC action.

power exchange. The TSO takes the necessary control actions in order to maintain or restore the power balance.

In large interconnected power systems, a frequency drop occurs throughout the system and is not a measure that can be used to indicate where in the system an imbalance occurred. In other words, from the value of the frequency alone, it cannot be determined which control area is responsible for the frequency drop. Additional information is necessary and can be obtained by determining the control area's imbalance, that is, a comparison of the actual and the scheduled power exchange of a control area to its neighboring control areas. Note that the effect of the primary control action in the control area must be subtracted from the control area's imbalance so that the secondary control action does not neutralize the primary one:

$$\text{ACE}_i = (P_{a,i} - P_{s,i}) + \lambda_i(f_a - f_s) = \Delta P_i + \lambda_i \Delta f \tag{5.26}$$

ACE_i the area control error of control area i [MW]
$P_{a,i}$ the actual power export of control area i [MW]
$P_{s,i}$ the scheduled power export of control area i [MW]
λ_i the network power frequency characteristic of control area i [MW/Hz]
f_a the actual frequency [Hz]
f_s the scheduled frequency [Hz]

The area control error (ACE) is an indication of the surplus or the lacking amount of power in a particular control area. A negative ACE shows that the control area generates too little power to exchange the scheduled amount. A positive ACE means that the control area produces excess power and exchanges more than the scheduled amount. The ACE must be brought back to zero in order to establish the scheduled power exchange at the nominal frequency.

Example 5.9 *Area control error (ACE)*
Consider a network that is divided in three control areas as drawn in Figure 5.8 (a) [1]. Control area C imports a scheduled amount of power (500 MW) from control area B; note that 100 MW flows via the intertie connections with area A, which has a zero-level scheduled import/export. The network power frequency characteristics of the three control areas are as follows:

Area A $\lambda_A = 13000\,\text{MW/Hz}$
Area B $\lambda_B = 16000\,\text{MW/Hz}$
Area C $\lambda_C = 11000\,\text{MW/Hz}$

For the whole system, the network power frequency characteristic amounts to

$$\lambda = -\frac{\Delta P}{\Delta f} = \sum_i \frac{1}{R_{gi}} \cdot \frac{P_{gi,r}}{f_r} = \lambda_A + \lambda_B + \lambda_C = 40000\,\text{MW/Hz} \tag{5.27}$$

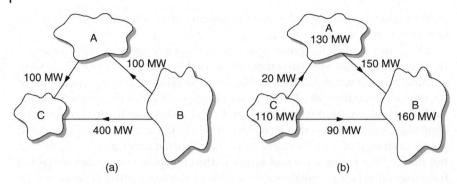

Figure 5.8 Power exchange between three control areas: (a) original (scheduled) situation; (b) incremental generation after losing 400 MW of generation in control area B and the resulting flows.

When suddenly a 400 MW generator is lost by a forced outage in control area B, the frequency in the whole system drops by 0.01 Hz:

$$\Delta f = -\frac{\Delta P}{\lambda} = -\frac{400}{40000} = -0.01\,\text{Hz} \tag{5.28}$$

Governed by the primary control, the generators in the three control areas increase their power output as follows (note that the sum of the increased generator power output equals the lost generating power of 400 MW in area B):

$$\Delta P_A = -\lambda_A \Delta f = 13000 \cdot 0.01 = 130\,\text{MW}$$
$$\Delta P_B = -\lambda_B \Delta f = 16000 \cdot 0.01 = 160\,\text{MW} \tag{5.29}$$
$$\Delta P_C = -\lambda_C \Delta f = 11000 \cdot 0.01 = 110\,\text{MW}$$

This effects the power flows through the intertie connections as depicted in Figure 5.8 (b), and the ACE for each area equals (Equation 5.26)

$$\text{ACE}_A = (130 - 0) + 13000(-0.01) = 0\,\text{MW}$$
$$\text{ACE}_B = (260 - 500) + 16000(-0.01) = -240 - 160 = -400\,\text{MW} \tag{5.30}$$
$$\text{ACE}_C = (-390 - (-500)) + 11000(-0.01) = 0\,\text{MW}$$

Based on these ACE values, only secondary control actions have to be taken in area B: the generation in area B is increased to compensate for the loss of 400 MW of power and to bring back the frequency to its original value. The areas A and C then automatically return to their original operating conditions.

5.4 Voltage Control and Reactive Power

The frequency is sometimes called the system frequency: in an interconnected power system, the frequency has the same value everywhere in the system; in

Figure 5.9 Simplified AVR diagram.

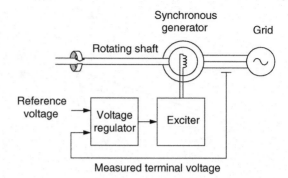

other words, it is independent of the location. A similar "system voltage" does not exist: the voltage amplitude depends strongly on the local situation in the system. As a consequence, voltages in the power system can only be controlled locally: either at generator buses by adjusting the generator voltage control or at fixed points in the system where tap-changing transformers, capacitor banks, or other reactive power consumers/producers are connected.

5.4.1 Generator Control (AVR)

The automatic voltage regulator (AVR) is the basis of the generator reactive power control. A simplified block diagram of an AVR is shown in Figure 5.9. The main task of the AVR is to keep the value of the voltage at the synchronous generator terminals at a specified level.

The principle of operation of the AVR is rather straightforward. When the terminal voltage of the generator decreases (increases), the voltage regulator magnifies (reduces) the excitation, which results in a higher (lower) internal EMF and terminal voltage. The same result can be obtained by increasing (decreasing) the reference voltage that is offered to the voltage regulator (see Figure 5.9). When we consider a generator with a fixed terminal voltage, the effect that the reactive power output rises when the internal EMF increases is easily understood from the phasor diagram as depicted in Figure 5.10 (see also Section 2.5; Figure 2.28).

The AVR can be tuned such that the generator terminal voltage remains constant and that it is independent of the generator loading as is shown in Figure 5.11 (a). But such a voltage control would cause problems when generators run in parallel for two reasons. When a voltage drop occurs, for instance, each controller will separately try to restore the terminal voltage and increases the excitation. As a result, the generators equipped with fast-operating controllers will supply more reactive power than the generators with slow-acting controllers. After a number of control actions, the reactive power is distributed randomly among the generators. Furthermore, if two generators run in parallel, one having a voltage controller setting that is slightly

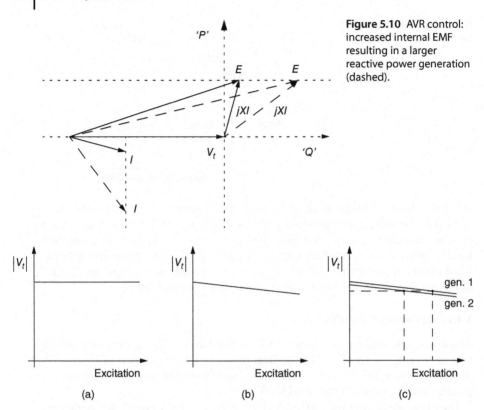

Figure 5.10 AVR control: increased internal EMF resulting in a larger reactive power generation (dashed).

Figure 5.11 AVR characteristics.

above the common setpoint and the other one with a setting that is slightly below the common setpoint, both the controllers will react in an opposite way. The busbar voltage will settle at a value between the two settings, which is, however, too low for the first and too high for the second controller. Now, one controller will try to increase its voltage and starts to boost the generator's reactive power output, but the other controller does just the opposite and consumes reactive power. In this way an unwanted reactive power exchange between the two generators is the result.

For this reason, the generator AVR is equipped with a voltage droop characteristic as shown in Figure 5.11 (b) in order to avoid these problems. The droop characteristic of the generator AVR should be declining in order to give a higher excitation when the voltage drops and a lower excitation when the terminal voltage increases.

In Figure 5.11 (c) two generators with slightly different AVR settings run in parallel and share the same terminal voltage; the controllers are in an equilibrium state. The AVR settings can be tuned in such a way that the reactive power

Figure 5.12 Tap-changing transformer.

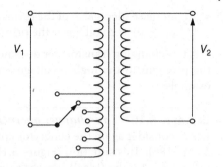

generation is proportional to the nominal power of the generators. For example, generator 1 in Figure 5.11 (c) supplies more reactive power to the system than generator 2 does. When the voltage declines, the controllers of both generators increase their excitation in order to adjust the terminal voltage to a (slightly lower) value as required by the droop characteristic.

5.4.2 Tap-Changing Transformers

Tap-changing transformers usually have a mechanically tapped high-voltage winding as is schematically shown in Figure 5.12. The high-voltage winding is chosen because less current flows through the high-voltage winding, compared with the low-voltage winding, and a lower current is switched by the tap changer under operating conditions. In this way the transformer ratio can be adjusted, and the voltage at the low-voltage terminals can be kept more or less constant. The voltage can usually be adjusted between $\pm10\%$ and $\pm20\%$ of the nominal rated values by using about 10–30 taps.

Let us examine a tap-changing transformer under no-load conditions. In this case the relation between the voltage at the primary side and the voltage at the secondary side is

$$|V_2| = \frac{|V_1|}{n} \tag{5.31}$$

n the turns ratio: the number of windings at the primary side divided by the number of windings at the secondary side

When we make a small change in the parameters of Equation 5.31, we get

$$\Delta|V_1| = n\Delta|V_2| + |V_2|\Delta n \tag{5.32}$$

Three cases can be distinguished:

- The turns ratio is a fixed number ($\Delta n = 0$): voltage changes are directly coupled to each other similar to the ideal transformer, without taps.
- An increase (decrease) of the turns ratio while the primary voltage is fixed ($\Delta|V_1| = 0$): the voltage at the secondary side will drop (rise).

- An increase (decrease) of the turns ratio while the secondary voltage is fixed ($\Delta|V_2| = 0$): the voltage at the primary side will rise (drop).

A tap-changing transformer does not solve a voltage problem; it only moves the problem to the higher voltage levels. This is illustrated by the following example.

Example 5.10 *Tap-changer operation*
Let us consider again the generator connected to a variable load from Example 5.2 (p. 186). This situation (Figure 5.1) is in fact similar to the network shown in Figure 5.13; the (ideal) tap-changing transformer has a turns ratio $n = 1$ in that case. When the reactive power consumption of the load increases from $Q \approx 1$ Mvar to $Q = 1.5$ Mvar, while the active power consumption remains fixed at $P = 2$ MW, we calculated already in Example 5.2 (p. 186) that the increased reactive power consumption causes the voltage at the load terminals to drop by almost 100 V. Now the tap-changing transformer can adjust the voltage level at the load terminals.

Seen from the terminals of the generator, the tap-changing transformer does not change the power flow: the same amount of power is consumed and therefore the generator terminal voltage and current stay the same: $V_t = 5676.6\angle 0°$ V and $I = 146.8\angle -36.9°$ A.

The relations between the voltages and currents at the primary side and at the secondary side of the ideal tap-changing transformer are (see also Appendix B.2)

$$|V_L| = \frac{|V_t|}{n}$$

$$|I_L| = n|I| \tag{5.33}$$

When the turns ratio of the transformer is set to $n = 0.983$, the voltage at the load terminals is brought back to the nominal value again: $|V_L| = 5676.6/0.983 = 5773.5$ V. The value of the current becomes $|I_L| = 0.983 \cdot 146.8 = 144.3$ A.

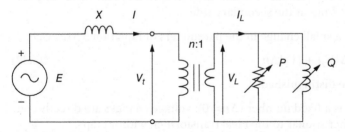

Figure 5.13 Generator connected to a variable load by means of a tap-changing transformer.

Figure 5.14 Capacitor banks. Reproduced with permission of TenneT TSO B.V.

5.4.3 Reactive Power Injection

At the beginning of this chapter, we already learned that there is a tight relation between reactive power exchange and voltage level (Examples 5.2 (p. 186) and 5.3 (p. 189)): reactive power consumption (by an inductive component) at a network node results in a lower node voltage, but reactive power injection (by a capacitive component) gives a higher node voltage.

Static shunt capacitors and reactors

Static shunt capacitors (capacitor banks), shown in Figure 5.14, are applied in the network to inject reactive power into systems with a lagging power factor, and inductors are applied to consume reactive power from the system when it is operated with a leading power factor, for instance, a system with lightly loaded cables. A drawback of static shunt capacitors is that their influence reduces when they are needed most, because the reactive power produced by a shunt capacitor becomes less when the node voltage drops:

$$Q = \frac{|V|^2}{-X} = \omega C |V|^2 \tag{5.34}$$

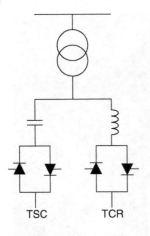

Figure 5.15 Static var compensator (SVC). TSC, thyristor-switched capacitor; TCR, thyristor-controlled reactor.

TSC   TCR

Synchronous compensators

A synchronous compensator is a synchronous machine running without a mechanical torque. By varying the excitation, the machine consumes or generates reactive power (see also Section 2.5). The excitation is controlled by means of an AVR. This method for voltage control is rather costly.

Static var compensator (SVC)

In Figure 5.15, the principle of a static var compensator (SVC) is shown. It is in fact a variable reactance connected to the system in shunt. The reactance consists of one or more capacitors and/or reactors; the control actions are done by power-electronic devices (thyristors). Depending on the switching mode of the thyristors, the SVC can inject reactive power into or absorb reactive power from the system.

The basis of an SVC is the thyristor-controlled reactor (TCR): an inductor in series with two back-to-back connected thyristors (a thyristor valve). A conventional thyristor is a power semiconductor that conducts current in one direction only; in the other direction it blocks the current. It is turned on and brought in the conductive state by applying a trigger signal to the gate of the thyristor, and it turns off (the blocking state) at the next current zero crossing. A back-to-back setup of the thyristors makes conduction possible in both the positive and negative half cycle of the supplying AC voltage. The current through the inductance can, in this way, be controlled from its maximum value to zero by changing the firing angle of the thyristor trigger signal between $\alpha = 90°$ and $\alpha = 180°$, as illustrated in Figure 5.16. It is obvious that the current flowing through the reactor does not have a nice sinusoidal-shaped waveform (this is only the case when the thyristors are fully conducting). This means that the current contains a first harmonic component or ground wave of 50 Hz and several higher-order components, the so-called higher harmonics. In practice, filters are connected to remove the harmonics. The fundamental component is

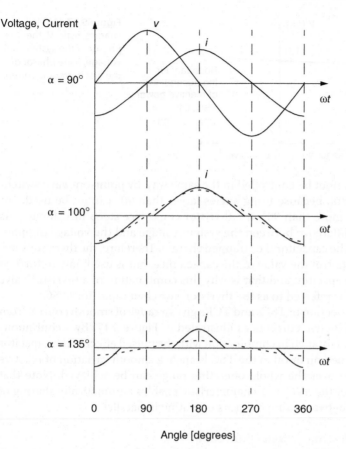

Figure 5.16 Current (solid line) through a TCR as a function of the delay angle α and its first harmonic component (dotted line). v, the system voltage; i, the lagging current through the reactor.

generally the dominant part of the current, and its amplitude gets smaller when the delay angle changes from 90° to 180°. This fundamental current component lags the voltage by 90°, and by varying the delay angle one controls in fact the value of the inductance:

$$L(\alpha) = \frac{\pi L}{2\pi - 2\alpha + \sin(2\alpha)} \tag{5.35}$$

α   the firing angle of the thyristor trigger signal, which can vary between π/2 and π radians (or, in electrical degrees, between 90° and 180°)

$L(\alpha)$ the effective inductance of the TCR [H]

L the nominal rated inductance of the TCR [H]

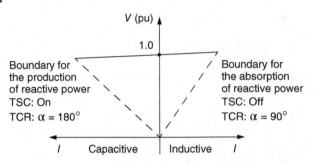

A capacitor cannot be controlled in the same way, by point-on-wave switching, as the reactor, because firing angles larger than 90° cannot be used. For firing angles of more than 90°, the thyristor could face steep inrush currents caused by the difference between the system voltage and the voltage of opposite polarity on the capacitor (i.e., trapped charge). Therefore, the thyristor valve is not used to control the value of the capacitance, but is used only to turn on or turn off the capacitor, and that is why this combination of a thyristor valve and a capacitor is referred to as the thyristor-switched capacitor (TSC).

A parallel connection of TSCs and TCRs gives a range of smooth control from capacitive to inductive currents as illustrated in Figure 5.17. By a continuous tuning of the TCR branch in combination with the pulsed effect of the capacitor banks being turned on or off in the TSC branch, a smooth variation of reactive power exchange over the whole operating range can be achieved. Note that the droop line of the SVC *V–I* characteristic enables automatically sharing of reactive power between compensators operating in parallel.

Static synchronous compensator (STATCOM)

A gate turnoff (GTO) thyristor is a more advanced type of thyristor: it not only can be turned on but also turned off by means of control signals. With GTOs it becomes possible to build a voltage source converter (VSC); VSCs are power-electronic devices that can serve as an interface between DC and AC systems. The principle layout of a single-phase VSC is shown in Figure 5.18. The power flow in the device is bidirectional: from DC to AC and also from AC to DC. The amplitude and the frequency of the AC output are controlled by means of pulse-width modulation (PWM).

A static synchronous compensator (STATCOM) is based on the operating principle of the VSC. The basic scheme of a STATCOM is shown in Figure 5.19. The exchange of reactive power between the STATCOM and the grid is controlled by the amplitude of the STATCOM output voltage, as illustrated in Figure 5.20. A STATCOM produces the reactive power by exchanging the instantaneous reactive power between the phases of the power system. The capacitance is needed to provide a circulating current path as well as a voltage source.

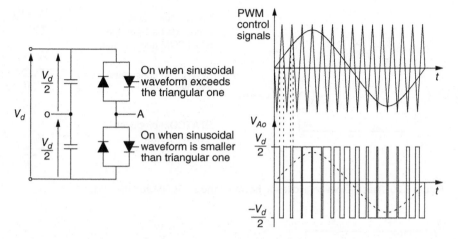

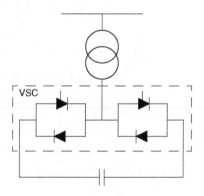

Figure 5.18 Voltage source (PWM) converter: upper graph, pulse-width modulation (PWM) control signals; lower graph, the AC voltage (solid line) at the output and its first harmonic component (dotted line).

Figure 5.19 Principle layout of a STATCOM.

5.5 Control of Transported Power

5.5.1 Controlling Active Power Flows

Apart from the phase shifter, which is described hereafter, there are other options to control the active power flows in the system. When HVDC links are part of the grid (see also Section 1.3.1), the active power flowing through the DC links can be influenced; this is an additional advantage of HVDC links. Another piece of equipment that is able to control active power flows is the unified power flow controller (UPFC); as this device can serve other purposes as well, it is treated separately in Section 5.5.3.

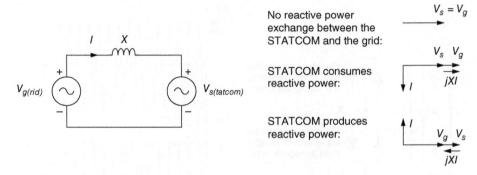

Figure 5.20 Reactive power exchange between the STATCOM and the grid.

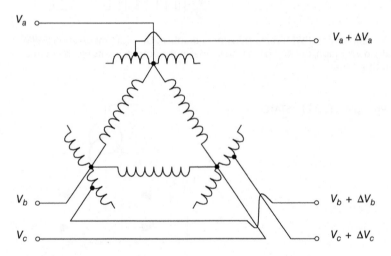

Figure 5.21 The phase shifter [1].

The phase shifter

The phase shifter or phase-shifting transformer is shown in Figure 5.21. The windings drawn in parallel to each other are on the same iron core and are magnetically coupled. Therefore, a tapped winding is on the same magnetic core as the winding whose voltage is 90° out of phase with the line-to-neutral voltage that is connected to the center of the tapped winding:

$$\Delta V_a \propto (V_b - V_c)$$
$$\Delta V_b \propto (V_c - V_a) \tag{5.36}$$
$$\Delta V_c \propto (V_a - V_b)$$

The result of this "phase shifting" is demonstrated by the phasor diagram in Figure 5.22. By altering the tap position, the length of the ΔV-phasor changes

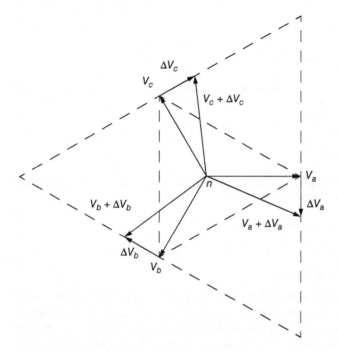

Figure 5.22 Phasor diagram of the phase shifter.

(and can become negative too), but the orientation of the phasors remains fixed. The resulting phasors, corresponding with the voltages at the transformer terminals, move along the dashed lines of the outer triangle and can be leading or lagging the primary voltage. Note that the voltage amplitude remains almost constant for small phase corrections.

In the following example, we see how a phase shifter can be applied to control the active power flow in the system.

Example 5.11 *Phase-shifter operation*
Let us take the situation that a generator is connected to a load via two parallel transmission lines and transformers as shown in Figure 5.23 (a). The current is $I = 1\angle-30°$ pu and the voltage $V_2 = 1\angle0°$ pu. The values of the per unit base quantities are chosen such that the ideal transformer in the single-phase representation of transformer T_1 can be left out from the equations and only the series reactance of the transmission line plays a role (see also Section 1.8). The single-phase equivalent circuit is shown in Figure 5.23 (b).

When transformer T_2 has the same turns ratio as T_1, the ideal transformer in the single-phase representation of transformer T_2 can be left out too, and only the two parallel reactances remain. The current flowing through these reactances is $I_1 = I_2 = I/2 = 0.433 - j0.25$ pu, and the complex power supplied

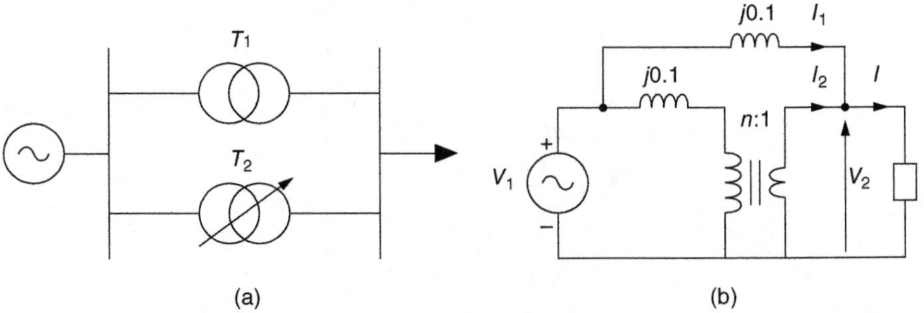

Figure 5.23 Phase-shifter application in case of two parallel connections: (a) the single-line diagram; (b) the single-phase equivalent circuit.

by each transformer is equal to

$$S_{T_1} = V_2 I_1^* = 0.433 + j0.25$$
$$S_{T_2} = V_2 I_2^* = 0.433 + j0.25 \tag{5.37}$$

In case the power rating (or the transportation capacity) of the two transmission lines is different, a phase shifter can be used to control the active power flows. Assume that transformer T_2 establishes a phase shift of 3° (it has a complex turns ratio of $n{:}1$ with $n = 1\angle{-3}°$; see also Appendix B.5) and that the current and voltage remain constant: $I = 1\angle{-30}°$ pu and $V_2 = 1\angle 0°$ pu. Then, the following equations must apply:

$$I_1 + I_2 = 1\angle{-30}°$$
$$V_1 = 1\angle 0° + j0.1 \cdot I_1 = 1\angle{-3}° + j0.1 \cdot I_2\angle{-3}° \tag{5.38}$$

The currents I_1 and I_2 can be calculated from these equations: $I_1 = 0.1646 - j0.2613$ pu and $I_2 = 0.7014 - j0.2387$ pu. The complex power supplied by each transformer is

$$S_{T_1} = V_2 I_1^* = 0.1646 + j0.2613$$
$$S_{T_2} = V_2 I_2^* = 0.7014 + j0.2387 \tag{5.39}$$

When we compare Equation 5.39 with Equation 5.37, we see that the active power flows through the transformers have changed drastically, while the reactive power flow remains more or less unchanged.

5.5.2 Controlling Reactive Power Flows

Reactive power flows in the system can be influenced by making use of series compensation: the effective series reactance of a transmission line is reduced in order to reduce the voltage drop across the line and the reactive power "loss" in the transmission line.

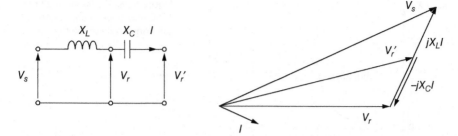

Figure 5.24 A transmission line with a series capacitor. V_r, the voltage at the receiving end of the line without a series capacitor; V_r', the voltage at the receiving end of the line with a series capacitor.

Static series capacitors

Capacitors can be put in series with the conductors of a long transmission line in order to reduce the line reactance:

$$Z = R + jX = R + j\omega L + \frac{1}{j\omega C} = R + j\left(\omega L - \frac{1}{\omega C}\right) \tag{5.40}$$

R the total series resistance of the transmission line [Ω]
L the total inductance of the transmission line [H]
C the series capacitance [F]

When the series impedance of the transmission line is reduced, the voltage drop between the receiving and the sending end of the line decreases, as is illustrated in Figure 5.24.

Thyristor-controlled series capacitor (TCSC)

The thyristor-controlled series capacitor (TCSC) consists of a series capacitance in parallel with a TCR, as shown in Figure 5.25. When the thyristor valve blocks the current, the TCSC is in fact only the series capacitor. When the thyristor valve is fired, the TCSC becomes a continuously controllable capacitor or a continuously controllable inductor. A smooth transition between the capacitive and the inductive mode is not allowed because of the resonance interval between the two modes, as shown in Figure 5.26 (the resonance interval lies around the 135°).

Figure 5.25 Thyristor-controlled series capacitor (TCSC).

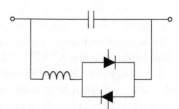

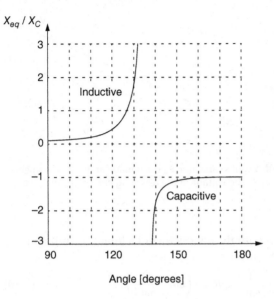

Figure 5.26 The variation of the TCSC reactance as a function of the firing angle of the thyristor.

This graph can be drawn when we realize that the equivalent impedance of the TCSC is equal to

$$Z_{eq} = \cfrac{1}{\cfrac{1}{1/(j\omega C)} + \cfrac{1}{j\omega L(\alpha)}} = -j\cfrac{1}{\omega C - \cfrac{1}{\omega L(\alpha)}} \qquad (5.41)$$

C   the fixed capacitor [F]
$L(\alpha)$ the effective inductance of the TCR [H]

The reactance of the capacitor is larger than that of the inductor (e.g., $X_C \approx 8 \cdot X_L$).

When $\alpha = 90°$, the effective inductance of the TCR reaches its minimum value, $L(\alpha) = L$, and the equivalent reactance of the TCSC is a parallel combination of the capacitor and the reactor. As the reactance of the capacitor is larger than that of the reactor, for example, $X_C \approx 8 \cdot X_L$, the current through the TCSC module is dominantly inductive. When $\alpha = 180°$, the effective inductance of the TCR becomes infinite, $L(\alpha) = \infty$, and the TCSC is now a series capacitor: the current through the TCSC module is capacitive.

Static synchronous series compensator (SSSC)

The static synchronous series compensator (SSSC) is depicted in Figure 5.27 and looks like the STATCOM (see Section 5.4.3). The SSSC is a device that can change the effective impedance of a transmission line by injecting a voltage ΔV that is in antiphase with the voltage drop across the transmission line. The output voltage ΔV is controlled such that it lags the current I by 90° and thus

Figure 5.27 Static synchronous series compensator (SSSC).

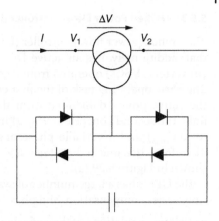

behaves as a series capacitor. As the injected voltage can have any magnitude between zero and maximum voltage, a continuous degree of series compensation is obtained. The phase of the injected voltage ΔV can be reversed too, thus increasing the overall line reactance, in order to limit fault currents.

When the SSSC is equipped with an energy-storage device of suitable capacity (in parallel to the capacitor), the active power exchange can be controlled by adjusting the angle between the SSSC output voltage and the current I (see also Section 5.5.3); only when the angle is 90° the active power exchange is zero.

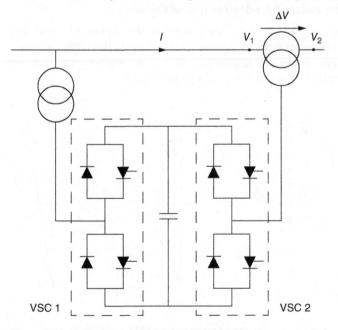

Figure 5.28 Unified power flow controller (UPFC).

5.5.3 Unified Power Flow Controller (UPFC)

The unified power flow controller (UPFC) is in fact an SSSC (see Section 5.5.2) that, additionally, has an active DC source. The UPFC is equipped with two converters (VSCs) operated from a common DC link as shown in Figure 5.28. The basic operating task of the first converter (VSC 1) is to supply and absorb the active power demanded from the second converter (VSC 2) at the DC link. The second converter (VSC 2) injects a voltage with a variable amplitude $(0-\Delta V_{max})$ and a variable phase angle $(0-360°)$ into the transmission line. Therefore, the injected voltage ΔV can take any position inside the circle shown in Figure 5.29 (a).

The UPFC has a large number of system applications, such as voltage regulation, series compensation, and phase shifting:

- Voltage regulation is achieved when the voltage ΔV is in phase with the voltage V_1, as shown in Figure 5.29 (b).
- In Figure 5.29 (c) a combination of voltage regulation and series compensation is demonstrated. The voltage ΔV consists of two components: one that is in phase with the voltage V_1 (voltage regulation) and another component that lags the line current I by 90° (series compensation).
- In Figure 5.29 (d) a combination of voltage regulation and phase-angle regulation is pictured. The voltage ΔV consists of two components: one that is in phase with the voltage V_1 (voltage regulation) and another component that shifts the resulting voltage by α degrees (phase shifting).

It is also possible to combine the three functions simultaneously, and this makes the UPFC a universal device for power flow control. The UPFC is, however, a rather expensive piece of equipment, which is the main reason that the UPFC has been installed in only a few places in the world.

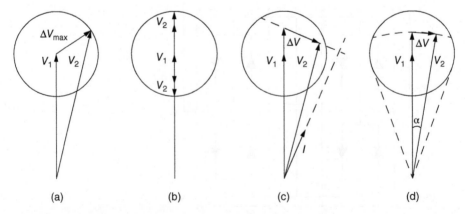

(a)   (b)   (c)   (d)

Figure 5.29 Phasor diagrams illustrating the operation of the UPFC.

5.6   Flexible AC Transmission Systems (FACTS)

In the power system, the power flow follows the laws of physics. In the past, the possibilities to control the power flows in the system were limited; they were mainly based on mechanical devices, such as transformer tap changers and turbine governors. These mechanisms automatically introduce a limitation to the speed of control. Nowadays, FACTS devices are available that enable a greater flexibility in the operation of AC power systems (FACTS is an acronym for flexible AC transmission systems) [2–6]. FACTS devices are large power-electronic controlled devices and can do control actions at a considerably higher speed. Some of these devices are operated in a shunt configuration for reactive power and voltage control, whereas others are put in series to control the power flow. This gives the system operator flexibility and a certain degree of freedom in operating the system, which is of great value in the present-day market environment (see also Chapter 7).

In the previous sections we explained the operating principles of the most common FACTS devices:

- SVC – static var compensator (Section 5.4.3)
- STATCOM – static synchronous compensator (Section 5.4.3)
- TCSC – thyristor-controlled series capacitor (Section 5.5.2)
- SSSC – static synchronous series compensator (Section 5.5.2)
- UPFC – unified power flow controller (Section 5.5.3).

Problems

5.1 A generator supplies a load of 100 MW at 50 Hz. The generator is not connected to the grid or to other generators. The nominal rated power of the generator is 400 MW and the speed governor has a droop of $R = 0.02$ pu.

a. What is the drop in frequency when the load suddenly increases from 100 to 200 MW?

b. The droop of the generator is changed to $R = 0.06$. What is the increase in active power when the frequency drops from 50 to 49.5 Hz?

5.2 Two generators run in parallel and are not connected to the grid. Generator 1 has a nominal rated power $P_{g1,r} = 400$ MW, and its speed governor has a droop $R_{g1} = 0.02$ pu. Generator 2 has a nominal rated power $P_{g2,r} = 400$ MW, and its speed governor has a droop $R_{g2} = 0.06$ pu. When the load suddenly consumes more active power ΔP, the frequency drops with an amount of Δf. The speed governors take action and divide the increase in power consumption over the two generators.

a. What is the ratio of the power division over the two generators?
b. What is the network power frequency characteristic of this small power system with only two generators?

5.3 In a hydropower plant the water turbine drives the synchronous generator. The rotation speed of the water turbine is 500 RPM. The generator is connected (via a step-up transformer) with an infinite bus having a power frequency of 50 Hz.

a. With the turbine rotating with 500 RPM, how can we deliver electricity with a frequency of 50 Hz?
b. When the hydropower plant comes into service, the generator should be connected to the grid. Which four conditions have to be met before a synchronous generator can be connected to the grid?
c. The generator is connected to an infinite bus: an ideal voltage source with a fixed voltage amplitude and frequency.
   How are the active power and reactive power output of the hydropower plant controlled?
d. The synchronous reactance of the generator is $2\,\Omega$. The copper losses can be neglected. The generator supplies a three-phase power of 5 MW to the infinite bus. The terminal voltage of the generator is 10 kV line-to-line, and the power factor is 0.9 leading. Draw the single-phase equivalent circuit of the synchronous generator and calculate the value of the internal EMF.
e. How can the infinite bus be modeled with two lumped elements in parallel: as $R//L$, $R//C$, or $L//C$?
   Calculate the component values.

5.4 The hydropower plant of Problem 5.3 has been extended and also supplies a small village across the mountains. For the transmission a short three-phase transmission line has been built with a series impedance of $3.3 + 5j\,\Omega$. At the village side (the receiving end of the line), a three-phase power of 2500 kW at a line-to-line voltage of 10 kV is consumed. The power factor of the load is 0.8 leading.

a. Draw the equivalent circuit of the short transmission line and calculate the line-to-neutral voltage at the generator terminals (the sending end of the line).
b. Calculate the three-phase active and reactive power supplied by the generator.

c. What is the three-phase active power loss in the line.

d. Draw the phasor diagram with the voltage at the generator side, the village side, the current through the line, and the voltage drop across the line.

5.5 When we consider the expression $S = 3VI^*$, which of the following statements is always true?

a. S is the three-phase active power.

b. V is the line-to-line voltage.

c. I lags V.

d. None of the answers is correct.

5.6 In a three-phase network the voltages of phase b and phase c are $V_b = 1\angle 0°$ and $V_c = 1\angle 120°$.What is the line-to-line voltage V_{bc}?

a. 1

b. $\left(\dfrac{1}{\sqrt{3}} \right) \angle -120°$

c. $\sqrt{3}\angle -30°$

d. $1\angle 120°$

5.7 A load consumes 7.5 MW active power and supplies 10 Mvar reactive power. The apparent power of the load is:

a. 12.5 MVA

b. 17.5 MVA

c. $-7.5 + j10$ MVA

d. $7.5 - j10$ MVA

5.8 What is the purpose of applying ground wires on an overhead transmission line?

a. Return path for the current

b. Giving more mechanical strength to the tower frame

c. Shielding the phase wires from a direct hit by lightning

d. Helping the three-phase system to maintain the balance

5.9 The three-phase base power is 100 MVA and the base voltage (line-to-line) is 100 kV. What is the base impedance?

a. 1000 Ω

b. 100 Ω

c. 0.01 Ω

d. 0.001 Ω

5.10 When a transformer faces a short circuit at its load side, the short circuit is limited by:
a. The magnetizing susceptance.
b. The core losses.
c. The leakage reactance.
d. None of the answers is correct.

References

1 Grainger, John J., and Stevenson, William D., Jr.: *Power System Analysis*, McGraw-Hill, Singapore, 1994, ISBN 0-07-113338-0.
2 Grünbaum, R., Noroozian, M., and Thorvaldsson, B.: 'FACTS – powerful systems for flexible power transmission', *ABB Review*, May 1999, pp. 4–17, www.abb.com/abbreview.
3 Moore, P., and Ashmole, P.: 'Flexible AC transmission systems', *Power Engineering Journal*, Vol. **9**, Issue 6, December 1995, pp. 282–6.
4 Moore, P., and Ashmole, P.: 'Flexible AC transmission systems – Part 2: Methods of transmission line compensation', *Power Engineering Journal*, Vol. **10**, Issue 6, December 1996, pp. 273–8.
5 Moore, P., and Ashmole, P.: 'Flexible AC transmission systems – Part 3: Conventional FACTS controllers', *Power Engineering Journal*, Vol. **11**, Issue 4, August 1997, pp. 177–83.
6 Moore, P., and Ashmole, P.: 'Flexible AC transmission systems – Part 4: Advanced FACTS controllers', *Power Engineering Journal*, Vol. **12**, Issue 2, April 1998, pp. 95–100.

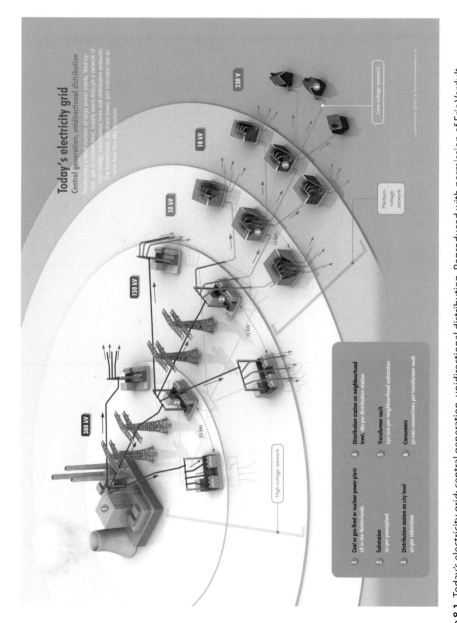

Figure 8.1 Today's electricity grid: central generation, unidirectional distribution. Reproduced with permission of Eric Verdult, www.kennisinbeeld.nl.

Electrical Power System Essentials, Second Edition. Pieter Schavemaker and Lou van der Sluis.
© 2017 John Wiley & Sons Ltd. Published 2017 by John Wiley & Sons Ltd.

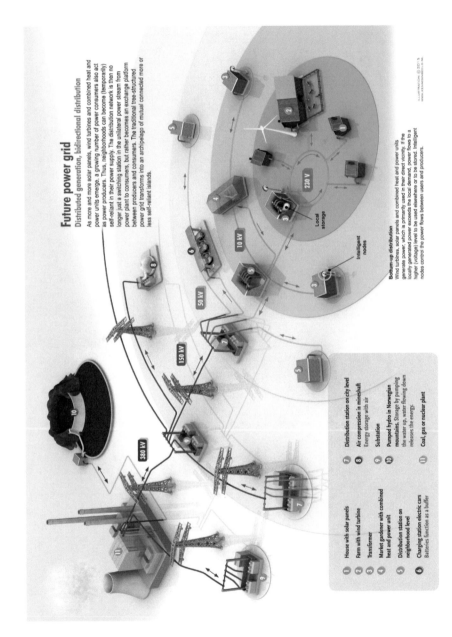

Future power grid

Distributed generation, bidirectional distribution

As more and more solar panels, wind turbines and combined heat and power units emerge, a growing number of power consumers also act as power producers. Thus, neighbourhoods can become (temporarily) self-reliant in their power supply. The distribution network is then no longer just a switching station in the unilateral power stream from power plant to consumers, but rather becomes an exchange platform between producers and consumers. The traditional tree-structured power grid transforms into an archipelago of mutual connected more or less self-reliant islands.

Bottom-up distribution

Wind turbines, solar panels and combined heat and power units generate power, which is primarily used in their direct vicinity. If the locally generated power exceeds the local demand, power flows to a higher (voltage) level to be used elsewhere or to be stored. Intelligent nodes control the power flows between users and producers.

1. House with solar panels
2. Farm with wind turbine
3. Transformer
4. Market gardener with combined heat and power unit
5. Distribution station on neighbourhood level
6. Charging station electric cars
   Batteries function as a buffer
7. Distribution station on city level
8. Air compression in mineshaft
   Energy storage with air
9. Substation
10. Pumped hydro in Norwegian mountains. Storage by pumping the water up, water flowing down releases the energy.
11. Coal, gas or nuclear plant

230 V · 10 kV · 50 kV · 150 kV · 380 kV

Local storage

Intelligent nodes

ILLUSTRATION © 2015,
WWW.KENNISINBEELD.NL

Figure 8.2 Future power grid: distributed generation, bidirectional distribution. Reproduced with permission of Eric Verdult, www.kennisinbeeld.nl.

6

Energy Management Systems

6.1 Introduction

In the control center, the transmission and distribution of electrical energy are monitored, coordinated, and controlled. The energy management system (EMS) is the interface between the operator and the actual power system.

An indispensable part of the EMS is the supervisory control and data acquisition (SCADA) system. The main functionalities of the SCADA system are to collect real-time measured data from the system and to present it to the computer screen of the operator and to control actual components in the network from the control center. The actual network status is stored in the "real-time telemetered database," which is used as input for the other EMS functions.

The EMS is in fact an extension of the basic functionality of the SCADA system. This includes tools for the analysis and the optimal operation of the power system. In the following sections some of these software tools are highlighted. The state estimator (Section 6.4) serves as a "filter" of the real-time database; it determines the state of the power system that matches best with the available measurements (some of them can be erroneous). The state estimator runs at regular time intervals (typically every 5 min) because its output is the necessary input for the other analysis programs, for example,

- Load flow (Section 6.2)
- Optimal power flow (Section 6.3)
- Controlled switching
- Contingency analysis
- Short-circuit computations
- Stability analysis

These analysis programs operate in the so-called "extended real-time" mode; this means that the actual power system state is the starting point from where (possible) future situations are investigated.

Electrical Power System Essentials, Second Edition. Pieter Schavemaker and Lou van der Sluis.
© 2017 John Wiley & Sons Ltd. Published 2017 by John Wiley & Sons Ltd.

6.2 Load Flow or Power Flow Computation

The word load flow or power flow computation is self-explanatory: the flows of power in the network are computed. Later on we see that the name is rather misleading as first the node voltages in the network are computed before the power flows can be calculated. The load flow is the most important network computation, as it allows insight into the steady-state behavior of the power system. In the following, some of the applications of the load flow are described.

When a transmission line is taken out of service temporarily, for planned maintenance, for example, the power originally flowing through the transmission line will find itself a new path to the loads. The operators want to be certain, in advance, that other transmission lines and/or cables in the vicinity are not overloaded after taking the particular line out of service. A load flow computation of the network configuration in which the transmission line is taken out of service gives insight into the new power flows and indicates possibly overloaded connections or components.

After a blackout has paralyzed a part of the network, the utilities put a lot of effort in finding the cause and want a reconstruction of the phenomena that eventually caused the blackout. A load flow calculation gives the state of the network for a certain steady-state situation. A blackout is a longer process – of multiple more or less steady-state situations – in the course of time, so a consecutive series of load flow computations gives information about the system behavior preceding to the blackout.

The node voltages in the network should be kept within close limits during normal operation between 1.1 and 0.9 pu. A load flow computes the voltages in the network and visualizes the effect of tap-changing transformers, capacitor banks, and load shedding on the voltage profile in the system.

The system operation should be robust and therefore the power system is operated $n-1$ secure. This means that a system component may fail without overloading other components or without violating the voltage limits. A list with transformers, transmission lines, cables, generators, and so on is available from which the components are taken out of service in a simulation one by one and each time a load flow is computed. When a load flow calculation shows an overloaded connection or transformer, preventive actions can be taken in the real network to prevent that particular situation. This analysis, based on a large number of load flow computations, is called the contingency analysis. The contingency analysis is performed in the control center of the utilities on a regular time basis.

6.2.1 Load Flow Equations

The load flow computation needs of course input data and, after performing the necessary calculations, generates the output as shown in Figure 6.1. The

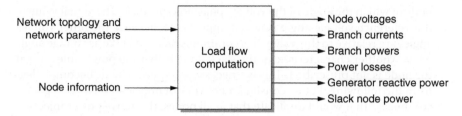

Figure 6.1 Load flow computation: input data and computational results.

Table 6.1 Network node types.

Node Type	Number of Nodes	Specified	Unknown
Slack node ($i = 1$)	1	$|V_i|, \delta_i$	P_i, Q_i
Generator bus ($i = 2 \ldots N_g + 1$)	N_g	$P_i, |V_i|$	Q_i, δ_i
Load bus ($i = N_g + 2 \ldots N$)	$N - N_g - 1$	P_i, Q_i	$|V_i|, \delta_i$

input data consists of the network topology and the parameters, so that the admittance matrix of the network can be built (see Example 6.2 (p. 223)), and node information. A network node is fully described (electrically) by four parameters:

- The voltage phasor magnitude: $|V|$
- The voltage phasor angle: δ
- The injected active power: P
- The injected reactive power: Q

Three types of network nodes can be distinguished, and only two of the four parameters are known for each node, as shown in Table 6.1. The following is a short explanation of the three bus types:

- The load bus
  Loads are modeled as constant power sinks instead of impedances. The reason for this is that the load flow is usually computed for the higher voltage levels (50 kV and above). The voltage deviations, which would result from changes in the load, are canceled out by changing the tap position of regulating transformers, and the system "sees" a load of constant power.
- The generator bus
  The generator has two controls: the active power control and the voltage control. A wind generator, however, does not have these controls and is treated in the load flow as a load bus.
- The slack node
  The load flow computation needs one network node to be addressed as slack node. The slack node serves as a reference for the other nodes: it is the only

node of which the angle of the voltage phasor is specified. The actual value of this reference voltage angle is not of importance, because the other voltage angles are relative to this value. It is common practice to take as voltage angle $\delta = 0$. Another characteristic of the slack node is that the power injection at the node is not prescribed. This is a necessary requirement as becomes clear from the following. Let us consider a network of only load buses and generator buses. That would implicate that at all nodes, the active power injection is known (see Table 6.1) and, as a consequence, that the active power losses ($|I|^2R$) in the network are prescribed. This is not possible: the load flow needs to compute the unknown node voltages first before the line currents and the losses can be calculated. The difference (the slack) between the total active power input and total active power output plus the computed total $|I|^2R$ losses is balanced by the slack node. Similarly for the reactive power, the difference (the slack) between the total reactive power input and total reactive power output plus the calculated total $|I|^2X$ "losses" is balanced by the slack node. In the considered network with only load buses and generator buses, one of the generator buses must be assigned as slack node.

Example 6.1   *Node types*
Nodes where no loads or generator stations are connected, for instance, substations in the system, are treated in the load flow as load buses with $P = Q = 0$.

A node can also have both a generator and a load connected. The (negative) power injected by the load (P_{load} and Q_{load}) is known as well as the P_{gen} and $|V|_{\text{gen}}$ of the generator. Because the voltage magnitude is a common parameter, the node can be modeled as a generator bus with $|V| = |V|_{\text{gen}}$ and $P = P_{\text{gen}} + P_{\text{load}}$.

In order to find a relation between the four quantities that electrically describe a network node, as specified in Table 6.1, the so-called load flow equations are derived in the following.

The admittance matrix Y describes the relation between the current injected at the network nodes and the node voltages:

$$I = YV \leftrightarrow \begin{bmatrix} I_1 \\ I_2 \\ \vdots \\ I_N \end{bmatrix} = \begin{bmatrix} Y_{11} & Y_{12} & \cdots & Y_{1N} \\ Y_{21} & Y_{22} & \cdots & Y_{2N} \\ \vdots & \cdots & \cdots & \vdots \\ Y_{N1} & Y_{N2} & \cdots & Y_{NN} \end{bmatrix} \begin{bmatrix} V_1 \\ V_2 \\ \vdots \\ V_N \end{bmatrix} \tag{6.1}$$

N the total number of network nodes

The admittance matrix Y can be built up as follows:

- The element on the main diagonal Y_{ii} equals the sum of all admittances directly connected to node i.
- The off-diagonal element Y_{ij} equals the negative value of the net admittance connected between node i and node j (note that the off-diagonal element Y_{ji}

has the same value, as the net admittance connected between node j and node i is the same as the net admittance connected between node i and node j).

This is illustrated in the following example.

Example 6.2 *Elements of the admittance matrix Y*

Consider the example admittance network shown in Figure 6.2. The node voltages are measured with respect to the reference node (node 0 in Figure 6.2). For all the nodes, apart from the reference node, we write down Kirchhoff's current law that states that the current that is injected into node i by a current source must equal the sum of all currents leaving node i:

$$I_1 = V_1 Y_a + (V_1 - V_2)Y_b + (V_1 - V_3)Y_d$$
$$0 = (V_2 - V_1)Y_b + (V_2 - V_3)Y_c \qquad (6.2)$$
$$I_3 = (V_3 - V_1)Y_d + (V_3 - V_2)Y_c$$

This equation can be written in the matrix notation $I = YV$:

$$\begin{bmatrix} I_1 \\ 0 \\ I_3 \end{bmatrix} = \begin{bmatrix} (Y_a + Y_b + Y_d) & -Y_b & -Y_d \\ -Y_b & (Y_b + Y_c) & -Y_c \\ -Y_d & -Y_c & (Y_c + Y_d) \end{bmatrix} \begin{bmatrix} V_1 \\ V_2 \\ V_3 \end{bmatrix} \qquad (6.3)$$

When we compare this admittance matrix with the circuit in Figure 6.2, we see that the admittance matrix Y can indeed be built up easily by looking at the circuit and applying the two abovementioned rules for the diagonal and off-diagonal elements.

The elements in the admittance matrix Y can be written as (see also Section 1.4)

$$Y_{ij} = |Y_{ij}|\angle\theta_{ij} = |Y_{ij}|(\cos\theta_{ij} + j\sin\theta_{ij}) = G_{ij} + jB_{ij} \qquad (6.4)$$

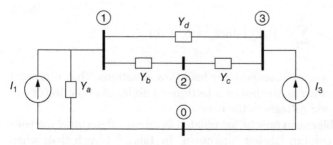

Figure 6.2 Example admittance network for building the admittance matrix Y.

G_{ij} the conductance
B_{ij} the susceptance

The voltages of the network nodes can be expressed as follows:

$$V_i = |V_i| \angle \delta_i = |V_i|(\cos \delta_i + j \sin \delta_i) \tag{6.5}$$

From Equation 6.1 the injected current at node i can be obtained:

$$I_i = Y_{i1}V_1 + Y_{i2}V_2 + \cdots + Y_{iN}V_N = \sum_{n=1}^{N} Y_{in}V_n \tag{6.6}$$

The complex power injected into node i is

$$S_i = V_i I_i^* = P_i + jQ_i \tag{6.7}$$

Substituting the equations for the voltage at node i (Equation 6.5) and the injected current at node i (Equation 6.6), gives us

$$S_i = |V_i| \angle \delta_i \sum_{n=1}^{N} Y_{in}^* V_n^* = \sum_{n=1}^{N} |V_i V_n Y_{in}| \angle(-\theta_{in} - \delta_n + \delta_i)$$

$$= \sum_{n=1}^{N} |V_i V_n Y_{in}|[\cos(-\theta_{in} - \delta_n + \delta_i) + j \sin(-\theta_{in} - \delta_n + \delta_i)] \tag{6.8}$$

$$= \sum_{n=1}^{N} |V_i V_n Y_{in}|[\cos(\theta_{in} + \delta_n - \delta_i) - j \sin(\theta_{in} + \delta_n - \delta_i)]$$

For the injected active and reactive power at node i, we can write

$$P_i = \sum_{n=1}^{N} |V_i V_n Y_{in}| \cos(\theta_{in} + \delta_n - \delta_i)$$

$$= |V_i|^2 G_{ii} + \sum_{\substack{n=1 \\ \neq i}}^{N} |V_i V_n Y_{in}| \cos(\theta_{in} + \delta_n - \delta_i) \tag{6.9}$$

$$Q_i = -\sum_{n=1}^{N} |V_i V_n Y_{in}| \sin(\theta_{in} + \delta_n - \delta_i)$$

$$= -|V_i|^2 B_{ii} - \sum_{\substack{n=1 \\ \neq i}}^{N} |V_i V_n Y_{in}| \sin(\theta_{in} + \delta_n - \delta_i) \tag{6.10}$$

Both Equations 6.9 and 6.10 compose the load flow equations. They represent the computed active and reactive power injection at a node i as a function of – in principle – all the node voltages in the network.

The load flow problem can now be formulated as follows: determine the voltages (V_i, δ_i) in the column labeled "unknown" in Table 6.1 such that, when all the voltage values are inserted into the load flow equations (Equations 6.9 and 6.10), the computed active and reactive power injections at the specific nodes correspond with the values as given in the column labeled "specified" in Table 6.1. Or in equations

$$\Delta P_i = P_{i,\text{specified}} - P_{i,\text{computed}} = 0 \tag{6.11}$$
$$\Delta Q_i = Q_{i,\text{specified}} - Q_{i,\text{computed}} = 0 \tag{6.12}$$

The load flow computation is demonstrated by the following numerical examples.

Example 6.3   *Load flow equations*

As an example the two-node network in Figure 6.3 is analyzed. Despite the fact that the network is rather simple, it is not possible to "see" or estimate what the value of the voltage at node 2 is. Therefore, a load flow computation has to be done. The network has two nodes: one with a generator connected to it, which is taken as the slack node, and a node with a load connected, which is a load bus. The nodes are interconnected by a short transmission line (see Appendix E.4).

In the example network (Figure 6.3), the voltage at node 2 is unknown, whereas the power injections P_2 and Q_2 are specified, and that is why only the corresponding load flow equations at node 2 need to be derived (Equations 6.9 and 6.10, with $N = 2$ and $i = 2$):

$$P_2 = |V_2|^2 G_{22} + |V_2 V_1 Y_{21}| \cos(\theta_{21} + \delta_1 - \delta_2) \tag{6.13}$$
$$Q_2 = -|V_2|^2 B_{22} - |V_2 V_1 Y_{21}| \sin(\theta_{21} + \delta_1 - \delta_2) \tag{6.14}$$

The admittance of the line is $Y = 1/Z = 0.396 - j3.96$, and the admittance matrix for the network of Figure 6.3 is

$$Y = \begin{bmatrix} 0.396 - j3.96 & -0.396 + j3.96 \\ -0.396 + j3.96 & 0.396 - j3.96 \end{bmatrix} \tag{6.15}$$

The elements in the admittance matrix can be denoted as

$$Y_{ij} = G_{ij} + jB_{ij} = |Y_{ij}| \angle \theta_{ij} \tag{6.16}$$

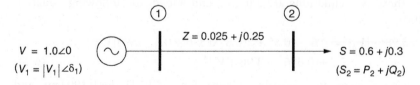

$$V = 1.0\angle 0$$
$$(V_1 = |V_1|\angle \delta_1)$$

$$Z = 0.025 + j0.25$$

$$S = 0.6 + j0.3$$
$$(S_2 = P_2 + jQ_2)$$

Figure 6.3 Example load flow network.

Thus, the following terms in the load flow equations are obtained from the admittance matrix (angles in radians):

$$Y_{21} = |Y_{21}|\angle\theta_{21} = 3.98\angle1.6705$$
$$Y_{22} = G_{22} + jB_{22} = 0.396 - j3.96 \tag{6.17}$$

Substitution of these admittance values and the voltage data at node 1 ($|V_1| = 1.0$ and $\delta_1 = 0$) into the load flow equations (Equations 6.13 and 6.14) results in the following set of equations:

$$P_2 = |V_2|^2 0.396 + 3.98|V_2|\cos(1.6705 - \delta_2) \tag{6.18}$$
$$Q_2 = 3.96|V_2|^2 - 3.98|V_2|\sin(1.6705 - \delta_2) \tag{6.19}$$

The unknown voltage quantities in the network, that is, $|V_2|$ and δ_2, must be determined such that the calculated power injections at node 2 equal the specified power injections at node 2:

$$P_2 = 0.396|V_2|^2 + 3.98|V_2|\cos(1.6705 - \delta_2) = -0.6 \tag{6.20}$$
$$Q_2 = 3.96|V_2|^2 - 3.98|V_2|\sin(1.6705 - \delta_2) = -0.3 \tag{6.21}$$

The minus sign of the specified power values is caused by the fact that in Figure 6.3 the power consumed at node 2 is given instead of the injected power.

The load flow problem of the example two-node network has been reduced to a set consisting of two nonlinear equations (Equations 6.20 and 6.21) and two unknowns ($|V_2|$ and δ_2). How this system can be solved is demonstrated in the following examples.

Example 6.4 *Solving the load flow equations directly*

Consider again the two-node network in Figure 6.3. Although the example network is simple, calculating the unknown voltage quantities $|V_2|$ and δ_2 is not so easy, as can be seen in the following direct calculation. First, Equations 6.20 and 6.21 are rewritten:

$$\cos(1.6705 - \delta_2) = \frac{-0.6 - 0.396|V_2|^2}{3.98|V_2|} \tag{6.22}$$

$$\sin(1.6705 - \delta_2) = \frac{0.3 + 3.96|V_2|^2}{3.98|V_2|} \tag{6.23}$$

When these two equations are squared and added, the following equation results:

$$3.98^2 \cdot |V_2|^2 = 0.6^2 + 0.3^2 + 2(0.6 \cdot 0.396 + 0.3 \cdot 3.96)|V_2|^2$$
$$+(0.396^2 + 3.96^2)|V_2|^4 \tag{6.24}$$

This system gives two possible values for $|V_2|$: $|V_2| = 0.1904\,\text{pu}$ and $|V_2| = 0.8853\,\text{pu}$ (see also Figure 8.5 in Section 8.6.1). The corresponding values for δ_2 can be obtained from Equations 6.22 and 6.23: $\delta_2 = -0.8458$ rad and $\delta_2 = -0.1617$ rad.

It becomes clear from this exercise that it is not so easy to solve the nonlinear load flow equations by means of straightforward arithmetic, not even for a two-node network. Therefore, an iterative method is applied to solve the nonlinear load flow equations as demonstrated with the Newton–Raphson method in the following example.

Example 6.5 *Solving the load flow equations iteratively*
Consider again the load flow equations as determined in Example 6.3 (p. 225). We call the exact solution of the voltage at node 2:

$$|V_2|^\dagger, \delta_2^\dagger \tag{6.25}$$

Because we do not know this exact solution yet, we start the calculation by making a guess:

$$|V_2|^{(0)}, \delta_2^{(0)} \tag{6.26}$$

The difference between the exact solution and our guess is

$$\Delta|V_2|^{(0)} = |V_2|^\dagger - |V_2|^{(0)}, \Delta\delta_2^{(0)} = \delta_2^\dagger - \delta_2^{(0)} \tag{6.27}$$

For the exact solution point, we can write

$$P_2(|V_2|^\dagger, \delta_2^\dagger) = P_2(|V_2|^{(0)} + \Delta|V_2|^{(0)}, \delta_2^{(0)} + \Delta\delta_2^{(0)}) = -0.6 \tag{6.28}$$

$$Q_2(|V_2|^\dagger, \delta_2^\dagger) = Q_2(|V_2|^{(0)} + \Delta|V_2|^{(0)}, \delta_2^{(0)} + \Delta\delta_2^{(0)}) = -0.3 \tag{6.29}$$

The next step is to determine the difference between the exact solution and our initial guess, being the corrections, by solving $\Delta|V_2|^{(0)}, \Delta\delta_2^{(0)}$ from the following first-order Taylor expansion of Equations 6.28 and 6.29:

$$P_2(|V_2|^\dagger, \delta_2^\dagger) = P_2(|V_2|^{(0)} + \Delta|V_2|^{(0)}, \delta_2^{(0)} + \Delta\delta_2^{(0)}) = -0.6$$
$$\approx P_2(|V_2|^{(0)}, \delta_2^{(0)}) + \Delta|V_2|^{(0)} \left.\frac{\partial P_2}{\partial |V_2|}\right|^{(0)} + \Delta\delta_2^{(0)} \left.\frac{\partial P_2}{\partial \delta_2}\right|^{(0)} \tag{6.30}$$

$$Q_2(|V_2|^\dagger, \delta_2^\dagger) = Q_2(|V_2|^{(0)} + \Delta|V_2|^{(0)}, \delta_2^{(0)} + \Delta\delta_2^{(0)}) = -0.3$$
$$\approx Q_2(|V_2|^{(0)}, \delta_2^{(0)}) + \Delta|V_2|^{(0)} \left.\frac{\partial Q_2}{\partial |V_2|}\right|^{(0)} + \Delta\delta_2^{(0)} \left.\frac{\partial Q_2}{\partial \delta_2}\right|^{(0)} \tag{6.31}$$

The notation $(\partial./\partial.)|^{(0)}$ indicates that the partial derivatives are evaluated based on our initial guess for the voltage values: $|V_2|^{(0)}, \delta_2^{(0)}$.

The terms of order higher than one in the Taylor expansion have been neglected: in this way a linear relation between the corrections and the other terms results, so that the corrections can be computed from the following matrix equation:

$$\begin{bmatrix} \dfrac{\partial P_2}{\partial \delta_2} & \dfrac{\partial P_2}{\partial |V_2|} \\[2mm] \dfrac{\partial Q_2}{\partial \delta_2} & \dfrac{\partial Q_2}{\partial |V_2|} \end{bmatrix}^{(0)} \underbrace{\begin{bmatrix} \Delta \delta_2^{(0)} \\[2mm] \Delta |V_2|^{(0)} \end{bmatrix}}_{\text{corrections}} = \begin{bmatrix} \Delta P_2^{(0)} \\[2mm] \Delta Q_2^{(0)} \end{bmatrix} = \underbrace{\begin{bmatrix} -0.6 - P_2(|V_2|^{(0)}, \delta_2^{(0)}) \\[2mm] -0.3 - Q_2(|V_2|^{(0)}, \delta_2^{(0)}) \end{bmatrix}}_{\text{mismatches}} \tag{6.32}$$

$$\underbrace{\phantom{\begin{bmatrix} \dfrac{\partial P_2}{\partial \delta_2} \end{bmatrix}}}_{\text{Jacobian}}$$

The mismatches are the difference between the prescribed power injected at node 2 and the injected power at node 2 as computed with the load flow equations based on the guess for the voltage at node 2.

The corrections can be calculated by inversion or factorization of the Jacobian matrix. When the corrections are added to the initial guess, we find an approximation of the exact solution:

$$\Delta |V_2|^{(0)} + |V_2|^{(0)} = |V_2|^{(1)} \neq |V_2|^{\dagger}, \quad \Delta \delta_2^{(0)} + \delta_2^{(0)} = \delta_2^{(1)} \neq \delta_2^{\dagger} \tag{6.33}$$

That we do not find the exact solution after one computational step is because we truncated the Taylor expansion in Equations 6.30 and 6.31 after the first-order term. Therefore, the iterative process has to be repeated until the mismatches become smaller than a certain predefined value.

Let us now make an initial guess for the unknown voltage at node 2. The easiest thing to do is to choose the voltage equal to that of node 1:

$$|V_2|^{(0)} = 1.0, \quad \delta_2^{(0)} = 0 \tag{6.34}$$

Substitution in the load flow equations (Equations 6.18 and 6.19) gives us

$$\begin{aligned} P_2 &= 0.396|V_2|^2 + 3.98|V_2|\cos(1.6705 - \delta_2) \\ &= 0.396 \cdot 1 + 3.98 \cdot 1 \cdot \cos(1.6705) = 0 \end{aligned} \tag{6.35}$$

$$\begin{aligned} Q_2 &= 3.96|V_2|^2 + 3.98|V_2|\sin(1.6705 - \delta_2) \\ &= 3.96 \cdot 1 - 3.98 \cdot 1 \cdot \sin(1.6705) = 0 \end{aligned} \tag{6.36}$$

For the elements in the Jacobian matrix, we find

$$\frac{\partial P_2}{\partial \delta_2} = |V_2 V_1 Y_{21}| \sin(\theta_{21} + \delta_1 - \delta_2)$$

$$\left. \frac{\partial P_2}{\partial \delta_2} \right|^{(0)} = 1 \cdot 1 \cdot 3.98 \cdot \sin(1.6705) = 3.96 \tag{6.37}$$

$$\frac{\partial P_2}{\partial |V_2|} = 2|V_2|G_{22} + |V_1 Y_{21}| \cos(\theta_{21} + \delta_1 - \delta_2)$$

$$\left. \frac{\partial P_2}{\partial |V_2|} \right|^{(0)} = 2 \cdot 1 \cdot 0.396 + 1 \cdot 3.98 \cdot \cos(1.6705) = 0.396 \tag{6.38}$$

$$\frac{\partial Q_2}{\partial \delta_2} = |V_2 V_1 Y_{21}| \cos(\theta_{21} + \delta_1 - \delta_2)$$

$$\left. \frac{\partial Q_2}{\partial \delta_2} \right|^{(0)} = 1 \cdot 1 \cdot 3.98 \cdot \cos(1.6705) = -0.396 \tag{6.39}$$

Table 6.2 Consecutive iterations with initial voltage: $V_2 = 1\angle 0$.

Iteration	$|V_2|$ [pu]	δ_2 [rad]	ΔP_2 [pu]	ΔQ_2 [pu]
0	1.0000	0.0000	−0.6	−0.3
1	0.9100	−0.1425	−0.0594	−0.0634
2	0.8863	−0.1610	−0.0021	−0.0024
3	0.8853	−0.1617	−0.3140e−5	−0.3754e−5

The shade indicates that a solution has been reached, where the power mismatches are sufficiently small.

$$\frac{\partial Q_2}{\partial |V_2|} = -2|V_2|B_{22} - |V_1 Y_{21}| \sin(\theta_{21} + \delta_1 - \delta_2)$$

$$\left.\frac{\partial Q_2}{\partial |V_2|}\right|^{(0)} = -2 \cdot 1 \cdot -3.96 - 1 \cdot 3.98 \cdot \sin(1.6705) = 3.96$$

(6.40)

Substitution of the obtained values in the matrix equation (Equation 6.32) gives us the following set of equations:

$$\begin{bmatrix} 3.96 & 0.396 \\ -0.396 & 3.96 \end{bmatrix} \begin{bmatrix} \Delta\delta_2^{(0)} \\ \Delta|V_2|^{(0)} \end{bmatrix} = \begin{bmatrix} -0.6 - 0 \\ -0.3 - 0 \end{bmatrix}$$

(6.41)

The corrections can be computed with Equation 6.41:

$$\begin{bmatrix} \Delta\delta_2^{(0)} \\ \Delta|V_2|^{(0)} \end{bmatrix} = \begin{bmatrix} -0.1425 \\ -0.09 \end{bmatrix}$$

(6.42)

Adding these corrections to our initial guess results in an approximation of the voltage at node 2:

$$\delta_2^{(1)} = 0 - 0.1425 = -0.1425 \ \text{rad}$$

(6.43)

$$|V_2|^{(1)} = 1 - 0.09 = 0.91 \,\text{pu}$$

(6.44)

These values are used as a new "guess" for the voltage at node 2 and the computational process is repeated with this new "guess."

The consecutive iterations are shown in Table 6.2. The iterative process (in this example) is stopped when the absolute values of the power mismatches become smaller than 1.e-3.

This answer corresponds to the one that we found earlier by solving the system directly in Example 6.4 (p. 226). In that example we also found a second solution, which we can calculate here as well, by choosing a very awkward initial guess for the value of the voltage at node two:

$$|V_2|^{(0)} = 0.1, \quad \delta_2^{(0)} = 0$$

(6.45)

The consecutive iterations are shown in Table 6.3.

Table 6.3 Consecutive iterations with initial voltage: $V_2 = 0.1\angle 0$.

Iteration	$|V_2|$ [pu]	δ_2 [rad]	ΔP_2 [pu]	ΔQ_2 [pu]
0	0.1000	0.0000	−0.5644	0.0564
1	0.1000	−1.4250	−0.2064	−0.3213
2	0.1572	−0.5295	−0.2416	0.1080
3	0.1761	−0.9242	−0.0136	−0.0583
4	0.1895	−0.8419	−0.0045	0.0016
5	0.1904	−0.8458	0.0319e−4	−0.1796e−4

The shade indicates that a solution has been reached, where the power mismatches are sufficiently small.

It is unnecessary to mention that this last solution is far from realistic, because the load is supplied at a very low voltage and, therefore, a high current (see also Section 8.6).

It is common practice to choose a flat start, that is, voltage magnitudes equal to one, while the angles are set to zero values (e.g., $|V_2|^{(0)} = 1.0, \delta_2^{(0)} = 0$), as initial guess for each load flow computation.

6.2.2 General Scheme of the Newton–Raphson Load Flow

The load flow computation is in fact the calculation of the voltage magnitude and angle at each bus of the power system under specified conditions of system operation. Other system quantities such as the current values, power values, and power losses can be calculated when the voltages are known. Speaking in mathematical terms, the load flow problem is nothing more than a system consisting of as many nonlinear equations as there are variables to be determined.

The unknown voltage quantities, or state variables, can be combined in a single vector:

$$x = \begin{bmatrix} \delta \\ |V| \end{bmatrix} = \begin{bmatrix} \delta_2 \\ \vdots \\ \delta_N \\ |V|_{N_g+2} \\ \vdots \\ |V|_N \end{bmatrix} \tag{6.46}$$

Note that the subscript of the voltage angles starts from 2, as only the voltage angle of the slack node, being node 1 in the network, is known. The subscript of the voltage magnitudes starts from $N_g + 2$, as only the voltage magnitudes of the load buses are unknown (see also Table 6.1).

The load flow equations at node i represent the computed active and reactive power injection (at node i) as a function of – in principle – all the node voltages in the network:

$$P_i = |V_i|^2 G_{ii} + \sum_{\substack{n=1 \\ \neq i}}^{N} |V_i V_n Y_{in}| \cos(\theta_{in} + \delta_n - \delta_i) \qquad (6.47)$$

$$Q_i = -|V_i|^2 B_{ii} - \sum_{\substack{n=1 \\ \neq i}}^{N} |V_i V_n Y_{in}| \sin(\theta_{in} + \delta_n - \delta_i) \qquad (6.48)$$

The unknown voltage magnitudes and angles must be determined such that, when all the voltage values are entered into the load flow equations (Equations 6.47 and 6.48), the computed active and reactive power injections at the specific nodes correspond with the specified power injections. In other words, the state variables must be determined such that the power mismatches, being the difference between the specified and computed power injections, are equal to zero:

$$\Delta P(x) = \begin{bmatrix} P_{2,\text{specified}} - P_2(x) \\ \vdots \\ P_{N,\text{specified}} - P_N(x) \end{bmatrix} = 0 \qquad (6.49)$$

$$\Delta Q(x) = \begin{bmatrix} Q_{N_g+2,\text{specified}} - Q_{N_g+2}(x) \\ \vdots \\ Q_{N,\text{specified}} - Q_N(x) \end{bmatrix} = 0 \qquad (6.50)$$

The power mismatches can also be combined in one vector:

$$h(x) = \begin{bmatrix} \Delta P(x) \\ \Delta Q(x) \end{bmatrix} = 0 \qquad (6.51)$$

When for a particular state vector x the power mismatches are not equal to zero ($h(x) \neq 0$), a correction Δx must be determined such that $h(x + \Delta x) = 0$. Application of a first-order Taylor approximation results in the Newton–Raphson iterative formula:

$$H \Delta x = -h(x) \qquad (6.52)$$

H the Jacobian of the power mismatches h

When J is defined as the Jacobian of the load flow equations, it holds that $J = -H$ and the Newton–Raphson iterative formula can be written as

$$J \Delta x = h(x) \qquad (6.53)$$

J the Jacobian of the load flow equations

The iteration process of the Newton–Raphson method is graphically illustrated in Figure 6.4 for a one-dimensional case. The general scheme of the Newton–Raphson load flow is listed in the flowchart in Figure 6.5. The

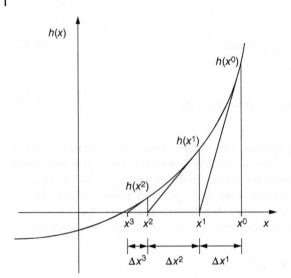

Figure 6.4 The Newton–Raphson method.

Jacobian terms, that is, the partial derivatives of the injected active and reactive power to the voltage angles and magnitudes, are found with Equations 6.47 and 6.48 as

$$\frac{\partial P_i}{\partial \delta_i} = \sum_{\substack{n=1 \\ \neq i}}^{N} |V_i V_n Y_{in}| \sin(\theta_{in} + \delta_n - \delta_i) \tag{6.54}$$

$$\frac{\partial P_i}{\partial \delta_j} = -|V_i V_j Y_{ij}|(\theta_{ij} + \delta_j - \delta_i) \tag{6.55}$$

$$\frac{\partial Q_i}{\partial \delta_i} = \sum_{\substack{n=1 \\ \neq i}}^{N} |V_i V_n Y_{in}| \cos(\theta_{in} + \delta_n - \delta_i) \tag{6.56}$$

$$\frac{\partial Q_i}{\partial \delta_j} = -|V_i V_j Y_{ij}| \cos(\theta_{ij} + \delta_j - \delta_i) \tag{6.57}$$

$$\frac{\partial P_i}{\partial |V_i|} = 2|V_i|G_{ii} + \sum_{\substack{n=1 \\ \neq i}}^{N} |V_n Y_{in}| \cos(\theta_{in} + \delta_n - \delta_i) \tag{6.58}$$

$$\frac{\partial P_i}{\partial |V_j|} = |V_i Y_{ij}| \cos(\theta_{ij} + \delta_j - \delta_i) \tag{6.59}$$

$$\frac{\partial Q_i}{\partial |V_i|} = -2|V_i|B_{ii} - \sum_{\substack{n=1 \\ \neq i}}^{N} |V_n Y_{in}| \sin(\theta_{in} + \delta_n - \delta_i) \tag{6.60}$$

$$\frac{\partial Q_i}{\partial |V_j|} = -|V_i Y_{ij}| \sin(\theta_{ij} + \delta_j - \delta_i) \tag{6.61}$$

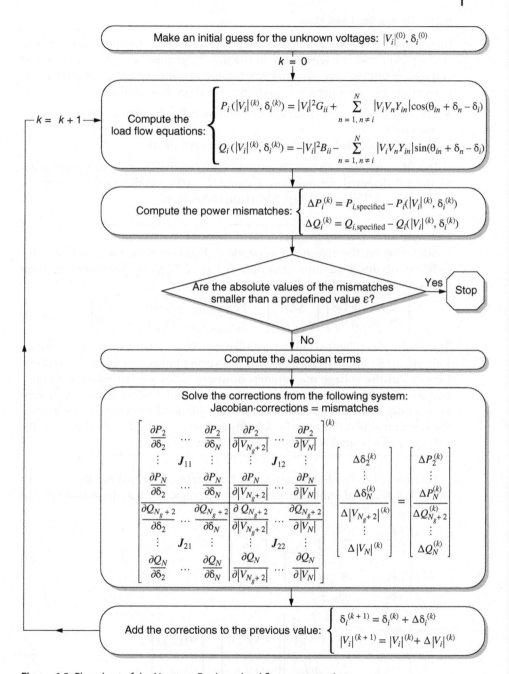

Figure 6.5 Flowchart of the Newton–Raphson load flow computation.

6.2.3 Decoupled Load Flow

In Section 5.2 we found out that there is a kind of "decoupling" between the active power and voltage angles on the one hand and the reactive power and voltage magnitudes on the other when some specific properties of power systems under normal operating conditions are taken into account:

- The resistance of the overhead transmission lines is much smaller than its reactance.
- The differences between the voltage angles are small.

This "decoupling" can be recognized in the Jacobian matrix, as the elements of the off-diagonal submatrices J_{12} and J_{21} (the off-diagonal elements in Equation 6.62 (Example 6.6 (p. 234))) are rather small.

Example 6.6 *Decoupling between active and reactive power*
Consider again the two-node network in Figure 6.3. In the iterative computation of this load flow (Example 6.5 (p. 227)), the following equation (Equation 6.41) was derived:

$$
\begin{bmatrix} 3.96 & 0.396 \\ -0.396 & 3.96 \end{bmatrix} \begin{bmatrix} \Delta\delta_2^{(0)} \\ \Delta|V_2|^{(0)} \end{bmatrix} = \begin{bmatrix} P_{2,\text{specified}} - P_2(|V_2|^{(0)}, \delta_2^{(0)}) \\ Q_{2,\text{specified}} - Q_2(|V_2|^{(0)}, \delta_2^{(0)}) \end{bmatrix}
\tag{6.62}
$$

The off-diagonal elements in the Jacobian matrix are relatively small, and therefore, changes in the voltage angles have a strong impact on the active powers. Changes in the voltage magnitudes, however, influence strongly the reactive powers. This "decoupling" between the active and reactive power equations occurs only when the resistance of the transmission link is much smaller than the reactance ($(R=0.025) \ll (X=0.25)$) and when the difference between the voltage angles is small ($\delta_1 - \delta_2 = 0.1617 \text{ rad} = 9.2647°$).

Therefore, the Jacobian matrix can be simplified by making the elements in the off-diagonal submatrices J_{12} and J_{21} zero, so that the following two equations result:

$$
\begin{bmatrix} \dfrac{\partial P_2}{\partial \delta_2} & \cdots & \dfrac{\partial P_2}{\partial \delta_N} \\ \vdots & J_{11} & \vdots \\ \dfrac{\partial P_N}{\partial \delta_2} & \cdots & \dfrac{\partial P_N}{\partial \delta_N} \end{bmatrix} \begin{bmatrix} \Delta\delta_2 \\ \vdots \\ \Delta\delta_N \end{bmatrix} = \begin{bmatrix} \Delta P_2 \\ \vdots \\ \Delta P_N \end{bmatrix}
\tag{6.63}
$$

$$
\begin{bmatrix} \dfrac{\partial Q_{N_g+2}}{\partial |V_{N_g+2}|} & \cdots & \dfrac{\partial Q_{N_g+2}}{\partial |V_N|} \\ \vdots & J_{22} & \vdots \\ \dfrac{\partial Q_N}{\partial |V_{N_g+2}|} & \cdots & \dfrac{\partial Q_N}{\partial |V_N|} \end{bmatrix} \begin{bmatrix} \Delta|V_{N_g+2}| \\ \vdots \\ \Delta|V_N| \end{bmatrix} = \begin{bmatrix} \Delta Q_{N_g+2} \\ \vdots \\ \Delta Q_N \end{bmatrix}
\tag{6.64}
$$

Those equations seem to be decoupled as the real power mismatches are used only to calculate the voltage-angle corrections, whereas the reactive power mismatches are applied only to calculate the voltage-magnitude corrections. But the elements in the matrices J_{11} and J_{22} still depend on both the voltage magnitude and the voltage angle. In order to obtain an efficient computational procedure, further simplifications can be made in the matrices J_{11} and J_{22} when we take into account that during normal system operation [1]:

- The line susceptances are much larger than the line conductances: $G_{ij}\sin(\delta_j - \delta_i) \ll B_{ij}\cos(\delta_j - \delta_i)$.
- The differences between the voltage angles are small: $\sin(\delta_j - \delta_i) = \delta_j - \delta_i$ and $\cos(\delta_j - \delta_i) = 1$.
- The reactive power injected into a node is much smaller than the reactive flow that would result if all lines connected to that bus were short-circuited to reference: $Q_i \ll |V_i|^2 B_{ii}$.

The elements in the matrix J_{11} can now be simplified as follows:

$$\frac{\partial P_i}{\partial \delta_i} = \sum_{\substack{n=1 \\ \neq i}}^{N} |V_i V_n Y_{in}|\sin(\theta_{in} + \delta_n - \delta_i)$$

$$= -Q_i - |V_i|^2 B_{ii} = -|V_i|^2 B_{ii} \tag{6.65}$$

$$\frac{\partial P_i}{\partial \delta_j} = -|V_i V_j Y_{ij}|\sin(\theta_{ij} + \delta_j - \delta_i)$$

$$= -|V_i V_j Y_{ij}|[\sin(\theta_{ij})\cos(\delta_j - \delta_i) + \cos(\theta_{ij})\sin(\delta_j - \delta_i)] \tag{6.66}$$

$$= -|V_i V_j|[B_{ij}\cos(\delta_j - \delta_i) + G_{ij}\sin(\delta_j - \delta_i)] = -|V_i V_j|B_{ij}$$

For the elements in the matrix J_{22}, it is convenient to multiply the partial derivatives with the voltage magnitude:

$$|V_i|\frac{\partial Q_i}{\partial |V_i|} = -2|V_i|^2 B_{ii} - \sum_{\substack{n=1 \\ \neq i}}^{N} |V_i V_n Y_{in}|\sin(\theta_{in} + \delta_n - \delta_i)$$

$$= Q_i - |V_i|^2 B_{ii} = -|V_i|^2 B_{ii} \tag{6.67}$$

$$|V_j|\frac{\partial Q_i}{\partial |V_j|} = -|V_i V_j||Y_{ij}|\sin(\theta_{ij} + \delta_j - \delta_i) = -|V_i V_j|B_{ij} \tag{6.68}$$

When we fill in those simplified partial derivatives in Equations 6.63 and 6.64, we obtain the following system of equations:

$$\begin{bmatrix} -|V_2|^2 B_{22} & \cdots & -|V_2 V_N|B_{2N} \\ \vdots & & \vdots \\ -|V_2 V_N|B_{N2} & \cdots & -|V_N|^2 B_{NN} \end{bmatrix} \begin{bmatrix} \Delta\delta_2 \\ \vdots \\ \Delta\delta_N \end{bmatrix} = \begin{bmatrix} \Delta P_2 \\ \vdots \\ \Delta P_N \end{bmatrix} \tag{6.69}$$

$$
\begin{bmatrix} -|V_{N_g+2}|B_{N_g+2,N_g+2} & \cdots & -|V_{N_g+2}|B_{N_g+2,N} \\ \vdots & & \vdots \\ -|V_N|B_{N,N_g+2} & \cdots & -|V_N|B_{N,N} \end{bmatrix} \begin{bmatrix} \Delta|V_{N_g+2}| \\ \vdots \\ \Delta|V_N| \end{bmatrix} = \begin{bmatrix} \Delta Q_{N_g+2} \\ \vdots \\ \Delta Q_N \end{bmatrix}
$$

(6.70)

When we divide each row in this system of equations by the voltage magnitude that is common in this row, and make the remaining voltage magnitudes in the left-hand side of Equation 6.69 equal to 1.0 per unit, we actually obtain a decoupled system of equations:

$$
\begin{bmatrix} -B_{22} & \cdots & -B_{2N} \\ \vdots & & \vdots \\ -B_{N2} & \cdots & -B_{NN} \end{bmatrix} \begin{bmatrix} \Delta\delta_2 \\ \vdots \\ \Delta\delta_N \end{bmatrix} = \begin{bmatrix} \frac{\Delta P_2}{|V_2|} \\ \vdots \\ \frac{\Delta P_N}{|V_N|} \end{bmatrix}
$$

(6.71)

$$
\begin{bmatrix} -B_{N_g+2,N_g+2} & \cdots & -B_{N_g+2,N} \\ \vdots & & \vdots \\ -B_{N,N_g+2} & \cdots & -B_{N,N} \end{bmatrix} \begin{bmatrix} \Delta|V_{N_g+2}| \\ \vdots \\ \Delta|V_N| \end{bmatrix} = \begin{bmatrix} \frac{\Delta Q_{N_g+2}}{|V_{N_g+2}|} \\ \vdots \\ \frac{\Delta Q_N}{|V_N|} \end{bmatrix}
$$

(6.72)

Both matrices are constant and depend on the grid parameters only. This means that the matrices have to be calculated and factorized only once, and this results in a faster load flow algorithm despite the fact that more iterations may be needed to reach zero mismatches. It is important to realize that all approximations are made in the Jacobian matrix only, while the power mismatches are evaluated without approximation. Therefore, only the speed of convergence is affected, but not the final result. This becomes clear from Figure 6.6, in which the iteration process of the Newton–Raphson method is visualized for a one-dimensional case in which the Jacobian matrix is kept constant: more iterations are needed than in Figure 6.4, but the final solution is the same.

The decoupled load flow is solved as follows. The procedure starts with an initial guess for the unknown voltages. Furthermore, the matrices in Equations 6.71 and 6.72 are built and inverted or factorized, followed by an iterative procedure:

- Calculate the active power mismatches: $\Delta P_i(|V_i|^{(k)}, \delta_i^{(k)})/|V_i|^{(k)}$.
- Solve the voltage-angle corrections from Equation 6.71.
- Update the voltage angles: $\delta_i^{(k+1)} = \delta_i^{(k)} + \Delta\delta_i^{(k)}$.
- Use the updated voltage angles to calculate the reactive power mismatches: $\Delta Q_i(|V_i|^{(k)}, \delta_i^{(k+1)})/|V_i|^{(k)}$.
- Solve the voltage-magnitude corrections from Equation 6.72.
- Update the voltage magnitudes: $|V_i|^{(k+1)} = |V_i|^{(k)} + \Delta|V_i|^{(k)}$.

Figure 6.6 The Newton–Raphson method with a fixed Jacobian.

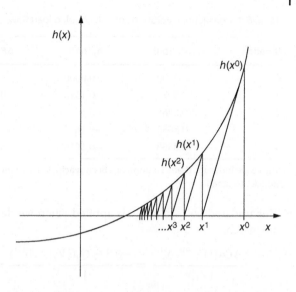

- Repeat this iterative procedure until the absolute values of the active and reactive power mismatches are smaller than a predefined value.

This procedure is illustrated in the following example by solving the load flow of the two-node network in Figure 6.3.

Example 6.7   *Decoupled load flow*

Consider again the two-node network in Figure 6.3. The load flow of this network will now be computed by a decoupled load flow.

We first choose the unknown voltage of node 2 to be equal to the voltage of node 1:

$$|V_2|^{(0)} = 1.0, \quad \delta_2^{(0)} = 0 \tag{6.73}$$

The active power mismatch amounts to

$$\frac{\Delta P_2(|V_2|^{(0)}, \delta_2^{(0)})}{|V_2|^{(0)}} = \frac{-0.6 - P_2(|V_2|^{(0)}, \delta_2^{(0)})}{|V_2|^{(0)}} = \frac{-0.6 - 0}{1} = -0.6 \tag{6.74}$$

Now we can calculate the voltage-angle correction (the value of B_{22} can be read from the admittance matrix in Equation 6.15):

$$-B_{22}\Delta\delta_2^{(0)} = \frac{\Delta P_2(|V_2|^{(0)}, \delta_2^{(0)})}{|V_2|^{(0)}} \rightarrow 3.96\Delta\delta_2^{(0)} = -0.6 \tag{6.75}$$

This gives us for the voltage-angle correction $\Delta\delta_2^{(0)} = -0.1515\,\text{rad}$, so that we can update the voltage angle:

$$\delta_2^{(1)} = \delta_2^{(0)} + \Delta\delta_2^{(0)} = 0 - 0.1515 = -0.1515\,\text{rad} \tag{6.76}$$

Table 6.4 Consecutive iterations of a decoupled load flow.

Iteration	$\lvert V_2 \rvert$ [pu]	δ_2 [rad]	ΔP_2 [pu]	ΔQ_2 [pu]
0	1.0000	0.0000		
1	0.8977	−0.1515	−0.6	−0.4051
2	0.8868	−0.1603	−0.0346	−0.0430
3	0.8855	−0.1615	−0.0048	−0.0053
4	0.8853	−0.1616	−6.3506e−4	−6.7891e−4

The shade indicates that a solution has been reached, where the power mismatches are sufficiently small.

This updated voltage angle is applied when we determine the reactive power mismatch:

$$\frac{\Delta Q_2(\lvert V_2 \rvert^{(0)}, \delta_2^{(1)})}{\lvert V_2 \rvert^{(0)}} = \frac{-0.3 - Q_2(\lvert V_2 \rvert^{(0)}, \delta_2^{(1)})}{\lvert V_2 \rvert^{(0)}} = \frac{-0.3 - 0.1051}{1} = -0.4051$$

(6.77)

Now we can calculate the voltage-magnitude correction from

$$-B_{22}\Delta \lvert V_2 \rvert^{(0)} = \frac{\Delta Q_2(\lvert V_2 \rvert^{(0)}, \delta_2^{(1)})}{\lvert V_2 \rvert^{(0)}} \rightarrow 3.96 \Delta \lvert V_2 \rvert^{(0)} = -0.4051 \qquad (6.78)$$

This gives us for the voltage-magnitude correction $\Delta \lvert V_2 \rvert^{(0)} = -0.1023$ pu, so that we can update the voltage magnitude:

$$\lvert V_2 \rvert^{(1)} = \lvert V_2 \rvert^{(0)} + \Delta \lvert V_2 \rvert^{(0)} = 1 + (-0.1023) = 0.8977 \, \text{pu} \qquad (6.79)$$

The iterative process (in this example) is stopped when the absolute values of the power mismatches become smaller than 1.e-3. As this is not the case yet, the updated voltage magnitude will now be applied when determining the active power mismatches, and so the procedure continues. The consecutive iterations are shown in Table 6.4.

A comparison with Table 6.2 shows that the same solution of the load flow problem is obtained.

In power system analysis, the so-called fast-decoupled load flow is frequently used [2, 3]. The fast-decoupled load flow is closely related to the algorithm that we derived here.

6.2.4 DC Load Flow

In situations where a lot of load flow computations have to be made, as is, for instance, the case for reliability computations or for security analysis, a linear approximation of the load flow problem can be made to save computation time: this is called the DC load flow. The DC load flow is principally different from the decoupled load flow. In the DC load flow, the nonlinear load flow equations are linearized to ease the calculation and to speed up the computation of the unknown voltages; this means that the actual model of the power system is altered, and this affects the final solution of the load flow. In the decoupled load flow, the nonlinear load flow equations are solved iteratively (the model of the power system remains unchanged), and approximations are made to the Jacobian matrix only; therefore only the speed of convergence is affected, but the final result remains the same.

The DC load flow can also be applied to find a fairly good approximation of the unknown voltages that can be used as initial values in a Newton–Raphson/decoupled load flow calculation.

Active power equations
The following approximations are made:

- The node voltage magnitudes are 1 pu.
- The resistances of the transmission lines are neglected: $Y_{ij} = G_{ij} + jB_{ij} = |Y_{ij}|\angle\theta_{ij}$ with $G_{ij} = 0$ and $\theta_{ij} = \pi/2\,\mathrm{rad}$.
- The differences between the voltage angles are small: $\sin(\delta_j - \delta_i) = \delta_j - \delta_i$ and $\cos(\delta_j - \delta_i) = 1$.

Under these assumptions, the active power load flow equation (Equation 6.47) can be written as

$$P_i = |V_i|^2 G_{ii} + \sum_{\substack{n=1 \\ \neq i}}^{N} |V_i V_n Y_{in}| \cos(\theta_{in} + \delta_n - \delta_i)$$

$$= \sum_{\substack{n=1 \\ \neq i}}^{N} |Y_{in}| \cos\left(\frac{\pi}{2} + \delta_n - \delta_i\right) = \sum_{\substack{n=1 \\ \neq i}}^{N} |Y_{in}| \sin(-\delta_n + \delta_i) \qquad (6.80)$$

$$= -\sum_{\substack{n=1 \\ \neq i}}^{N} |Y_{in}|(\delta_n - \delta_i)$$

Now, a linear relation is obtained between the active power injections and the busbar voltage angles.

Example 6.8 *DC load flow computation of the voltage angle*
Consider again the two-node network in Figure 6.3. In the DC load flow approximation, the resistance of the transmission line is neglected, so that the admittance of the line equals $Y = 1/Z = 1/j0.25 = -j4\,\mathrm{S}$, and the admittance matrix of the network can be written as

$$Y = \begin{bmatrix} -j4 & j4 \\ j4 & -j4 \end{bmatrix} \qquad (6.81)$$

The unknown voltage angle δ_2 will be approximated by means of the DC load flow computation (Equation 6.80):

$$P_2 = -|Y_{21}|(\delta_1 - \delta_2) = 4\delta_2 = -0.6 \tag{6.82}$$

$$\delta_2 = -0.15 \, \text{rad} \tag{6.83}$$

The exact value is $\delta_2 = -0.1617 \, \text{rad}$ (Example 6.4 (p. 226)).

Reactive power equations

The following approximations are made:

- The resistances of the transmission lines are neglected: $Y_{ij} = G_{ij} + jB_{ij} = |Y_{ij}|\angle\theta_{ij}$ with $G_{ij} = 0$ and $\theta_{ij} = \pi/2 \, \text{rad}$.
- The differences between the voltage angles are small: $\sin(\delta_j - \delta_i) = \delta_j - \delta_i$, and $\cos(\delta_j - \delta_i) = 1$.

With these assumptions, the reactive power load flow equation (Equation 6.48) can be written as

$$
\begin{aligned}
Q_i &= -|V_i|^2 B_{ii} - \sum_{\substack{n=1 \\ \neq i}}^{N} |V_i V_n Y_{in}| \sin(\theta_{in} + \delta_n - \delta_i) \\
&= -|V_i|^2 B_{ii} - \sum_{\substack{n=1 \\ \neq i}}^{N} |V_i V_n Y_{in}| \sin\left(\frac{\pi}{2} + \delta_n - \delta_i\right) \\
&= -|V_i|^2 B_{ii} - \sum_{\substack{n=1 \\ \neq i}}^{N} |V_i V_n Y_{in}| \cos(-\delta_n + \delta_i) \\
&= -|V_i|^2 B_{ii} - \sum_{\substack{n=1 \\ \neq i}}^{N} |V_i V_n Y_{in}|
\end{aligned}
\tag{6.84}
$$

The relation between the voltage magnitudes and the reactive power injections is still nonlinear. This equation can be linearized when, instead of the actual voltage magnitudes, the variations of the magnitudes around a 1 pu voltage are taken into account. In that case the voltage magnitudes can be approximated by [4]

$$|V| = 1 + \Delta|V| \quad \text{and} \quad \frac{1}{1 + \Delta|V|} \approx 1 - \Delta|V| \tag{6.85}$$

Rewriting Equation 6.84 gives

$$\frac{Q_i}{|V_i|} = -|V_i|B_{ii} - \sum_{\substack{n=1 \\ \neq i}}^{N} |V_n Y_{in}| \tag{6.86}$$

Substitution of Equation 6.85 leads to the following expression:

$$Q_i(1 - \Delta|V_i|) = -(1 + \Delta|V_i|)B_{ii} - \sum_{n=N_g+2, \neq i}^{N} |Y_{in}|(1 + \Delta|V_n|) - \sum_{n=1, \neq i}^{N_g+1} |V_n Y_{in}|$$

(6.87)

The summation has been split in two parts: the first summation is taken over all the load buses, of which the voltage magnitudes are unknown, and the second summation is done over all the generator buses and the slack node, of which the voltage magnitudes are known. Bringing all the knowns to the left-hand side and the unknowns to the right-hand side gives

$$Q_i + B_{ii} + \sum_{n=N_g+2, \neq i}^{N} |Y_{in}| + \sum_{n=1, \neq i}^{N_g+1} |V_n Y_{in}| = \Delta|V_i|(Q_i - B_{ii}) - \sum_{n=N_g+2, \neq i}^{N} |Y_{in}|\Delta|V_n|$$

(6.88)

We now have obtained a linear relation between the reactive power injections and the deviations of the unknown voltage magnitudes from a 1 pu voltage level.

Example 6.9   *DC load flow computation of the voltage magnitude*
We take again the two-node network in Figure 6.3, of which the admittance matrix under the DC load flow approximations is given in Example 6.8 (p. 239), Equation 6.81. The unknown voltage magnitude $|V_2|$ will be approximated by means of the DC load flow computation (Equation 6.88):

$$Q_2 + B_{22} + |V_1 Y_{21}| = \Delta|V_2|(Q_2 - B_{22})$$
$$-0.3 - 4 + 4 = \Delta|V_2|(-0.3 + 4)$$

(6.89)

$$\Delta|V_2| = -0.0811 \, \text{pu}$$
$$|V_2| = 1 + \Delta|V_2| = 0.9189 \, \text{pu}$$

(6.90)

The exact value is $|V_2| = 0.8853$ pu (Example 6.4 (p. 226)).

6.3 Optimal Power Flow

The load flow or power flow computation solves the node voltages in a given network under specified load conditions and a selected generation. Besides the generator power, also the injected power (e.g., reactive power from capacitor banks), the transformer tap ratios, and the phase-shifter settings can be chosen. It is this freedom of choice that makes it possible to optimize the system for certain criteria. An optimal power flow is a load flow computation, in which the calculation of the node voltages and the optimization of the controllable variables are carried out simultaneously.

The optimization problem can be defined as follows:

$$\begin{aligned}
\text{Minimize} \quad & f(x, u) \\
\text{subject to} \quad & h(x, u) = 0 \\
& g^{\min} \le g(x, u) \le g^{\max} \\
& u^{\min} \le u \le u^{\max}
\end{aligned}$$

(6.91)

x the vector with the state variables (this is the vector with the unknown voltage quantities for the load flow computation (see Equation 6.46))

u the vector with the controllable variables (these are the injected power, the transformer tap ratios, and the phase-shifter settings)

$f(x, u)$ the function to be minimized (the goal of the optimization can serve many purposes: minimization of the generating costs, minimization of active power loss or reactive power loss, and so on)

$h(x, u)$ the vector with the equality constraints (these are the power mismatches from the load flow computation (see Equation 6.51))

$g(x, u)$ the vector with the inequality constraints (these equations contain all limitations of technical or institutional kind other than the load flow equations, such as maximum values for node voltages/power flows/currents, restrictions regarding power exchanges, and fuel restraints; the constraints imposed on the controllable variables are treated separately)

In mathematical terms this is a constrained optimization problem. For a more extensive treatment of these kind of problems, the reader is referred to the mathematical literature on this topic, for example [5].

6.4 State Estimator

The operator of a power system must be able to observe every part of the system at all times. The system view should be coherent; there should be no gaps, no incongruities, and no misinformation. The SCADA system, which collects real-time measurements from the power system, cannot fulfill all of these requirements. Some measurements may not be available, other measurements may be corrupted because of hardware failures or communication problems, redundant measurements of the same quantity are rarely the same, and even some measurements could be totally wrong. The combination with the state estimator overcomes the deficiencies of the SCADA system. In effect, the state estimator enhances and maintains the integrity of the real-time database, thus making it possible for the control center software to support the operator with a complete, consistent, and accurate system overview.

The state estimator requires input data and, based on this, generates the output as shown in Figure 6.7. The input data consists of the network topology

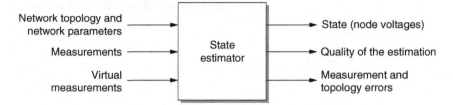

Figure 6.7 State estimator: input data and computational results.

and the values for the network parameters. Furthermore, measurement data is supplied. The measurements can be divided into three classes:

- Telemetered measurements: Online telemetered data such as bus voltages, branch currents and current injections, real power flows and injections, and reactive power flows and injections.
- Pseudo-measurements: Manufactured data such as guessed MW generation or substation load demand based on historical data.
- Virtual measurements: Not real measurements, but data generated in the control center; a substation with only one incoming feeder and two outgoing feeders, for instance, has a zero injected active and reactive power; this is "measurement" data that we can generate in the control center and that we are absolutely certain of.

The state of the power system is described by the bus voltage magnitudes and phase angles. The number of state variables amounts to $2N - 1$ (with N the number of nodes in the network), as one phase angle serves as a reference and is not included in the state vector:

$$x = \begin{bmatrix} \delta_2 \\ \vdots \\ \delta_N \\ |V|_1 \\ \vdots \\ |V|_N \end{bmatrix} \tag{6.92}$$

In practical state estimation, the measurement set is redundant: the number of measurements (m) is much larger than the number of state variables that has to be determined ($m > 2N - 1$). Therefore, the state estimation problem is an optimization problem: determine the unknown state variables such that they correspond best with the redundant measurement set. Or in mathematical terms, solve a system consisting of more equations than there are variables to be determined ($m > 2N - 1$). In the case that there is no redundancy, that is, the number of measurements equals the number of state variables ($m = 2N - 1$), the state estimation transfers into a load flow problem – a system with as many equations as there are variables to be determined. When the number of measurements is

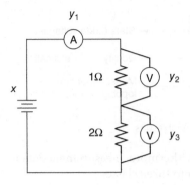

Figure 6.8 DC circuit with two voltmeters and one ammeter.

smaller than the number of state variables ($m < 2N - 1$), the system cannot be solved.

Example 6.10 *State estimation in a DC network*

Take as an example the DC network that is depicted in Figure 6.8. The voltage x of the DC source is not measured and must be estimated based on the three measurements: $y_1 = 2$ A, $y_2 = 3$ V, and $y_3 = 4$ V. What we do not know is that measurement device number 2 is malfunctioning and gives an incorrect reading of $y_2 = 3$ V instead of $y_2 = 2$ V (which would correspond to the true value of the state variable $x = 6$ V).

Based on the current measurement y_1 only, a voltage of $x = 2 \cdot 3 = 6$ V is computed, while based on the voltage measurement y_2 only, a voltage of $x/3 = 3 \rightarrow x = 9$ V results. The state estimator calculates the voltage x such that this value corresponds best to all of the measurements. The difference between the measured value and the computed value of the same quantity, as a function of the voltage x, is called the residue. The residues of the three measurements in this example are

$$r_1 = 2 - x/3$$
$$r_2 = 3 - x/3 \tag{6.93}$$
$$r_3 = 4 - 2x/3$$

The voltage x must be calculated such that the residues are as small as possible. This can be done by applying the least-squares method. The function to be minimized is the sum of squared residues:

$$J(x) = \sum_{i=1}^{3} r_i^2 = \frac{2}{3}x^2 - \frac{26}{3}x + 29 \tag{6.94}$$

The value of the voltage x where the sum of squared residues J is at its minimum can be found by differentiating the function J to x and determining for what value of x the $\partial J/\partial x$ equals zero:

$$\frac{\partial J}{\partial x} = \frac{4}{3}x - \frac{26}{3} = 0 \rightarrow x = 6.5\,\text{V} \tag{6.95}$$

A voltage $x = 6.5\,\text{V}$ corresponds best with the three measurements in the circuit.

Besides an estimate of the state of the system, the state estimator provides an indication on the quality of the estimation too. This quality can be read from the value of the sum of squared residues for the estimated voltage x. In the ideal case, that is, no erroneous measurements, the value of the sum of squares equals zero. This value deviates from zero as soon as the residues have a value other than zero:

$$J(6.5) = \frac{2}{3} \cdot (6.5)^2 - \frac{26}{3} \cdot (6.5) + 29 = \frac{5}{6} \tag{6.96}$$

Inspection of the residue values at the estimated voltage $x = 6.5\,\text{V}$ makes it clear that the second measurement shows the largest deviation from the zero value:

$$\begin{aligned} r_1(6.5) &= 2 - (6.5)/3 = -1/6 \\ r_2(6.5) &= 3 - (6.5)/3 = 5/6 \\ r_3(6.5) &= 4 - 2 \cdot (6.5)/3 = -1/3 \end{aligned} \tag{6.97}$$

From these residue values, it is obvious why the sum of squared residues is minimized instead of a sum of residues. A sum of residues could equal zero, while the individual residues are far from zero, because positive and negative residues cancel each other out.

6.4.1 General Scheme of the State Estimator

Between the true values of the measurements and the true values of the state variables, the following relation exists:

$$y_t = f(x_t) \tag{6.98}$$

y_t the true values of the measurements ($m \times 1$)
x_t the true state variables ($2N - 1 \times 1$)
f the measurement functions ($m \times 1$)
m the number of measurements
N the number of network nodes

As instead of the true values of the measurements, only with random noise corrupted values are available, the relation is rewritten:

$$y = f(x_t) + \varepsilon \tag{6.99}$$

y the measurements ($m \times 1$)
ε the measurement noise terms ($m \times 1$) (each term ε_i has an expectation $E(\varepsilon_i) = 0$ and a variance $E(\varepsilon_i^2) = \sigma_i^2$)

Because the noise terms are known by their statistical quantities only, this model cannot be used to obtain the true state variables, and the model is adjusted:

$$y = f(x) + r \tag{6.100}$$

x the state variables to be estimated $(2N - 1 \times 1)$
r the residues $(m \times 1)$

The state is determined by solving the unconstrained weighted least-squares minimization problem of the form

$$J(x) = \sum_{i=1}^{m} w_i r_i^2 = r^T W r \tag{6.101}$$

J the function to be minimized
W the diagonal matrix with the measurements weight factors w_i $(m \times m)$ (each weight factor is taken as the inverse of the measurement noise variance $w_i = 1/\sigma_i^2$)

Example 6.11 *Weighting factors*

It is logical to relate the weighting factors to the accuracy of the measurements. An accurate measurement gets a high weighting factor, while a less reliable measurement gets a lower weighting factor. The relation between an analog measurement and the true measurement value y_t is

$$y_{\text{analog}} = (1 \pm \phi) y_t \tag{6.102}$$

Analog-to-digital (A/D) conversion also introduces a small error:

$$y_{\text{digital}} = (1 \pm \psi) y_{\text{analog}} \tag{6.103}$$

The deviation of the analog measurement is, for instance, 1%, while the error introduced by an eight-bit A/D converter amounts $\psi = 1/2^8 \approx 0.004$. A weighting factor that corresponds with the accuracy of the measurement can be defined as follows:

$$w_i = \frac{1}{\sigma_i^2} = \frac{1}{\phi^2 + \psi^2} \tag{6.104}$$

As virtual measurements are data generated in the control center of which we are absolutely certain, the weighting factors of these measurements can be set to high values.

The minimum of Equation 6.101 is found when the gradient equals zero:

$$J_x(x) = 0 \tag{6.105}$$

J_x the gradient of the function J $(2N - 1 \times 2N - 1)$

This set of nonlinear equations is solved in the same way as we did with the load flow calculation (Section 6.2.2). When for a certain state vector x

the gradient is not equal to zero ($J_x(x) \neq 0$), a correction Δx must be found such that $J_x(x + \Delta x) = 0$. A first-order Taylor approximation results in the Newton–Raphson iterative formula:

$$J_{xx}(x)\Delta x = -J_x(x) \tag{6.106}$$

J_{xx} the Hesse matrix of the function J ($2N - 1 \times 2N - 1$)

The element expression and the matrix expression of the gradient J_x are given by

$$\frac{\partial J}{\partial x_j} = 2\sum_{i=1}^{m} w_i r_i \frac{\partial r_i}{\partial x_j} = -2\sum_{i=1}^{m} w_i r_i \frac{\partial f_i}{\partial x_j} \tag{6.107}$$

$$J_x(x) = -2A^T W r \tag{6.108}$$

A the Jacobian matrix of the measurement functions f ($m \times 2N - 1$)

The element expression and the matrix expression of the Hesse matrix J_{xx} are given by

$$\frac{\partial^2 J}{\partial x_j \partial x_k} = 2\sum_{i=1}^{m} w_i \frac{\partial f_i}{\partial x_j} \frac{\partial f_i}{\partial x_k} - 2\sum_{i=1}^{m} w_i r_i \frac{\partial^2 f_i}{\partial x_j \partial x_k} \tag{6.109}$$

$$J_{xx}(x) = 2A^T W A - 2\sum_{i=1}^{m} w_i r_i H_i \tag{6.110}$$

H_i the Hesse matrix of the measurement function f_i ($2N - 1 \times 2N - 1$)

When this is substituted, the Newton–Raphson iterative formula (Equation 6.106) becomes

$$\left(A^T W A - \sum_{i=1}^{m} w_i r_i H_i \right) \Delta x = A^T W r \tag{6.111}$$

Instead of this iterative equation, in practice often the following equation, in which the Hesse matrices are neglected, is used:

$$A^T W A \Delta x = A^T W r \tag{6.112}$$

Solving this iterative formula is basically the same as solving the iterative equation of the Newton–Raphson load flow.

6.4.2 Bad Data Analysis

In the context of power system state estimation, bad data are measurements that are much more inaccurate than was assumed when the measurement errors were modeled. Bad data are caused by a variety of factors, such as failing communication links, defective meters, and so on. Flagrantly erroneous data are rejected by prefiltering the measurements.

The presence of bad data among the observations processed by a least-squares estimator is, as a rule, detrimental to the performance of the estimator and usually results in poor state estimates. Therefore, a procedure is needed to check whether abnormally erroneous measurements are present in the measurement set. Moreover, it is also necessary to identify the faulty observations so that they can be removed from the measurement set.

Bad data analysis is normally done in three steps:

1) The detection procedure to determine whether bad data are present
2) The identification procedure to determine which data are bad
3) The elimination procedure to eliminate the influence of bad data on the state estimate

The statistical properties of the measurement errors facilitate the detection and identification of bad data. These properties are described in Section 6.4.3.

Statistical theory shows that the weighted sum of squared residues has a chi-square distribution with $m - (2N - 1)$ degrees of freedom (as demonstrated in the following section). However, this is no longer the case if a measurement that is erroneous enough to violate the normality assumption of the measurement noise vector is present. Therefore, bad data can be detected by means of a chi-square test, which can be outlined as follows. After each state estimation run, the weighted sum of squared residues is computed. This value is compared with a critical value from a chi-square distribution with $m - (2N - 1)$ degrees of freedom and a specified probability α, which is the probability that the sum of weighted squared residues exceeds the critical value (see Figure 6.9). If the weighted sum of squared residues is larger than this critical value, one concludes that bad data are present, and an identification procedure can be invoked to find out which measurements are erroneous. Otherwise, the state estimates are accepted on the ground that there is not enough evidence to indicate the presence of bad data.

Locating the bad data requires the individual examination of the estimation residues. A possible identification strategy could be to find the maximum residue and then to conclude that the corresponding measurement is the faulty one. However, this is not necessarily true for two reasons:

- The residues are, in general, correlated among themselves so that an error associated with a measurement can spread over other residues.
- Meters (for different quantities) can have different accuracies and the variances of the corresponding measurements can be significantly different.

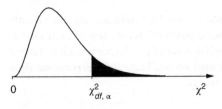

Figure 6.9 Probability density function of the chi-square distribution for *df* degrees of freedom. The black area is equal to α. $\chi^2_{df,\alpha}$ indicates the value from a chi-square distribution (with *df* degrees of freedom), which has a given area α above it; those critical values can be read from Table 6.5.

Table 6.5 χ^2 critical values.

df	α = 0.05	α = 0.025	α = 0.01	α = 0.005
1	3.841	5.024	6.635	7.879
2	5.991	7.378	9.210	10.597
3	7.815	9.348	11.345	12.838
4	9.488	11.143	13.277	14.860
5	11.070	12.833	15.086	16.750
6	12.592	14.449	16.812	18.548
7	14.067	16.013	18.475	20.278
8	15.507	17.535	20.090	21.955
9	16.919	19.023	21.666	23.589
10	18.307	20.483	23.209	25.188
11	19.675	21.920	24.725	26.757
12	21.026	23.337	26.217	28.300
13	22.362	24.736	27.688	29.819
14	23.685	26.119	29.141	31.319
15	24.996	27.488	30.578	32.801
16	26.296	28.845	32.000	34.267
17	27.587	30.191	33.409	35.718
18	28.869	31.526	34.805	37.156
19	30.144	32.852	36.191	38.582
20	31.410	34.170	37.566	39.997
21	32.671	35.479	38.932	41.401
22	33.924	36.781	40.289	42.796
23	35.172	38.076	41.638	44.181
24	36.415	39.364	42.980	45.559
25	37.652	40.646	44.314	46.928
26	38.885	41.923	45.642	48.290
27	40.113	43.195	46.963	49.645
28	41.337	44.461	48.278	50.993
29	42.557	45.722	49.588	52.336
30	43.773	46.979	50.892	53.672
40	55.758	59.342	63.691	66.766
50	67.505	71.420	76.154	79.490
60	79.082	83.298	88.379	91.952
70	90.531	95.023	100.425	104.215
80	101.879	106.629	112.329	116.321
90	113.145	118.136	124.116	128.299
100	124.342	129.561	135.807	140.169

As different types of meters can have different variances, a residual value that is an outlier for a specific measurement could be very well acceptable for another one. Therefore, a normalization of the residues is necessary to make up for that imbalance. A convenient and simple way to make a comparison of residues is to normalize them with respect to their standard deviations. After this normalization, the measurement that corresponds to the maximum normalized residue is, most likely, the bad measurement.

The general scheme of the state estimation is illustrated by the flowchart in Figure 6.10.

Example 6.12 *State estimation and bad data detection in a DC network*
Consider again the DC network that is depicted in Figure 6.8. The voltage x of the DC source is not measured and must be estimated based on the three measurements: $y_1 = 2$ A, $y_2 = 3$ V, and $y_3 = 4$ V. What we do not know is that measurement device number 2 is malfunctioning and gives an incorrect reading $y_2 = 3$ V instead of $y_2 = 2$ V (which corresponds to the true value of the state variable $x = 6$ V).

Measurement device 1 has an inaccuracy of $\sigma_1 = 0.14$ A, measurement device 2 has an inaccuracy of $\sigma_2 = 0.082$ V, and measurement device 3 has an inaccuracy of $\sigma_3 = 0.14$ V. As it is beneficial to give the more accurate measurements a higher weighting factor in the estimation, the weighting factors are defined as follows: $w_i = 1/\sigma_i^2$. Therefore, the matrix with the weighting factors is

$$W = \begin{bmatrix} 50 & 0 & 0 \\ 0 & 150 & 0 \\ 0 & 0 & 50 \end{bmatrix} = 50 \cdot \begin{bmatrix} 1 & 0 & 0 \\ 0 & 3 & 0 \\ 0 & 0 & 1 \end{bmatrix} \tag{6.113}$$

The measurement functions, as a function of the voltage x, are

$$\begin{aligned} y_1 &= 2 = x/3 + r_1 \\ y_2 &= 3 = x/3 + r_2 \\ y_3 &= 4 = 2x/3 + r_3 \end{aligned} \tag{6.114}$$

The Jacobian matrix of the measurement functions A is obtained by differentiating the measurement functions to the state variable ($\partial y_i / \partial x$):

$$A = \frac{1}{3} \cdot \begin{bmatrix} 1 \\ 1 \\ 2 \end{bmatrix} \tag{6.115}$$

The residue vector can be easily obtained from Equation 6.114:

$$r = \begin{bmatrix} 2 - x/3 \\ 3 - x/3 \\ 4 - 2x/3 \end{bmatrix} \tag{6.116}$$

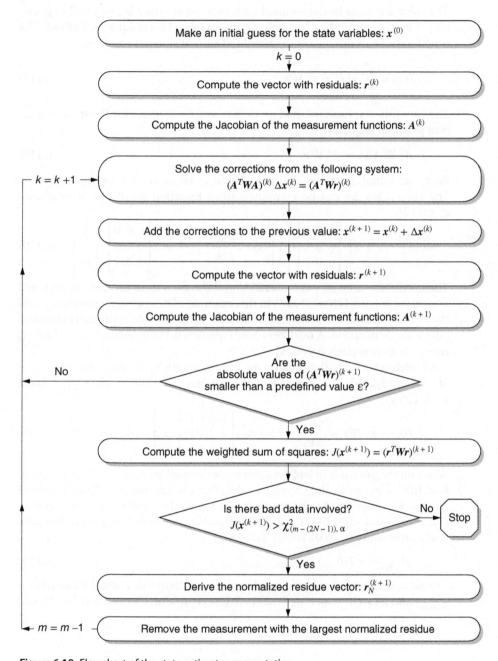

Figure 6.10 Flowchart of the state estimator computation.

The voltage x must be determined such that the residues are as small as possible. This is accomplished by applying the weighted least-squares method. The function to be minimized is the weighted sum of squared residues:

$$J(x) = \sum_{i=1}^{m} w_i r_i^2 = r^T W r \tag{6.117}$$

The Newton–Raphson method gives the iterative formula for the state estimation problem (Equation 6.112):

$$A^T W A \Delta x = A^T W r \tag{6.118}$$

First, an initial guess of the state variable is made, for example, $x = 1$ V. The correction Δx can be computed from Equation 6.118 that is evaluated at $x = 1$ V:

$$\frac{50}{9} \cdot \begin{bmatrix} 1 & 1 & 2 \end{bmatrix} \cdot \begin{bmatrix} 1 & 0 & 0 \\ 0 & 3 & 0 \\ 0 & 0 & 1 \end{bmatrix} \cdot \begin{bmatrix} 1 \\ 1 \\ 2 \end{bmatrix} \cdot \Delta x = \frac{50}{9} \cdot \begin{bmatrix} 1 & 1 & 2 \end{bmatrix} \cdot \begin{bmatrix} 1 & 0 & 0 \\ 0 & 3 & 0 \\ 0 & 0 & 1 \end{bmatrix} \cdot \begin{bmatrix} 5 \\ 8 \\ 10 \end{bmatrix} \tag{6.119}$$

The correction amounts $\Delta x = 6.125$ V and the final state estimate is obtained $x = x + \Delta x = 1 + 6.125 = 7.125$ V. In this example a DC system is considered, and the equations involved are linear. Therefore, the final state estimate is obtained after a single iteration. A next iteration, Equation 6.118 evaluated at $x = 7.125$ V, results in a correction $\Delta x = 0$ V.

An indication on the quality of the estimation is given by the value of the weighted sum of squared residues for the estimated voltage $x = 7.125$ V:

$$J(x) = r^T W r = \frac{50}{64} \cdot \begin{bmatrix} -3 & 5 & -6 \end{bmatrix} \cdot \begin{bmatrix} 1 & 0 & 0 \\ 0 & 3 & 0 \\ 0 & 0 & 1 \end{bmatrix} \cdot \begin{bmatrix} -3 \\ 5 \\ -6 \end{bmatrix} = \frac{375}{4} = 93.75 \tag{6.120}$$

A test on the presence of bad data can be performed by means of the chi-square test. In this example, the chi-square distribution has two degrees of freedom, namely, the number of measurements (3) minus the number of state variables (1). The critical value of the chi-square distribution with 2 degrees of freedom and a probability $\alpha = 0.01$ is (see Table 6.5)

$$\chi^2_{2,0.01} = 9.210 \tag{6.121}$$

The weighted sum of squared residues is larger than this critical value ($J(x) > \chi^2_{2,0.01}$) and it can be concluded that bad data is present. Identification of the bad data based on the residues only (Equation 6.116 with $x = 7.125$ V) could give an incorrect identification:

$$r = \begin{bmatrix} -3/8 \\ 5/8 \\ -3/4 \end{bmatrix} \tag{6.122}$$

In this case the measurement with the largest (absolute value of the) residue, that is, measurement 3, would be falsely identified as being bad data.

Instead of the residues, the normalized residues are applied for bad data identification. The normalized residue can be calculated (see also Equation 6.149 in Section 6.4.3) from

$$r_N = \frac{r}{\sqrt{\text{diag}(\text{cov}(r))}} = \frac{r}{\sqrt{\text{diag}(W^{-1} - A(A^T W A)^{-1} A^T)}} \tag{6.123}$$

The normalized residues are

$$r_N = \begin{bmatrix} -2.8347 \\ 9.6825 \\ -7.5000 \end{bmatrix} \tag{6.124}$$

The measurement with the largest absolute value of the normalized residue is identified as bad data: measurement 2. This is indeed the faulty measurement and it will be removed from the measurement set (or the weighting factor is set to zero). The state estimation procedure is now repeated with measurements 1 and 3 only. The new state estimate is $x = 6$ V and the weighted sum of squared residues equals $J = 0$. Despite the fact that the weighted sum of squared residues equals zero and no bad data can be present, again a test on the presence of bad data is performed by means of the chi-square test. In this example, the chi-square distribution has one degree of freedom, namely, the number of measurements (2) minus the number of state variables (1). The critical value of the chi-square distribution with 1 degree of freedom and a probability $\alpha = 0.01$ is (see Table 6.5)

$$\chi^2_{1,0.01} = 6.635 \tag{6.125}$$

The weighted sum of squared residues is smaller than this critical value ($J(x) < \chi^2_{1,0.01}$), and it can be concluded that no bad data is present. Now, the state estimation procedure is ready: the state variable has been determined ($x = 6$ V), and measurement 2 has been identified as bad data.

6.4.3   Statistical Analysis of the State Estimator

Between the measurements and the state variables, the following relations exist (estimates of quantities are indicated with a circumflex):

$$y_t = f(x_t) \tag{6.126}$$

$$y = f(x_t) + \varepsilon \tag{6.127}$$

$$y = f(\hat{x}) + r \tag{6.128}$$

$$\hat{y} = f(\hat{x}) \tag{6.129}$$

y_t the true values of the measurements $(m \times 1)$

y the measurements $(m \times 1)$

$\hat{y}$ the estimated measurement values $(m \times 1)$

x_t the true state variables $(2N - 1 \times 1)$

$\hat{x}$ the estimated state variables $(2N - 1 \times 1)$

f the measurement functions $(m \times 1)$

ε the measurement noise terms $(m \times 1)$

r the residues $(m \times 1)$

m the number of measurements

N the number of network nodes

The statistical analysis is based on the assumption that the measurement noise terms are independent, normally distributed random variables $\varepsilon_i \sim N(0, \sigma_i^2)$. Therefore, the mean value equals $E(\varepsilon_i) = 0$, the variance equals $E(\varepsilon_i^2) = \sigma_i^2$, and the covariance equals $E(\varepsilon_i \varepsilon_j) = 0$.

Properties of the estimates

The differences between the true and estimated values of both the state vector and the measurement vector are analyzed. For the estimated state vector, the following equation holds:

$$J_x(\hat{x}) = -2A^T W(y - f(\hat{x})) = 0 \tag{6.130}$$

Linearization of $f(x_t)$ around the value of the estimated state vector gives

$$f(x_t) = f(\hat{x}) + A(x_t - \hat{x}) \tag{6.131}$$

Using Equations 6.127 and 6.131, Equation 6.130 can be written as follows:

$$\begin{aligned} J_x(\hat{x}) &= -2A^T W(f(x_t) + \varepsilon - f(x_t) + A(x_t - \hat{x})) \\ &= -2A^T W(\varepsilon + A(x_t - \hat{x})) = 0 \end{aligned} \tag{6.132}$$

As a result, the difference between the estimated and the true state vector is

$$\hat{x} - x_t = (A^T W A)^{-1} A^T W \varepsilon = B\varepsilon \tag{6.133}$$

The expected value of the difference between the estimated and the true state vector is

$$E(\hat{x} - x_t) = E(\hat{x}) - E(x_t) = E(\hat{x}) - x_t = E(B\varepsilon) = BE(\varepsilon) = 0 \tag{6.134}$$

Thus, the expected value of the estimated state vector is equal to the true state vector:

$$E(\hat{x}) = x_t \tag{6.135}$$

The difference between the estimated and the true measurement vector is

$$\hat{y} - y_t = f(\hat{x}) - f(x_t) = A(\hat{x} - x_t) \tag{6.136}$$

The expected value amounts to

$$E(\hat{y} - y_t) = E(\hat{y}) - E(y_t) = E(\hat{y}) - y_t = AE(\hat{x} - x_t) = 0 \tag{6.137}$$

Thus, the expected value of the estimated measurement vector equals the true measurement vector:

$$E(\hat{y}) = y_t \tag{6.138}$$

Bad data detection

To analyze the properties of the weighted sum of squared residues, first the properties of the residues need to be examined. The residues, evaluated at the estimated state, can be written as

$$r = y - f(\hat{x}) = f(x_t) + \varepsilon - f(\hat{x}) = A(x_t - \hat{x}) + \varepsilon = (I - AB)\varepsilon = R\varepsilon \tag{6.139}$$

R the residual sensitivity matrix ($m \times m$)

The residual sensitivity matrix is idempotent ($R^2 = R$):

$$\begin{aligned} R^2 &= (I - AB)^2 = (I - A(A^T WA)^{-1}A^T W)(I - A(A^T WA)^{-1}A^T W) \\ &= I - 2A(A^T WA)^{-1}A^T W + A(A^T WA)^{-1}A^T WA(A^T WA)^{-1}A^T W \\ &= I - A(A^T WA)^{-1}A^T W = I - AB = R \end{aligned} \tag{6.140}$$

Using Equation 6.139, the weighted sum of squared residues evaluated at the estimated state can be written as

$$J(\hat{x}) = r^T Wr = \varepsilon^T R^T WR\varepsilon \tag{6.141}$$

The necessary and sufficient conditions for a quadratic form in independent standard normals to have a chi-square distribution are as follows. Let $u^T = (u_1, u_2, \ldots, u_k)$, where the u_i are independent standard normal variables. Let $V = V^T$ denote a symmetric matrix with real constants as entries. Then the nonnegative quadratic form $u^T Vu$ has a chi-square distribution if and only if $V^2 = V$. In this case the number of degrees of freedom equals rank(V) = trace(V).

Rewriting the weighted sum of squared residues gives

$$\begin{aligned} J(\hat{x}) &= \varepsilon^T R^T WR\varepsilon = \varepsilon^T \sqrt{W}\left(\sqrt{W}W^{-1}R^T \sqrt{W}\right)\left(\sqrt{W}RW^{-1}\sqrt{W}\right)\sqrt{W}\varepsilon \\ &= u^T V^T Vu = u^T Vu \end{aligned} \tag{6.142}$$

First, let us verify that the vector u is standard normal distributed. The matrix W is a diagonal matrix with the measurements weight factors w_i:

$$W = \begin{bmatrix} 1/\sigma_1^2 & 0 & \cdots & 0 \\ 0 & 1/\sigma_2^2 & \cdots & 0 \\ \vdots & & \cdots & \vdots \\ 0 & 0 & \cdots & 1/\sigma_m^2 \end{bmatrix} \tag{6.143}$$

As the noise terms $\varepsilon_i \sim N(0, \sigma_i^2)$ are normal distributed, the variables $u_i = \varepsilon_i/\sigma_i$ are standard normal distributed $u_i \sim N(0, 1)$.

Now, we need to verify that the matrix V in Equation 6.142 has the following properties: $V = V^T$ (symmetric) and $V^2 = V$ (idempotent):

$$\begin{aligned} V^T &= \sqrt{W} W^{-1} R^T \sqrt{W} = \sqrt{W} W^{-1} \left(I - B^T A^T \right) \sqrt{W} \\ &= \sqrt{W} \left(W^{-1} - A \left(A^T W A \right)^{-1} A^T \right) \sqrt{W} \\ &= \sqrt{W} \left(I - A \left(A^T W A \right)^{-1} A^T W \right) W^{-1} \sqrt{W} \\ &= \sqrt{W} \left(I - AB \right) W^{-1} \sqrt{W} = \sqrt{W} R W^{-1} \sqrt{W} = V \end{aligned} \tag{6.144}$$

$$\begin{aligned} V^2 &= \sqrt{W} R W^{-1} \sqrt{W} \sqrt{W} R W^{-1} \sqrt{W} \\ &= \sqrt{W} R^2 W^{-1} \sqrt{W} = \sqrt{W} R W^{-1} \sqrt{W} = V \end{aligned} \tag{6.145}$$

Thus, the weighted sum of squared residues is chi-square distributed with $m - (2N - 1)$ degrees of freedom $\chi^2_{(m-(2N-1)),\alpha}$. As this is no longer the case if a measurement is present that is erroneous enough to violate the normality assumption of the measurement noise vector, bad data can be detected by means of a chi-square test.

Bad data identification

For bad data identification, the normalized residues have to be computed:

$$r_{i,N} = \frac{r_i}{\sqrt{\text{var}(r_i)}} = \frac{r_i}{\sqrt{D_{ii}}} \tag{6.146}$$

$r_{i,N}$ the normalized residue of measurement i
D the diagonal matrix with the variances of the residues $(m \times m)$, the diagonal elements of the covariance matrix of the residue vector:
     $D = \text{diag}(\text{cov}(r))$

In order to compute the normalized residue, the variance of the residues must be analyzed. For the residues, the following expression was derived (Equation 6.139):

$$r = (I - AB)\varepsilon = R\varepsilon \tag{6.147}$$

The covariance matrix of the residues equals

$$\text{cov}(r) = E\{(r - E(r))(r - E(r))^T\} = E\{(R\varepsilon - E(R\varepsilon))(R\varepsilon - E(R\varepsilon))^T\}$$
$$= RE(\varepsilon\varepsilon^T)R^T = RW^{-1}R^T = RW^{-1} \tag{6.148}$$

Therefore, the normalized residues are defined as

$$r_N = \frac{r}{\sqrt{D}} = \frac{r}{\sqrt{\text{diag}(RW^{-1})}} \tag{6.149}$$

Problems

6.1 a. The load flow computation is important for a safe and secure operation of the system. Why is the load flow computation performed?
  b. To solve the mathematical equations in the load flow computation,
numerical iterative solvers are applied. What is the reason for this?

6.2

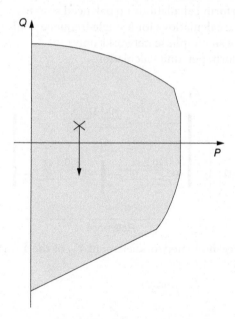

This is the loading capability diagram of a synchronous generator. The
operating point of the machine is at X. To move the operating point in
the direction of the arrow:

a. The excitation current has to increase.
b. The excitation current has to be reduced.
c. The torque of the prime mover has to increase.
d. The torque of the prime mover has to be reduced.

6.3 The complex power consumed by a single-phase load is $S = 100\angle-30°\,VA$. The power factor is:
   a. -0.707
   b. 0.707
   c. -0.866
   d. 0.866

6.4 The complex power consumed by a single-phase load is $S = 100\angle-30°\,VA$. The load is:
   a. An inductor and a resistor.
   b. A resistor.
   c. A capacitor and a resistor.
   d. None of the answers (a), (b), and (c) is correct.

6.5 When we perform network calculations in the phasor domain, we can:
   a. Only perform calculations on balanced systems
   b. Apply the calculations for a single frequency only
   c. Calculate single-phase networks only
   d. Not perform per-unit calculations

6.6

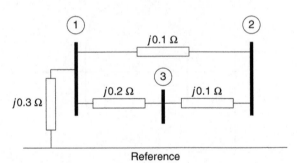

For this three-node network, element Y_{33} of the Y-matrix is:
   a. $-j2.86\,S$
   b. $-j15\,S$
   c. $j4\,S$
   d. $j0.3\,S$

6.7 A network has three load buses, three generator buses, and one slack node. What is the number of unknown voltage quantities (amplitudes and angles) that is calculated with a load flow computation?
   a. 6
   b. 9
   c. 10
   d. 11

6.8

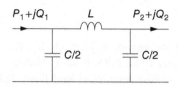

For this medium-length transmission line, the power at the sending end and the receiving end is as shown in the picture. The resistance of the line is neglected. Which of the following statements is *not* true?
a. Q_2 cannot be larger than Q_1.
b. $P_1 = P_2$.
c. When the line is unloaded, the line has a capacitive behavior.
d. For a medium-length line, the shunt capacitances can be omitted.

6.9

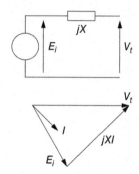

This is a phasor diagram of a synchronous machine. The resistance of the stator winding is neglected. E_i is the internal EMF of the machine, X the synchronous reactance, and V_t the voltage at the terminal of the machine.
We are dealing with:
a. An overexcited motor
b. An underexcited motor
c. An overexcited generator
d. An underexcited generator

6.10 The three-phase base power is 10 MVA and the base voltage (line-to-line) is 1 kV. What is the base impedance?
a. $10 \, \Omega$
b. $1 \, \Omega$
c. $0.1 \, \Omega$
d. $\sqrt{3} \, \Omega$

References

1 Grainger, John J., and Stevenson, William D., Jr.,: *Power System Analysis*, McGraw-Hill, Singapore, 1994, ISBN 0-07-113338-0.

2 Stott, B., and Alsaç, O.: 'Fast-decoupled load flow', *IEEE Transactions on Power Apparatus and Systems*, Vol. **PAS-93**, 1974, pp. 859–69.

3 Van Amerongen, R.A.M.: 'A general-purpose version of the fast-decoupled loadflow', *IEEE Transactions on Power Systems*, Vol. **PWRS-4**, 1989, pp. 760–70.

4 El-Hawary, Mohamed E.: *Electrical Power Systems – Design and Analysis*, IEEE Press, New York, 1995, ISBN 0-7803-1140-X.

5 Momoh, James A.: *Electric Power System Applications of Optimization*, CRC Press, Boca Rotan, 2009, ISBN 978-1-4200-6586-2.

7

Electricity Markets

7.1 Introduction

The concept behind the world of competition, and the free choice of the consumer to choose its electricity supplier, is that it is possible and even desirable to separate transportation as a means from the commodity that is transported. In other words, electric energy as a product can be commercially separated from transmission as a service. The drivers for restructuring the electricity sector in the different countries, traditionally under the rule of federal and state governments, are:

- The introduction of competition to reach more efficiency in the operation of the power-producing industry
- The possibility for the customers to choose a supplier

The main measures that are taken to achieve this are:

- The vertical unbundling to separate monopolistic activities from competitive ones, such as the separation of generation from the transmission and distribution of electricity
- The horizontal unbundling to stimulate competition in competitive fields, for instance, the creation of different generation companies

In essence, the electricity market is similar to any other economic market: buyers and sellers agree on a price. But there is one big difference: electricity cannot be stored in large quantities. That means that the cost conditions for the provision of electric supply vary constantly under the influence of the continuously changing demand and the possible dropout of generating units. In other markets similar phenomena do play a role, but then the ability to store the traded product mitigates this effect. There are some factors that complicate "electricity as a product." First of all, electric energy cannot be labeled or traced back with regard to its source or sink; electricity produced by a windmill, for example, cannot be distinguished from electricity that is produced by

Electrical Power System Essentials, Second Edition. Pieter Schavemaker and Lou van der Sluis.
© 2017 John Wiley & Sons Ltd. Published 2017 by John Wiley & Sons Ltd.

a nuclear power station. Furthermore, power flows cannot be controlled by financial instruments; they obey the laws of physics.

In this chapter, the focus is on the wholesale electricity market only, where the production companies trade their electricity with large consumers and retailers. The retailers offer their products on the retail electricity market, where the end consumers can choose their electricity supplier.

7.2 Electricity Market Structure

The "electricity market" is a rather broad concept that covers all aspects between the production/import and the consumption/export of electricity. Electric energy as a product can be separated commercially from transmission as a service, and this leads to a "market" structure as shown in Figure 7.1: production and supply are organized as a market, whereas the transmission and distribution of electricity are monopolistic activities. The square entitled "market" at the left-hand side of Figure 7.1 refers to a market in the economical context: a "marketplace" where trading takes place.

7.2.1 Transmission and Distribution

The entities involved in the transmission and distribution of electricity (the square at the right-hand side of Figure 7.1) are now briefly described.

Grid companies own parts of the transmission and/or distribution networks and facilitate the transmission and/or distribution of electricity efficiently, at low operating cost and with a high reliability. They have the duty to operate,

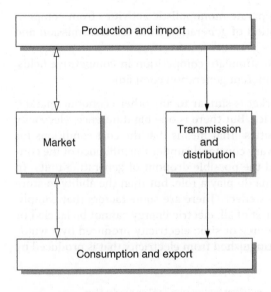

Figure 7.1 Organization of the electricity market; solid arrows, power flows; open arrows, commercial relations.

maintain, renew, and extend the network if this is necessary. Also, the handling of congestions and taking care of the power quality are included in their service. The transmission and distribution of electricity are monopolistic activities: there is only one (interconnected) electric infrastructure, of which the various parts are owned by different network companies. This makes a regulatory authority (being a governmental body) necessary to watch over the independence of the grid companies and to protect the customers. This body approves the tariffs of the grid companies as well.

Maintaining a continuous power balance and constant frequency, assuring adequate voltage support, and keeping operations within rigid security limits are essential to maintain a safe and reliable system. This requires a centralized coordination. All modern power systems, whether they support competitive markets or not, have a central "grid operator" to perform these duties: they have either an independent system operator (ISO) or a transmission system operator (TSO). The main difference between those two is that the ISO does not own network assets.

7.2.2 Market Architecture

The objective of the electricity market design is an efficient and reliable production of electricity to satisfy the demand. This objective includes both short-term (least-cost daily production) and long-term elements (induce efficient investment decisions) [1]. In this framework, where the TSO is responsible to operate the grid within certain boundaries, the proper incentives are required to have the market participants operate the supply side (and to a certain extent the demand, although most part of the demand is not flexible) in line with the system needs.

There are two basic ways to arrange trades between buyers and sellers [2]. They can make direct bilateral trades among themselves, or the trade can be arranged through an intermediary. This is elaborated in Table 7.1.

A bilateral market is highly flexible, as buyer and seller – often facilitated by a broker – can agree upon any (nonstandardized) contractual arrangement that they would like to. Flexibility and nonstandardized products/contracts are costly though, and the solvency of the counterparty needs to be assessed before

Table 7.1 Bilateral versus mediated market arrangements.

Arrangement	Type of Market				
Bilateral	Search	Bulletin Board	Brokered		
Mediated				Dealer	Exchange	Pool

Less organized ← → More centralized

entering into contractual arrangements. A bilateral market is also referred to as over-the-counter (OTC) market.

In a mediated market, it can be a dealer that matches a certain demand and supply by selling the energy at a higher price than what he or she bought it for. At an exchange, an auction is organized (see Section 7.3) where the exchange acts as the counterparty for all traders, taking away the solvency risks. Compared with a bilateral market, an exchange offers standard products, and as such lower costs and more participants, anonymous trading, and a transparent price formation and price signal. Where participation to an exchange is voluntary, it is mandatory in a pool. The bid structure in a pool is often more complex to allow for a more detailed proxy of the economic description of the generation process.

Trading is organized on different time frames, as illustrated in Figure 7.2.

The day-ahead market (DAM) is the market with the biggest volumes traded. Although the electricity is only physically delivered the day after, it is often referred to as being a spot market (which is characterized by trading for immediate delivery). It is on the DA market where the market actors take their positions, which can be fine-tuned during the intraday (ID) market, where – on most exchanges – can be traded till an hour before physical delivery. Indeed, on the intraday market, there is a better view on the wind production and solar infeed, as well as unforeseen outages of production facilities.

Both the long-term (e.g., year(s)-ahead) markets and the short-term (e.g., day-ahead) markets are usually organized as bilateral as well as mediated markets. For the long-term products, futures are traded on an exchange, whereas forwards are traded OTC. Long-term contracts are interesting for both buyers and sellers, as they allow them to lock in a price and avoid the price volatility of the spot market. This comes at a cost though: forward prices can be less competitive than spot prices, and the spot market provides more flexibility for the trading activities. An example of a future of forward contract may cover the supply of base electricity, that is, during all hours of the month, or at peak times, that is, from 8.00 a.m until 8.00 p.m from Monday to Friday.

Balancing refers to the process through which TSOs manage the physical equilibrium between injections (generation) and withdrawals (consumption) on the grid. Today, the balancing market in most of Europe – or maybe it is better to speak of a balancing mechanism – is not a real market where multiple buyers and sellers are active. Indeed it is a single-buyer market where the TSO is the only buyer of the balancing energy.

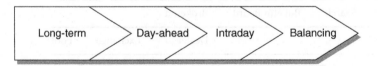

Figure 7.2 Market time frames.

Worldwide experience has shown that there is no market that functions exclusively through spot trading without a form of bilateral trading arrangements. Neither has a strictly bilateral model been implemented. Every market provides for both spot market trading coordinated by a grid/market operator and for bilateral contract arrangements scheduled through the same entity. Therefore we can state that, at a broad conceptual level, such a thing as a "common market model" exists. This is because of the universal desire to promote competitive principles such as efficiency, transparency, ease of entry, nondiscrimination, effective price signals, and so on.

7.3 Market Clearing

In most electricity markets a DAM is implemented, where hourly contracts for physical delivery on the following day are traded.

The DAM operated by power exchanges (PXs) is based on a two-sided auction model: both the supply and demand bids are gathered by the PX. Market equilibrium occurs when the aggregated supply and aggregated demand curve intersect; this intersection represents the market clearing price (MCP) and the market clearing volume (MCV). This market clearing is shown in Figure 7.3. The demand curve has a negative slope, indicating that as prices fall, the demand increases ("the first law of demand"). The supply curve has a positive slope. It illustrates the increased costs of providing an additional unit of electricity: from cheap units (wind, solar, hydro, nuclear, lignite) to more expensive production facilities (coal, gas, oil). Sale bids that are less than or

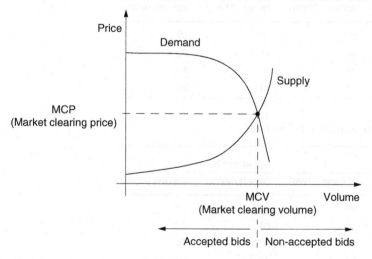

Figure 7.3 Market clearing algorithm.

equal to the MCP will be accepted. The settlement price will be the same as the MCP for the bidding quantity. Purchase bids that are higher than or equal to the MCP will be accepted. The settlement price will be the same as the MCP for the bidding quantity.

Example 7.1 *Market clearing*

Consider a market with two producers and two consumers. The two producers submit sale bids to the market for one specific hour of the day ahead as listed in Table 7.2. The first bid of producer 1 indicates that he or she is willing to sell a volume of 20 MWh for a minimum amount of 10 €/MWh.

The two consumers submit purchase bids for the same hour of the day ahead as listed in Table 7.3. The first bid of consumer 1 indicates that he or she is willing to buy a volume of 40 MWh for a maximum amount of 60 €/MWh.

In order to clear the market, the PX determines the aggregated supply and demand curves (see Table 7.4). The aggregated supply and demand curves are depicted in Figure 7.4. The MCP and the MCV can be obtained from the graph easily: MCP = 45 €/MWh and the MCV = 150 MWh. According to this market clearing, the two producers and two consumers are assigned trading volumes and have the ensuing revenues/expenses as given in Table 7.5.

Producer 1 only supplies 50 MWh of the 150 MWh (MCV) totally traded. As the MCP amounts to 45 €/MWh and two bids of producer 1 are above this

Table 7.2 The producers' sale bids.

Producer I		Producer 2	
Price [€/MWh]	Volume [MWh]	Price [€/MWh]	Volume [MWh]
10	20	5	15
30	30	15	30
50	40	30	40
70	60	45	50

Table 7.3 The consumers' purchase bids.

Consumer I		Consumer 2	
Price [€/MWh]	Volume [MWh]	Price [€/MWh]	Volume [MWh]
60	40	80	45
55	30	60	35
40	20	40	25
35	10	20	15

Table 7.4 Aggregated supply and demand.

Aggregated Supply		Aggregated Demand	
Price [€/MWh]	Volume [MWh]	Price [€/MWh]	Volume [MWh]
5	0–15	80	0–45
10	15–35	60	45–120
15	35–65	55	120–150
30	65–135	40	150–195
45	135–185	35	195–205
50	185–225	20	205–220
70	225–285		

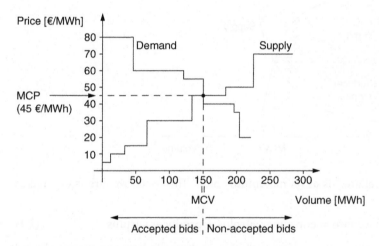

Figure 7.4 Market clearing example.

price, these bids are not accepted. Only the bids below or equal to the MCP are accepted. The market price equals the MCP = 45 €/MWh, and therefore the revenues of producer 1 amount to $50 \cdot 45 = 2250$ €.

7.4 Social Welfare

In the two-sided auction, market equilibrium occurs when the aggregated supply and aggregated demand curve intersect; this intersection represents the MCP and the MCV. At this market equilibrium, the benefits for society, called

Table 7.5 Trading volumes and revenues/expenses.

Market Actor	Trading Volume [MWh]	Revenues (Producers) Expenses (Consumers) [€]
Producer 1	50	2250
Producer 2	100	4500
Consumer 1	70	3150
Consumer 2	80	3600

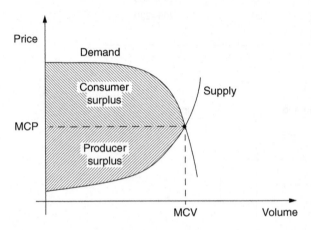

Figure 7.5 Consumer and producer surplus.

"the social welfare," is at its maximum value. The social welfare is defined as follows:

$$\text{Social welfare} = \text{consumer surplus} + \text{producer surplus} \qquad (7.1)$$

In Figure 7.5, the social welfare is the hatched area in between the supply and the demand curve. After all, the consumers are prepared to pay the price illustrated by the demand curve, whereas they only need to pay the MCP. The area between the MCP and the demand curve is the consumer surplus. The producers are willing to sell for the price given by the supply curve, but they receive the MCP. The area between the MCP and the supply curve is the producer surplus. Social welfare is the sum of both the consumer and producer surplus.

The market equilibrium represents the point where the economic balance among all participants is satisfied. To illustrate the fact that this is indeed an equilibrium, consider a trading volume that is below the MCV. In this case, the producers are prepared to sell for a price that is lower than the price the consumers are willing to pay for, and an increase in the trading volume will occur till the MCV is reached. The trading volume will not grow higher than the MCV, as no suppliers can be found that are prepared to sell for the price that the consumers are willing to pay. To summarize, there are no incentives for a seller or a buyer to deviate from the market equilibrium characterized by the MCP and MCV.

Example 7.2 *Social welfare*

Let us derive the social welfare of the two-sided auction, of which the aggregated demand and supply are tabulated in Table 7.4 and visualized in Figure 7.4. The consumer surplus can be calculated by determining the area in between the demand characteristic and the MCP for a volume from zero up to the MCV:

$$
\begin{aligned}
\text{Consumer surplus} = {}& 45 \cdot (80 - 45) + (120 - 45) \cdot (60 - 45) \\
& + (150 - 120) \cdot (55 - 45) = 3000\,€
\end{aligned} \tag{7.2}
$$

The producer surplus is the area in between the MCP and the supply curve for a volume from zero up to the MCV:

$$
\begin{aligned}
\text{Producer surplus} = {}& 15 \cdot (45 - 5) + (35 - 15) \cdot (45 - 10) \\
& + (65 - 35) \cdot (45 - 15) + (135 - 65) \cdot (45 - 30) \\
& + (150 - 135) \cdot (45 - 45) = 3250\,€
\end{aligned} \tag{7.3}
$$

The social welfare is the sum of the consumer surplus and the producer surplus: social welfare $= 3000 + 3250 = 6250\,€$.

7.5 Market Coupling

The "electricity market," of which a general structure was depicted in Figure 7.1, can be organized on a state level, a national level, or an international level, which we refer to as a market area. The MCP in such an area depends on a number of factors, one of which is the fuel mix that is used by the bulk of the production facilities. An area with a lot of nuclear power plants will have a lower MCP than an area with mainly gas turbines installed. But the power system is a large interconnected system, and these areas – "electricity markets" with their own characteristics, specific implementations, and MCPs – are physically interconnected with others as illustrated schematically in Figure 7.6. This facilitates the export of electricity from a low-price area to a high-price area.

Consider two areas named A and B. Each area has a PX that operates a DAM. The aggregated demand and supply curves in both areas are shown in Figure 7.7. The MCP in area B (MCP_B) is higher than in area A (MCP_A). Because the two areas are interconnected, import and export between the two areas are possible and desirable. The consumers in area B like to buy power from area A, as the price in area A is lower than what they are prepared to pay for. Practically, this

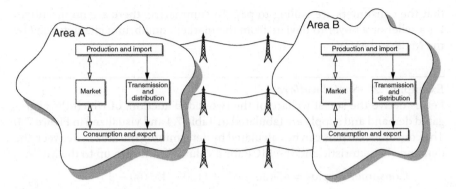

Figure 7.6 Two interconnected areas.

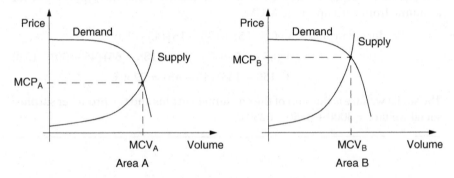

Figure 7.7 A low-price and a high-price area.

means that the demand in area A increases (i.e., the already existing demand in area A plus the demand coming from area B) and that the demand characteristic in area A shifts to the right as shown in Figure 7.8. Because of the higher price in area B, the producers in area A are willing to sell power to the consumers in area B. In essence, the supply in area B increases (i.e., the already existing supply in area B plus the imported power from area A) so that the supply characteristic in area B shifts to the right as shown in Figure 7.8. As a result of the power exchange between the two areas, the price goes up in the low-price area (area A) and goes down in the high-price area (area B). The new market equilibrium is the point where the economic balance among all participants is satisfied; this is the case when one common price (MCP* in Figure 7.8) is reached in both areas. There are no incentives for market players to deviate from this new market equilibrium.

In the following text, we make clear that the total social welfare increases when import and export take place between two areas with different prices. To

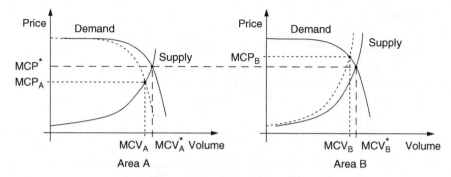

Figure 7.8 Import (area B) and export (area A). The dotted lines indicate the situation without import/export between the two areas.

do so, we first introduce the net export curve (NEC), which relates the market price in an area to the import and export volumes.

Example 7.3 *Net export curve (NEC)*

In this example we want to investigate what the impact of import and export is on the market clearing price that we calculated in Example 7.1 (p. 266). Let us focus first on a situation where the area under consideration exports 50 MW. Practically, the demand in this area increases with the amount that is demanded by the area to which the export takes place. In the case of the 50 MW export situation, the demand in the area under study becomes 50 MW higher, and the aggregated supply (unaltered) and demand (previous volume + 50 MW) as shown in Table 7.6 result. The aggregated supply and demand curves are depicted in Figure 7.9. The MCP and the MCV can be obtained from the graph easily: MCP = 50 €/MWh and the MCV = 200 MWh.

Table 7.6 Aggregated supply and demand.

Aggregated Supply		Aggregated Demand	
Price [€/MWh]	Volume [MWh]	Price [€/MWh]	Volume [MWh]
5	0–15	80	0–95
10	15–35	60	95–170
15	35–65	50	170–200
30	65–135	40	200–245
45	135–185	30	245–255
50	185–225	20	255–270
70	225–285		

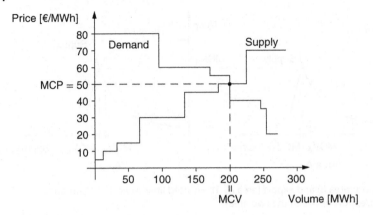

Figure 7.9 Market clearing example with 50 MW export.

Again we can calculate the social welfare in this 50 MW export situation by adding the consumer surplus and the producer surplus:

$$\text{Consumer surplus} = 95 \cdot (80 - 50) + (170 - 95) \cdot (60 - 50)$$
$$+ (200 - 170) \cdot (55 - 50) = 3750 \,€ \tag{7.4}$$

$$\text{Producer surplus} = 15 \cdot (50 - 5) + (35 - 15) \cdot (50 - 10)$$
$$+ (65 - 35) \cdot (50 - 15) + (135 - 65) \cdot (50 - 30) \tag{7.5}$$
$$+ (185 - 135) \cdot (50 - 45) = 4175 \,€$$

The social welfare = 4175 + 3750 = 7925 € and has increased compared to the non-export situation.

We can find the NEC of the area that we consider by altering the amount of export (shifting the demand characteristic) or import (shifting the supply characteristic) and by connecting the resulting MCP–MCV values by a line. At zero import/export, we found an MCP = 45 €/MWh, whereas in the 50 MW export situation an MCP = 50 €/MWh was the result; those are already two points (both are indicated by a dot) on the NEC shown in Figure 7.10.

The NECs of a low-price area A and a high-price area B are given in Figure 7.11. In this two-area model an export from area A is equivalent to an import into area B, so that we can put the NECs of the two areas in a single figure. We can achieve this by swapping the x-axis (and NEC) of area B in Figure 7.11. In this way we have obtained the same variable along the x-axes of the two NECs. Both NECs can now be combined into a single figure as shown in Figure 7.12.

The point where the two NECs intersect indicates the market equilibrium of the two coupled market areas; a common area price P^* results when a volume Q^* is exported from area A to area B as is shown in Figure 7.12. The hatched area

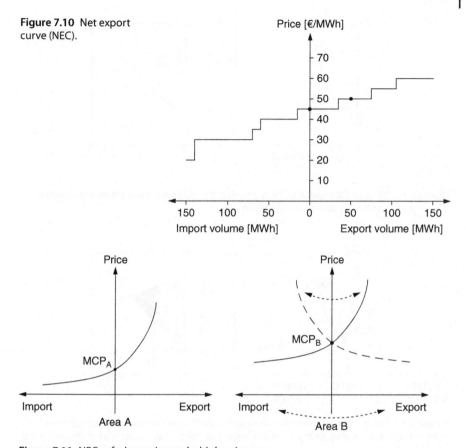

Figure 7.10 Net export curve (NEC).

Figure 7.11 NECs of a low-price and a high-price area.

in the figure, enclosed in between the two NECs and the price (zero volume) axis, is the increase in social welfare when power exchange takes place between the high-price and the low-price area. This situation holds as long as the physical interconnection capacity of the transmission links (the transmission lines and tower structures in Figure 7.6) between the two areas is sufficient. We have already noticed in Section 3.1 that the capacity of the tie lines is "limited" as they were originally designed for mutual support only and not for the exchange of large volumes of power between areas. Congestion of the interconnection capacity is therefore not unlikely, and a price difference between the two areas will remain. This is illustrated in Figure 7.13, in which ATC stands for the available transmission capacity between the two areas. The hatched area represents the market surplus. The gray area in the figure is the utility surplus or congestion rent; it is the cost that the market actors have to pay so that they can make

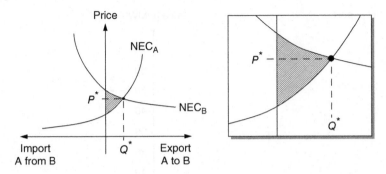

Figure 7.12 NECs of a low-price and a high-price area combined in a single graph and a close-up. Hatched area: gain in social welfare.

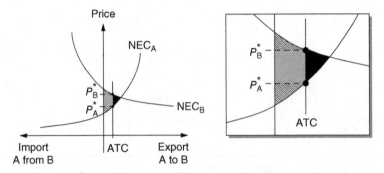

Figure 7.13 Congestion of the interconnection capacity and a close-up. Hatched area: market surplus. Gray area: utility surplus (congestion rent). Black area: market efficiency loss (deadweight loss).

use of the scarce transmission capacity. The total gain in social welfare obtained by the market coupling is given by the hatched and gray area together:

Social welfare = consumer surplus + producer surplus

$$+ \text{congestion rent} \tag{7.6}$$

The black area indicates the market efficiency loss, or deadweight loss; it is the difference between the social welfare that would exist in the uncongested case and the social welfare that exists in the congested case.

7.6 Allocation Mechanism and Zonal/Nodal Markets

In the previous section, two markets could be coupled as there was cross-border capacity (on the transmission lines interconnecting the two market areas) available between the two market areas to facilitate the physical transport. It is a so-called capacity allocation mechanism – where the order books of both market areas are available, as well as the available cross-border capacity as computed by the TSOs of both market areas – that matches the demand and supply in both market areas, such that the social welfare is maximized, while respecting the cross-border capacity made available. In essence, an allocation mechanism boils down to the following constrained optimization problem:

Objective: Maximize social welfare
Control variables: Import/export positions of the market areas involved
Subject to: Sum of import/export positions of market areas involved equals zero
                     Grid constraints

In an allocation mechanism, all commercial exchanges (i.e., trades from one market area to another) that are subject to the allocation mechanism are competing for the scarce capacity made available by the TSOs within the allocation mechanism.

The question "what is the proper size of a market area" is linked to congestion management in the grid and is not straightforward to answer. In essence, three systems may be distinguished in this respect [3]:

- The uniform system, where no exchanges are subject to an allocation mechanism (the copper-plate grid). This implies that all exchanges in the grid are feasible; in the case of possible congestions in the grid, the TSOs need to apply re-dispatch (increase the generation at one side of the congestion and reduce it at the other in order to induce a flow that relieves the congestion).
- The nodal system, where exchanges between all nodes are subject to an allocation mechanism. Indeed, this can be seen as the other extreme of the uniform system. All exchanges in the grid are now competing for the scarce transmission capacity. This system is usually applied in weaker power systems, being systems that are characterized by many (structural) congestions.
- The zonal system, where only exchanges between zones are subject to an allocation mechanism (i.e., a copper plate within the zone). Basically, this can be seen as the intermediate approach between the two extremes. There is no (structural) scarce transmission capacity within the market area, and, as such, the exchanges within the market area do not have to compete for scarce transmission capacity; possible (nonstructural) congestions need to be resolved by the TSO of the market area by means of re-dispatch. The cross-border exchanges are competing for the scarce transmission capacity and are subject to an allocation mechanism.

Nodal systems have been mainly implemented in the United States (Southwest Power Pool, California, New England, New York, PJM, and Texas), Argentina, Chile, New Zealand, Russia, and Singapore.

The European energy market is based on the zonal system. The zones (market areas) correspond in most of the cases to national borders (before the liberalization generation, grid and supply were organized mainly on a national basis

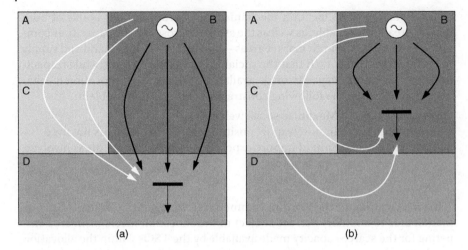

Figure 7.14 Transit flows (a) and loop flows (b) are indicated by the white arrows.

by integrated companies) and consequently vary largely in size. The Scandinavian market is an exception to that, as well as the Italian market; they both have multiple market areas within their national borders.

In Europe, the question "what is the proper size of a market area" is currently discussed intensively; the discussion does not focus on the question zonal or nodal though, but rather on the proper definition of the zones (market areas). Indeed, exchanges that are subject to the allocation mechanism are all competing for the scarce capacity made available within the allocation mechanism. Exchanges that are outside the allocation mechanism are all exchanges of which the impact is taken into account before the allocation mechanism itself, that is, exchanges that receive a "priority access" to the grid and that are exempted from the competition element within the allocation mechanism. Thus, the key question when asking "what is the proper size of a market area" is which exchanges need to be subject to an allocation mechanism and which exchanges can be left outside the allocation mechanism.

The grid interconnects the sources and the sinks in the system; the flows fan out in accordance to Kirchhoff's laws. A geographical concentration of sources and sinks in a market area can have a strong impact on the physical flows as is illustrated in Figure 7.14. The flows represented by the white arrows are often labeled "transit flows" if the source and the sink are in two different market areas and referred to as "loop flows" when the source and the sink are within the same market area.

A transit flow can be either an allocated or an unallocated flow:

- A transit flow is an unallocated flow when the exchange, causing the transit flow, is not subject to the same cross-border allocation mechanism as the

zone facing the transit flow (e.g., market area A and C in Figure 7.14 (a) are subject to one allocation mechanism and market area B and D to another).
- A transit flow is an allocated flow when the exchange, causing the transit flow, is subject to the same cross-border allocation mechanism as the zone facing the transit flow (e.g., market area A, B, C, and D in Figure 7.14 (a) are subject to one single allocation mechanism).

A loop flow is by definition an unallocated flow, as the source and the sink are located within the same market area (market area B in Figure 7.14 (b)) and the intrazonal exchange is not subject to an allocation mechanism.

Intrazonal exchanges can be made subject to the allocation mechanism by means of splitting a market area. For example, the north–south exchange within market area B will be subject to the allocation mechanism when the market area is split in two parts (being north and south). The splitting of the market area will make the former intrazonal exchange subject to the allocation mechanism and have it compete with other cross-border exchanges to make use of the scarce transmission capacity. The same holds the other way around: exchanges can be withdrawn from the allocation mechanism by merging two (or more) market areas into one.

References

1 Crampton, P.: Electricity Market Design: The Good, the Bad, and the Ugly, Hawaii International Conference on System Sciences, January 2003.
2 Stoft, S.: *Power System Economics, Designing Markets for Electricity*, IEEE Press and Wiley-Interscience, New York, 2002, ISBN 978-0-471-15040-4.
3 Schavemaker, P.H.: 'Flow-based market coupling and bidding zone delimitation: key ingredients for an efficient capacity allocation in a zonal system', EEM13 (European Energy Market Conference), Stockholm, Sweden, May 28–30, 2013.

8

Future Power Systems

8.1 Introduction

Electrical power systems can be regarded as one of the most complex systems designed, constructed, and operated by man. Of course it is a fact that the majority of the hardware that makes the generation, transmission, and distribution of electricity possible has not changed in essence since their first appearance more than a hundred years ago. In the design and construction of transformers, motors, generators, cables, and transmission lines, it is better to speak of evolution rather than revolution. But a lot of advanced technologies and techniques are applied in today's power system. Some examples of recent developments are described in the following text.

The fundamental design of the transformer has not altered much over the years, but the efficiency of the larger modern power transformers is now better than 99%! Anti-sound is sometimes applied to reduce the noise level of power transformers in densely populated areas. The mechanical tap changer for adjusting the voltage level will be gradually replaced by power-electronic voltage controllers.

Developments in the field of generators can also be mentioned, as one of the leading multinationals in power technology introduced the "powerformer" to the market [1]. This generator unit is in fact a combination of a generator and a transformer. The generator produces power at a higher voltage level, making a step-up transformer unnecessary.

For the transmission and distribution of electricity in urban areas, the application of underground cables is the trend. In particular in larger cities such as New York and Tokyo, the demand for electricity is steadily increasing, and the transmission and distribution voltages should be raised to higher voltage levels to prevent high ohmic losses. Underground cables, with voltage ratings up to the 500 kV level, are applied, and field experiments with superconductive cables are carried out.

In most countries the demand for electricity grows a few percent annually. In 2015, the ENTSO-E (European network of transmission systsem operators for

Electrical Power System Essentials, Second Edition. Pieter Schavemaker and Lou van der Sluis.
© 2017 John Wiley & Sons Ltd. Published 2017 by John Wiley & Sons Ltd.

electricity) consumption reached 3,278 TWh, which represents a 1.4 % increase compared to the previous year [2]. Every year the power system is operated closer to its limits, and FACTS (see Section 5.6) devices will play a more dominant role in the future in order to maintain stability [3]. The application of FACTS devices, however, does increase the complexity of the power system even further, and one cannot exclude the possibility that chaotic phenomena might occur in the near future (see also Section 8.6).

Apart from technical developments, economical changes also take place. In the majority of the developed countries, the generation companies and the utilities are now restructured under the legal force of deregulation and liberalization. The market is the playing field, and the utilities have to operate on an international scale and make strategic alliances to survive these open markets. A parallel with the earlier liberated telecom market may be considered but with the big difference that the mobile telephone infrastructure has no counterpart in the electrical power system. The daily practice is nowadays that industrial companies and individual customers can choose their supplier of electricity.

Power systems evolve continuously. This is not only driven by technical developments but also by politics: governments have a steering role in certain tendencies/trends within the sector. Of course, the liberalization of the electricity companies was initiated by the government. But also in the field of alternative energy sources, and emission/pollution restrictions, the government sets out the direction in which the sector has to develop. In addition, public opinion and rejection/acceptance of certain technologies, such as nuclear energy, have their impact on the system and its operation too. Action committees, raised by environmentalists, villagers, and/or other involved parties, cause tremendous delays in the case of network expansion or installation of power plants and have quite some influence on the network operation and planning.

In the following sections some of the foreseen developments, which originate from the complex technological, ecological, sociological, and political playing field and their possible consequences on the power system, are highlighted. But, of course, nobody knows what the future and the future power systems will look like.

8.2 Renewable Energy

Concern on the change in the Earth's climate, formalized in the Kyoto Protocol in 1997, stimulates research, promotion, development, and increased use of renewable energy. In power systems, the application of electricity generation based on renewable sources is not new, but a future large-scale implementation of these renewable energy sources will cause structural changes in the existing distribution and transmission networks. This is for the following reasons:

- Most renewable energy generators are connected to the distribution network, for example, the solar panels on the roofs of houses, small wind farms, and individual windmills. This is in contrast with the current layout of the system, where most of the generation is connected to the transmission system (see also Section 8.3).

- Most renewable energy generators are connected to the grid by means of power-electronic interfaces. This gives quite a different behavior compared with synchronous generators (see also Section 8.4).

- Most renewable energy generators depend on natural and uncontrollable sources, such as the wind and the sun, and the electrical power output cannot be controlled. If there is no wind, the windmill does not deliver energy. However, most renewable energy generators are connected to the grid by power-electronic interfaces, which offer the possibility and flexibility to control the power output – given a certain power input – in some way. In the case of photovoltaic systems, for example, the converter of a solar panel is programmed to maximize the energy yield (by means of a maximum power point tracker).

- Many renewable energy generators, for example, wind turbines and photovoltaic systems, have an intermittent character. A large-scale implementation of this type of generators can lead to strong power fluctuations in the grid.

8.3 Decentralized or Distributed Generation

Most of the power plants are large industrial sites located at strategic locations, nearby a river or a lake for cooling water, and close to energy resources or supply routes. These large power plants are connected to the transmission network by step-up transformers and are controlled in order to take care of the voltage and frequency stability of the power system. This is what we call "centralized generation." Until now, the power system is for the greater part supplied by this centralized generation, and we therefore say that the system is "vertically" operated, as illustrated by Figure 8.1. We can see from the system layout that there is a "vertical" power flow in the system: at the top, power is generated by a (relatively small) number of large power plants, and, via the transmission and distribution systems, power finds its way down to the consumers connected at the lowest voltage levels.

Nowadays, the trend is to integrate more and more decentralized generation (DG, also called distributed or dispersed generation) into the system by means of connecting small-scale generators at the lower voltage levels. Examples of DG units are windmills, solar panels, or combined heat and power units (producing steam for industrial processes and electricity as a by-product). When this trend continues, a large-scale implementation of these DG units

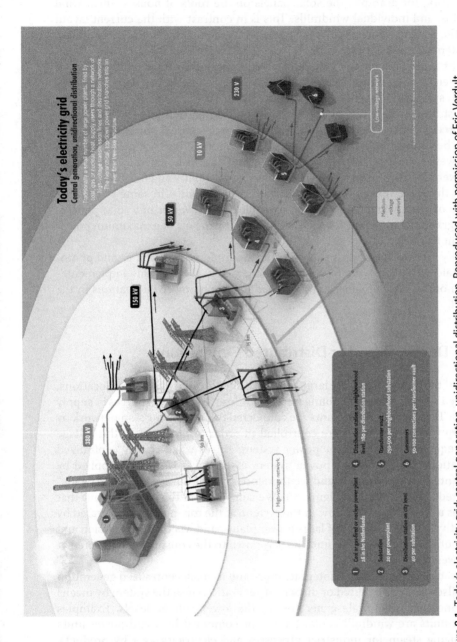

Figure 8.1 Today's electricity grid: central generation, unidirectional distribution. Reproduced with permission of Eric Verdult, www.kennisinbeeld.nl. (*See color plate section for the color representation of this figure.*)

will lead to a transition from the current "vertically operated power system" (Figure 8.1) into a "horizontally operated power system" in the future, as shown in Figure 8.2. Because of the increasing amount of DG, the most uneconomical and/or aged power plants are taken out of service, and this leaves a power system with the bulk of the consumption and the generation connected to the distribution networks so that a more or less "horizontal power flow" through the system results.

Possible developments and/or consequences of the transition from the current "vertically operated power system" into a future "horizontally operated power system" for the power system are [4] as follows:

- When most of the large power stations have vanished and the "horizontal" power system is a fact, the transmission network has lost one of its main purposes, namely, to facilitate the bulk transport of electrical energy from the centralized generators to the distribution networks. Although the import and export functionality is still present, a rather "empty" transmission network is the result, the main purpose of which is now to interconnect the various "active" distribution networks. Such a situation requires different control and operating strategies in order to keep the system operation within safe margins. The voltage, for example, will no longer be imposed by the large (centralized) power stations, and voltage stability of the system becomes an issue.
- Traditionally, the distribution network is a passive network, which depends totally on the transmission network for energy delivery, frequency control, and voltage regulation. In a future "horizontally operated power system," the power is not only consumed by but also generated in the distribution network. Therefore, the distribution network needs to change into an active and intelligent network, which is able to control and regulate the system parameters, without strong support from the transmission network.
- In the "vertical" power system, the direction of the power flow is more or less predictable: centralized generation → transmission network → distribution network → consumers. In the "horizontal" power system, with its active distribution networks, the situation is different: (centralized generation →) transmission network ↔ distribution network ↔ consumers. The direction of the power flow in the network is not predictable any more: one-way traffic becomes two-way traffic. This has a fundamental impact on the protection of the system.
- Autonomous networks could be developed. When the total amount of power generated in a certain part of the distribution network is sufficient to supply the local loads, the network could be operated autonomously by disconnecting it from the rest of the grid. From an operational point of view, this gives a system that can be controlled more easily. However, we saw in

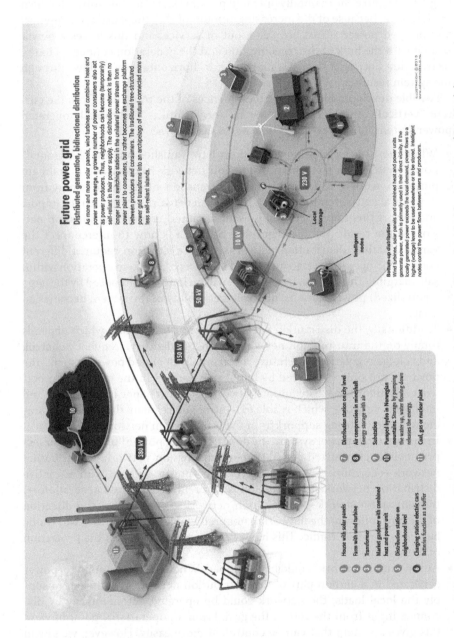

Future power grid

Distributed generation, bidirectional distribution

As more and more solar panels, wind turbines and combined heat and power units emerge, a growing number of power consumers also act as power producers. Thus, neighborhoods can become (temporarily) self-reliant in their power supply. The distribution network is then no longer just a switching station in the unilateral power stream from power plant to consumers, but rather becomes an exchange platform between producers and consumers. The traditional tree-structured power grid transforms into an archipelago of mutual connected more or less self-reliant islands.

Bottom-up distribution

Wind turbines, solar panels and combined heat and power units generate power, which is primarily used in their direct vicinity. If the locally generated power exceeds the local demand, power flows to a higher (voltage) level to be used elsewhere or to be stored. Intelligent nodes control the power-flows between users and producers.

① House with solar panels
② Farm with wind turbine
③ Transformer
④ Market gardener with combined heat and power unit
⑤ Distribution station on neighborhood level
⑥ Charging station electric cars. Batteries function as a buffer
⑦ Distribution station on city level
⑧ Air compression in mineshaft. Energy storage with air
⑨ Substation
⑩ Pumped hydro in Norwegian mountains. Storage by pumping the water up, water flowing down releases the energy.
⑪ Coal, gas or nuclear plant

ILLUSTRATION © 2013 ©
WWW.KENNISINBEELD.NL

Figure 8.2 Future power grid: distributed generation, bidirectional distribution. Reproduced with permission of Eric Verdult, www.kennisinbeeld.nl. (*See color plate section for the color representation of this figure.*)

Section 3.1 that interconnection of networks offers quite some advantages too. An important one is that the operation can be supported by others if there is a problem, for example, with the local supply because of unexpected loss of generation. Therefore, such autonomous operating systems will have to be equipped with a connection to the main transmission grid, or to neighboring autonomous distribution systems, to safeguard the supply in case of a sudden emergency situation. This is not as easy as it looks: both systems operate in practice at a slightly different frequency and usually have a different voltage angle at the instant in time we would like to make the interconnection. Another issue is that both systems have their own frequency control, and the question arises: which system will act as the master and which system will be the slave? These problems can be avoided by having a DC link connecting the systems; in this way, the systems have the possibility for power exchange but are frequency-wise and voltage-wise uncoupled.

8.4 Power-Electronic Interfaces

There are a number of different DG technologies, based on distinct energy sources, such as cogenerating plants, wind turbines, small hydro plants, photovoltaic systems, fuel cells, and microgeneration. DG units, especially the ones powered by renewable energy sources with intermittent characteristics (such as wind and solar), are quite often connected to the distribution network by means of power-electronic interfaces. When a renewable energy generator produces DC output power, as is the case with photovoltaics and fuel cells, the main task of the power-electronic interface is to do the DC to AC conversion. An additional task of the power-electronic converters is to maximize the energy yield, as accomplished by the maximum power point tracker in photovoltaic systems. In variable speed wind turbines, the converter is necessary to make variable speed operation of the electrical machine possible as it leads to a higher energy output from the wind. When a DG unit is connected to the power system through a power-electronic interface, no extra inertia is added to the system when:

- The DG unit itself has no rotating mass and generates DC power, as is the case with photovoltaic systems and fuel cells (schematically drawn in Figure 8.3 (a)).
- The inertia of the DG unit of which the prime mover does have rotating masses is separated from the grid by the power-electronic interface as it decouples the mechanical rotor speed of the DG unit from the grid frequency; this is, for example, the case for wind turbines with either a doubly fed induction generator or a direct drive synchronous generator (schematically drawn in Figure 8.3 (b)).

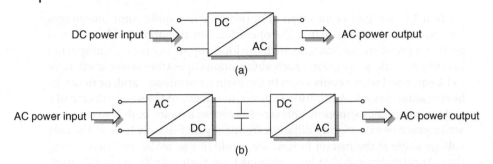

Figure 8.3 Power-electronic interfaces.

Thus, in the aforementioned cases, the DG does not contribute to the frequency oscillations of the synchronous generators in the grid in the case of a disturbance in the balance between generation and load in the system. So when the grid frequency drops, the mechanical frequency of a variable speed wind turbine with a doubly fed induction generator does not change because the power-electronic interface is in between and no energy stored in the rotating mass is supplied to the grid (as would be in the case of a conventional synchronous generator coupled to the grid via a transformer).

8.5 Energy Storage

When (renewable) DG comprises a substantial part of the power production, energy storage can give us an extra and valuable degree of freedom for the operation and control of the system and is sometimes even indispensable. In a traditional vertical power system, the inertia of the rotors of the synchronous generators covers a possible mismatch between power production and consumption. In a horizontal power system, with a relatively small amount of centralized generation and a considerable contribution of (renewable) DG, the total inertia of the rotating mass in the system is strongly reduced. This is because of the lower inertia of the DG units and/or because generators are separated from the grid by power-electronic interfaces. In consequence, the total inertia in the system may become insufficient to compensate adequately for disturbances of the power balance, resulting in relatively large frequency deviations and in a dynamically less stable system in the case of system disturbances and faults. Energy storage can help to compensate for this and help to maintain the power balance; energy can be stored when there is a surplus of energy, and subsequently it can be delivered in the case of a temporary deficiency. Energy storage can also be of help to level out large power fluctuations when the power is generated by renewable energy sources.

Large-scale energy storage is still behind the horizon, but some promising techniques have already been developed: superconducting magnetic energy storage (SMES, electrical storage of energy) systems, battery energy storage (BES, electrochemical storage of energy) systems, fuel cells (electrochemical storage of energy), flywheels (electromechanical storage of energy), and compressed air energy storage (CAES) systems.

There are already several utility-scaled battery plants in operation, and they are applied for load leveling and dynamic applications; one of the largest is a 10 MW, 40 MWh plant operated by Southern California Edison in Chino (California, United States).

8.6 Blackouts and Chaotic Phenomena

Expansion of the power system is expensive and takes a long time; the erection of a new overhead transmission line requires years of negotiation with the landowners or discussions with environmentalists, and – last but not least – it takes time to actually build it. This time-consuming process, in combination with the annual growth of the electricity consumption, means that the power system is operated closer to its limits. Network congestion becomes an important issue, and FACTS (see Section 5.6) devices are installed to keep the operation of the system within safe margins. FACTS devices, in turn, increase the complexity of the power system. The same applies to the growing penetration level of (power-) electronic interfaces in the system; many domestic loads are connected to the supply by converters (take a PC, for instance), and distributed generators, especially the ones powered by primary energy sources with a rather unpredictable behavior, are generally connected to the grid by a power-electronic interface. The converters introduce harmonics in the system and resonance at certain harmonic frequencies can happen. Power electronics also introduce nonlinear behavior in the system, and chaotic phenomena might occur in the near future. Accordingly, large system blackouts will probably happen more often.

8.6.1 Nonlinear Phenomena and Chaos

The alteration of the number of solutions of a dynamic system, as a result of a parameter change or a change of an initial condition, is called a bifurcation. There are two types of bifurcations. It is called a static bifurcation when there is a change in the number of equilibrium points. The other category is called the dynamic bifurcations, such as the Hopf bifurcation, the cyclic fold bifurcation, and the period doubling bifurcations. The static bifurcation and the Hopf bifurcation are local bifurcations; they always take place in the vicinity of an equilibrium. Cyclic fold and period doubling bifurcations are global bifurcations that occur outside the boundaries of the equilibria. Chaos emerges from a global bifurcation. Chaos is aperiodic time-asymptotic behavior in a deterministic system with a sensitive dependency on the initial system conditions. Or in other words, a small deviation of a certain state can cause the system to show a completely different behavior. Chaos can (in theory) occur in a dynamic system that is described by a system of first-order differential equations, when the following necessary conditions are met:

- The system is described by at least three independent variables.
- The equations have nonlinear terms that relate at least two of the independent variables.
- The trajectories described by the system are unambiguous.

For three or more independent variables ($n \geq 3$) and at least one nonlinear function f_i, a system like this can be written as

$$
\begin{aligned}
\frac{dx_1}{dt} &= f_1(x_1, x_2, \dots, x_n) \\
\frac{dx_2}{dt} &= f_2(x_1, x_2, \dots, x_n) \\
&\vdots \\
\frac{dx_n}{dt} &= f_n(x_1, x_2, \dots, x_n)
\end{aligned}
\tag{8.1}
$$

In this book, we focus on steady-state behavior of the power system only; a dynamic analysis of the system and the possible occurrence of chaos is beyond our scope (interested readers are referred to the literature in this field; e.g., Ref. [5]). However, we encountered an example of a static bifurcation earlier in this book when we treated the power flow equations. In Example 6.4 (p. 226), we learned that the power flow equations have multiple solutions, and this is a typical example of a static bifurcation. The existence of this static bifurcation makes that a small deviation of the working point of the system can lead to a voltage collapse, as explained in the following paragraphs.

In Figure 8.4, the two-node network used to demonstrate the load flow computation in Section 6.2 is shown. In Example 6.4 (p. 226), we calculated that the voltage at node two can have two values: $|V_2| = 0.1904$ pu and $|V_2| = 0.8853$ pu. In Figure 8.5, the voltage at node two is drawn as a function of the active power

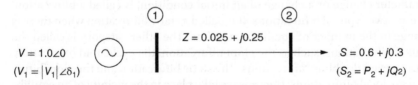

$V = 1.0\angle0$
$(V_1 = |V_1|\angle\delta_1)$

$Z = 0.025 + j0.25$

$S = 0.6 + j0.3$
$(S_2 = P_2 + jQ_2)$

Figure 8.4 Example network.

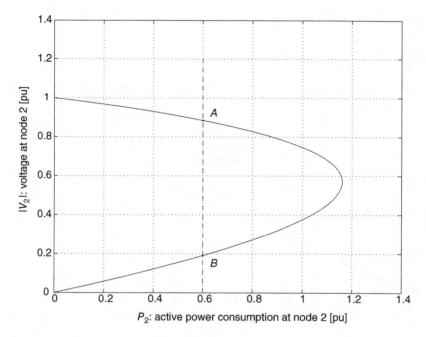

Figure 8.5 *PV* curve and load characteristic.

consumption, and the two possible working points of the system for a power consumption of $S_2 = 0.6 + j0.3$, being the situation as shown in Figure 8.4, are A and B. To draw this so-called *PV* curve, we assumed that the voltage at node one remains constant and that the load has a constant power factor $\cos(\varphi) = 0.6/\sqrt{0.36 + 0.09} \approx 0.9$. The *PV* curve has two traces: an upper trace starting at a voltage $|V_2| = 1.0 \text{ pu}$ (the no-load situation) and a lower trace that has its origin at a voltage $|V_2| = 0.0 \text{ pu}$ (the short-circuit state). When the power consumption increases, the upper and lower parts of the characteristic meet each other at a point where only one solution exists; this is the static bifurcation point.

In the case of a static load, both the working points A ("high" voltage and "small" current) and B ("low" voltage and "high" current) are stable working points, but point B is most probably nonviable because of the low voltage and the high current. Point B can therefore become an unstable working point; this is, for instance, the case when the load is supplied by a tap-changing transformer. Assume that the system operates in working point B. The tap changer will try to raise the voltage at the load side, which leads to an increased line current and an even higher voltage drop across the line. This leads to a progressive reduction of the voltage, which is called voltage collapse. In such a case, the static bifurcation separates the stable solutions from the unstable ones.

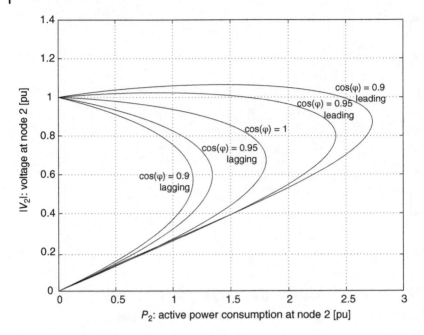

Figure 8.6 *PV* curves at various power factors.

A necessary condition for a stable system operation is the existence of an equilibrium point, a point where the *PV* curve and the load characteristic intersect. When this equilibrium point disappears, voltage collapse becomes a fact. An equilibrium point can disappear when the load demand increases such that the dashed line in Figure 8.5 shifts to the right, passing the "nose" of the *PV* curve. Also a change in the power factor can be a cause for an equilibrium point to disappear. In Figure 8.6, the *PV* curves are drawn for a number of different power factors. When we operate the system at full load and at a power factor $\cos(\varphi) = 1$, a sudden change to a power factor of $\cos(\varphi) = 0.95$ lagging will cause the voltage to collapse.

The voltage stability problem, as described earlier, is not an example of what we call chaos, but it illustrates one of the mechanisms that we encounter with chaos: a small deviation of a certain state can cause a system to show totally different behavior.

8.6.2 Blackouts

Major blackouts are usually caused by cascading contingencies, such as a short circuit, an overloaded component, and a generator outage, with complicated interactions. The vulnerability of the system to (in itself) low-probability incidents that expand to a cascading outage (which is also called the domino effect) increases when the system is already stressed by other causes, such as congested transmission corridors when there is a bulk exchange of power between parts of the system. Quite often, a cascading outage is initiated by forces of nature or by weather conditions: (thunder) storms, extreme temperatures, geomagnetic storms, forest fires, and so forth. The sequence of events leading to a blackout are usually diverse, but the result is always the same: an interruption of the power supply for a certain period of time.

In the next examples, two major blackouts are analyzed to show what can cause such a disastrous event and on what time scale it takes place. The first example describes the major blackout in the northeastern part of the United States and Canada on August 14, 2003; the blackout affected approximately 50 million people, and it took more than 24 hours to restore the power supply in New York City and other areas. The second example describes the sequence of events leading to the blackout of Italy on September 28, 2003. The blackout affected approximately 57 million people, and it took 5–9 hours to restore the power to Rome and the major cities of the country.

Example 8.1 *Blackout in Northeastern United States and Canada (August 14, 2003)*

The blackout in Northeastern United States and Canada has been investigated and documented by the North American Electric Reliability Council [6] and the US–Canada Power System Outage Task Force [7]. Here follows a short description of the sequence of events that led to the blackout in Northeastern United States and Canada (see Figure 8.7).

On August 14, 2003, there were voltage problems in the border area between the United States and Canada. The scenario that led to the actual blackout in New York and wide surroundings at 4.13 p.m. started around noon. Shortly after 12 o'clock, a 375 MW unit in the Conesville power station in mid-Ohio was disconnected from the grid, followed by 785 MW of the Greenwood power station in northern Detroit (Michigan) 1 hour 10 minutes later and 597 MW of the Eastlake power station in northern Ohio 18 minutes later. The loss of these three generating units caused a change in the power flow in the power pool around the great lakes, the so-called Eastern Interconnection.

At 2.02 p.m., an important 345 kV transmission link from the southwest to the north of Ohio was taken out of service. There was a forest fire close to the transmission lines, and there was fear that the heat would ionize the air surrounding the lines and would cause a short circuit. Between 3.00 p.m. and a 3.45 p.m., three other 345 kV links were taken out of service, and an important connection in the transport capacity between east and north Ohio had at that moment disappeared. The power flow divided up over the remaining connections, such as connections at the 138 kV level. These connections got overloaded. Because of the large voltage drop, 600 MW industrial consumption was disconnected in Ohio, as well as consumers at the 138 and 69 kV network.

Figure 8.7 The area around the great lakes.

Because of the loss of the 345 kV connections Canton Central–Tidd and Sammis–Star, the power could flow to the north of Ohio by means of three routes only (see Figure 8.8, I). The north of Ohio, which normally provided the east of Michigan with energy, became a weak point; the large industrial town Detroit depended on southeast Michigan for its power supply. Subsequently, the connections Galion–Ohio, Central–Muskingum (Figure 8.8, II), and East Lima–Fostoria Central (Figure 8.8, III), all 345 kV lines, were taken out of service, and the link between north Ohio and the east of Michigan was formed mainly by two 345 kV connections at the south side of Lake Erie. The power flows from Indiana and the lines in southwest Michigan in the direction of east Michigan and north Ohio increased but were insufficient to meet the demand, so that the voltage in north Ohio dropped.

Shortly after that, a number of large power plants were disconnected from the grid in north Ohio and west Michigan. It was 4.10 p.m. The increased power flows overloaded the remaining connections in service, and they were taken out of service (Figure 8.8, IV). The problem got worse, and when, at a certain moment in time, the 345 kV line Perry–Ashtabula–Erie West was taken

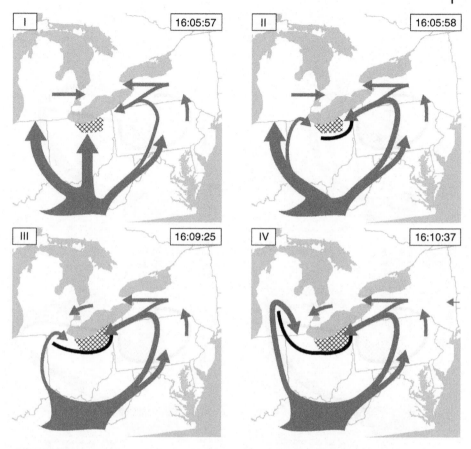

Figure 8.8 Cascade sequence [7]. The arrows indicate the power flows. Black lines represent separations between areas within the Eastern Interconnection. Regions affected by the blackout are highlighted by the dashed areas.

out of service, the area covering east Michigan and north Ohio had hardly any production left, and the voltage collapsed.

Subsequently, the line at the south side of Lake Erie was taken out of service, and the power flow changed its direction and flows, with a wide curve, from Pennsylvania to New York to Ontario and Michigan, around the north side of Lake Erie (Figure 8.8, V). This also drew Pennsylvania, New York, Ontario, and Quebec into the sequence of events. Two 345 kV and two 220 kV lines between Pennsylvania and New York were disconnected within 4 seconds of each other (Figure 8.8; VI and VII) due to the suddenly increased power flows. As a result, Pennsylvania and New York were no longer interconnected in a direct way, and large power flows appeared on the New York–New Jersey

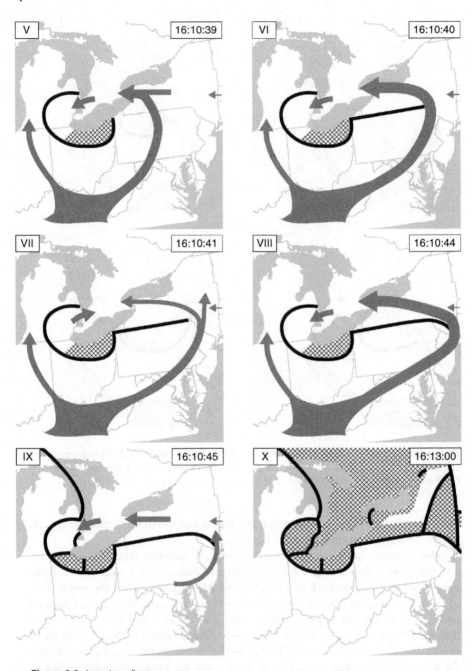

Figure 8.8 *(continued)*

connection (Figure 8.8; VIII). At the same time, more generation was lost in north Ohio, and also the Fostoria Central–Galion 345 kV connection was idle. The disconnection of the Beaver–Davis–Besse 345 kV line cut Cleveland (Ohio) off from the Eastern Interconnection. Michigan was still connected to Ontario in Canada, but two 230 kV connections became inactive, and Ontario was connected to Manitoba and Minnesota only (Figure 8.8; IX).

Around that time, the last link between the Eastern Interconnection and the area of New Jersey, the 500 kV connection Branchburg–Ramapo, went out of operation. As a consequence, the supply area around Greater New York was split in two parts: New England (except for southwest Connecticut) and the Maritimes. Large areas were islanding now and tried to reestablish the power balance. Also Ontario tried to balance the power and shed 2500 MW of load. The direct interconnection via the Niagara Falls between Ontario and New York was taken out of service, and the power flows were now heading for New York from Ontario via Quebec through the 765 kV lines. The recovery of the connection between Ontario and New York failed, and 4500 MW of generation was disconnected in Ontario.

Now, the power supply in the largest part of Ontario was interrupted. Problems in maintaining the power balance in the islanded parts of New York had as a result that the power supply also failed there. A major part of the Eastern Interconnection, the area in the United States around the great lakes, was now without electricity by which time it was 4.13 p.m. (Figure 8.8, X).

Example 8.2 *Blackout in Italy (September 28, 2003)*

The blackout in Italy has been investigated and documented by an investigation committee of the UCTE [8]. There follows a short description of the sequence of events that led to the blackout of the entire Italian peninsula. An overview of the Italian cross-border connections is shown in Figure 8.9.

The blackout in Italy was initiated by the loss of the Swiss 380 kV transmission line connection between Mettlen and Lavorgo. This high-voltage overhead line was 85% loaded at that time. The line was taken out of service after a short circuit between the line conductors and the branches of a tree that came too close to the line.

Such a short circuit could disappear spontaneously when the arc (a lightning-like conducting path, in this case between the conductor and a branch of the tree) extinguished by itself because of the cooling by the surrounding air. Probably this was not the case here, as the automatic reclosure of the high-voltage transmission line failed a number of times. Also by means of a manual reclosure command, seven minutes after the disturbance, the line could not be brought back into service. The phase angle between the voltages at the ends of the line was too large: 42°, whereas 30° is the maximum allowed phase difference between two points that are to be connected.

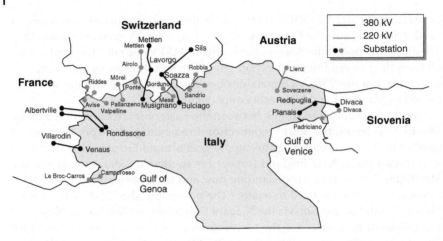

Figure 8.9 The Italian interconnectors [8].

When this happened, the Italian electricity import from Switzerland was 540 MW more than scheduled and the import from France 440 MW less than scheduled. At the same time, the remaining transmission links took care of the power transport, as dictated by the laws of physics; the major part of the transport capacity was taken over by the lines nearby. This resulted in a 110% loading of the Swiss 380 kV link Sils–Soazza, the so-called San Bernardino link. This overload is permitted, but may not last longer than 15 minutes.

Ten minutes after the start of the problems with the line Mettlen–Lavorgo, a discussion by phone was held between the ETRANS control center in Laufenburg, which is located at the border between Germany and Switzerland, and the GRTN control center in Rome, Italy. GRTN was asked to take measures in the Italian area in order to reduce the overload on the Swiss transmission lines and bring back a safe system operation. The measure that had to be taken was in fact the reduction of the Italian import by 300 MW, because that was the surplus amount of imported power at the time of the disturbance. The 300 MW import reduction was a fact 10 minutes after the telephone conversation, and the power balance in Italy was restored again. Also the Swiss took their measures, and, together with the Italian import reduction of 300 MW, the transmission lines in Switzerland were normally loaded again.

But four minutes later, another short circuit occurred between a treetop and a transmission line conductor – this time on the circuit Sils–Soazza, the connection that had been overloaded for 10 minutes. Because of the overload, the aluminum conductors were heated and the line sag became larger. As a result, the line conductors came closer to the tops of the trees, as illustrated in Figure 8.10. The connection Sils–Soazza also had to be taken out of service. The loss of these two interconnections resulted in dangerous overloads on the other system components and connections in the system (4 seconds after the disappearance of the line Sils–Soazza, the 220 kV line from

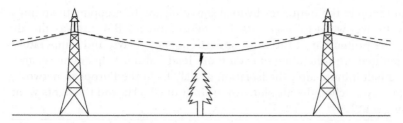

Figure 8.10 Increased sag of overloaded lines can cause a short circuit; dashed, the "normally" loaded line; solid, the overloaded line.

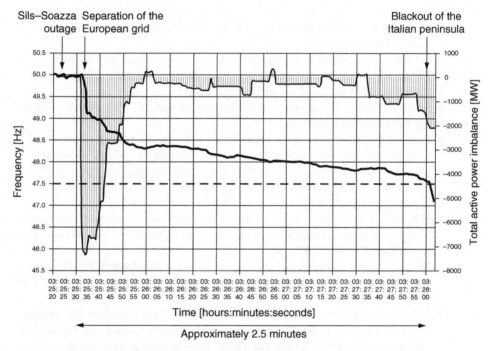

Figure 8.11 The Italian frequency (line only) and total active power imbalance (line + area) [8]; the dotted line at a frequency of 47.5 Hz is the critical threshold.

Airolo to Mettlen in Switzerland was overloaded and taken out of service), and therefore, 12 seconds after putting the line Sils–Soazza idle, the Italian network was isolated from the rest of Europe in order to reduce the affected area. During these 12 seconds, large power fluctuations in combination with severe transient voltage instabilities occurred.

The voltage drop in the northern part of Italy caused a number of power plants to disconnect automatically from the grid. The power balance was disturbed again, and, because Italy was now isolated from the rest of Europe, the frequency dropped to 49 Hz (see Figure 8.11). The primary frequency control halted the drop in the frequency by load shedding and by stopping the pumps that elevate river water to reservoirs (for energy storage). But this was already too late: the protection of turbines, underfrequency relays, and temperature relays operated. The shedding of even more load did not help anymore, and, two and a half minute after the isolation of Italy from the European network, the frequency reached the absolute lower limit of 47.5 Hz and the lights went out throughout Italy.

References

1 'Powerformer, record performer', *Modern Power Systems*, Vol. **19**, Issue 6, June 1999, pp. 57–8.
2 ENTSO-E: *Electricity in Europe 2015*, https://www.entsoe.eu/Documents/ Publications/Statistics/electricity_in_europe/entsoe_electricity_in_europe_ 2015_web.pdf (Accessed on 30 December 2016).
3 Edris, A.: 'FACTS technology development: an update', *IEEE Power Engineering Review*, March 2000, pp. 4–9.
4 Reza, M., Schavemaker, P.H., Kling, W.L., and van der Sluis, L.: 'A research program on intelligent power systems: self-controlling and self-adapting power systems equipped to deal with the structural changes in the generation and the way of consumption', CIRED 2003 Conference, Barcelona, Spain, May 12–15, 2003.
5 Hill, D.J., et al.: 'Special issue on nonlinear phenomena in power systems: theory and practical implications', *Proceedings of the IEEE*, Vol. **83**, No. 11, November 1995, pp. 1439–587.
6 NERC: *Technical Analysis of the August 14, 2003, Blackout: What Happened, Why, and What Did We Learn?*, July 13, 2004, North American Electric Reliability Council.
7 U.S.–Canada Power System Outage Task Force: *Final Report on the August 14, 2003 Blackout in the United States and Canada: Causes and Recommendations*, April 2004, U.S.–Canada Power System Outage Task Force.
8 UCTE: *Final Report of the Investigation Committee on the 28 September 2003 Blackout in Italy*, April 2004, Union for the Coordination of Transmission of Electricity.

A

Maxwell's Laws

A.1 Introduction

All macroscopic electromagnetic phenomena are described by Maxwell's equations. The equations express the distributed nature of the electromagnetic fields; the field quantities are functions of space as well as time.

Faraday's law relates the electromotive force, generated around a closed contour C, to the time rate of change of the total magnetic flux through the open surface S bounded by that contour. Or in other words, Faraday's law shows that a time-changing magnetic flux can induce an electric field:

$$\oint_C E \cdot dl = -\frac{d}{dt}\int_S B \cdot ds \qquad (A.1)$$

E the electric field intensity vector [V/m]
B the magnetic flux density vector [Wb/m^2]

Ampère's law states the opposite: a time-changing electric flux can induce a magnetic field:

$$\oint_C H \cdot dl = \int_S J \cdot ds + \frac{d}{dt}\int_S D \cdot ds \qquad (A.2)$$

H the magnetic field intensity vector [A/m]
J the current density vector [A/m^2]
D the electric flux density vector [C/m^2]

Gauss' law for the electric field states that the net flux of the electric flux density vector out of the closed surface S is equivalent to the net positive charge enclosed by the surface:

$$\oint_S D \cdot ds = \int_V \rho\, dv \qquad (A.3)$$

ρ the volume-free charge density [C/m^3]

Sections A.2–A.5 are based on Chapter 9 of Ref. [1].

Electrical Power System Essentials, Second Edition. Pieter Schavemaker and Lou van der Sluis.
© 2017 John Wiley & Sons Ltd. Published 2017 by John Wiley & Sons Ltd.

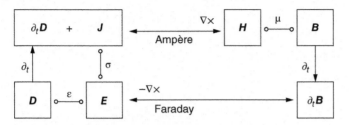

Figure A.1 Schematic outline of the Maxwell relations.

Gauss' law for the magnetic field outlines that all magnetic field lines form closed paths and that there are no isolated sources for the magnetic field – or in other words, that there are no magnetic monopoles:

$$\oint_S \boldsymbol{B} \cdot d\boldsymbol{s} = 0 \tag{A.4}$$

In differential form, Maxwell's equations are described in the following paragraphs and visualized in Figure A.1:

$$\nabla \times \boldsymbol{E} = -\frac{\partial \boldsymbol{B}}{\partial t} \tag{A.5}$$

$$\nabla \times \boldsymbol{H} = \boldsymbol{J} + \frac{\partial \boldsymbol{D}}{\partial t} \tag{A.6}$$

$$\nabla \cdot \boldsymbol{D} = \rho \tag{A.7}$$

$$\nabla \cdot \boldsymbol{B} = 0 \tag{A.8}$$

Circuit theory can be regarded as describing a restricted class of solutions of Maxwell's equations. In the following sections, power series approximations will be applied to describe the electromagnetic field. It is shown that the zero- and first-order terms in these approximations (i.e., the quasi-static fields) form the basis for the lumped-circuit theory. By means of the second-order terms, the validity of the lumped-circuit theory at various frequencies can be estimated.

A.2 Power Series Approach to Time-Varying Fields

All electromagnetic fields will, in general, alter their shape and their behavior as a function of the frequency ω. The following series expansion can be made for all the field quantities (such as $\boldsymbol{E}, \boldsymbol{H}, \boldsymbol{J}$, and ρ):

$$\boldsymbol{E}(x, y, z, \tau, \omega) = \boldsymbol{e}_0(x, y, z, \tau) + \omega \boldsymbol{e}_1(x, y, z, \tau) + \omega^2 \boldsymbol{e}_2(x, y, z, \tau) + \cdots \tag{A.9}$$

where τ is defined as $\tau = \omega t$. The coefficients of the expansion terms are independent of ω, and the coefficient of the kth term is given by

$$e_k(x, y, z, \tau) = \frac{1}{k!} \left(\frac{\partial^k E(x, y, z, \tau, \omega)}{\partial \omega^k} \right)_{\omega=0} \tag{A.10}$$

Substitution of the power series into the field laws gives

$$\nabla \times E = -\frac{\partial B}{\partial t} = -\frac{\partial B}{\partial \tau} \cdot \frac{\partial \tau}{\partial t} = -\omega \frac{\partial B}{\partial \tau} \tag{A.11}$$

$$\nabla \times E = ((\nabla \times e_0) + \omega(\nabla \times e_1) + \omega^2(\nabla \times e_2) + \cdots)$$

$$= -\omega \frac{\partial B}{\partial \tau} = -\left(\omega \left(\frac{\partial b_0}{\partial \tau} \right) + \omega^2 \left(\frac{\partial b_1}{\partial \tau} \right) + \omega^3 \left(\frac{\partial b_2}{\partial \tau} \right) + \cdots \right) \tag{A.12}$$

Combining terms leads to

$$\left((\nabla \times e_0) + \omega \left(\nabla \times e_1 + \frac{\partial b_0}{\partial \tau} \right) + \omega^2 \left(\nabla \times e_2 + \frac{\partial b_1}{\partial \tau} \right) + \cdots \right) = 0 \tag{A.13}$$

This equation must be valid for every possible frequency ω. Therefore, the coefficients of all the powers of ω are each equal to zero, and this gives us the following general equation:

$$\nabla \times e_k = -\frac{\partial b_{k-1}}{\partial \tau} \tag{A.14}$$

Similar results follow from the corresponding equations obtained from each of the other basic field laws.

The power series notation can be simplified by absorbing the various powers of ω directly into each of the terms of the power series expansion:

$$E(x, y, z, \tau, \omega) = e_0(x, y, z, \tau) + \omega e_1(x, y, z, \tau) + \omega^2 e_2(x, y, z, \tau) + \cdots$$

$$= E_0(x, y, z, \tau, \omega) + E_1(x, y, z, \tau, \omega) + E_2(x, y, z, \tau, \omega) + \cdots \tag{A.15}$$

The kth term of the series is taken as follows:

$$E_k(x, y, z, \tau, \omega) = \omega^k e_k(x, y, z, \tau) \tag{A.16}$$

The field laws can be written in terms of only E_k and B_k, without explicit appearance of ω^k:

$$\nabla \times E_k = \nabla \times (\omega^k e_k) = -\omega^k \frac{\partial b_{k-1}}{\partial \tau} = -\omega \frac{\partial B_{k-1}}{\partial \tau} = -\frac{\partial B_{k-1}}{\partial t} \tag{A.17}$$

The resulting set of equations is shown in Table A.1.

The series expansions are solved by determining the zero-order terms first. When the zero-order terms are known, the first-order terms can be calculated. This step-by-step calculation continues for the higher-order terms until the required degree of accuracy in the solution is reached.

The static field is described by the zero-order terms only, meaning that the time variation of the electromagnetic field is neglected. The quasi-static field

Table A.1 Electromagnetic field relations expressed in zero-, first- and k^{th}-order terms.

Order	Field Laws
Zero order	$\nabla \times E_0 = 0$
	$\nabla \times H_0 = J_0$
	$\nabla \cdot D_0 = \rho_0$
	$\nabla \cdot B_0 = 0$
	$\nabla \cdot J_0 = 0$
First order	$\nabla \times E_1 = -\dfrac{\partial B_0}{\partial t}$
	$\nabla \times H_1 = J_1 + \dfrac{\partial D_0}{\partial t}$
	$\nabla \cdot D_1 = \rho_1$
	$\nabla \cdot B_1 = 0$
	$\nabla \cdot J_1 = \dfrac{\partial \rho_0}{\partial t}$
k^{th} order	$\nabla \times E_k = -\dfrac{\partial B_{k-1}}{\partial t}$
	$\nabla \times H_k = J_k + \dfrac{\partial D_{k-1}}{\partial t}$
	$\nabla \cdot D_k = \rho_k$
	$\nabla \cdot B_k = 0$
	$\nabla \cdot J_k = \dfrac{\partial \rho_{k-1}}{\partial t}$

is described by the zero-order and first-order terms. In the next section we see that the fundamentals of circuit theory follow directly from the quasi-static approximations of Maxwell's equations.

A.3 Quasi-static Field of a Parallel-plate Capacitor

Consider the parallel-plate capacitor in air as shown in Figure A.2. The plates consist of perfectly conducting material ($\sigma = \infty$), and the length and width of the plates are much larger than their separation distance, so that all fringing in the resulting fields can be neglected. Therefore, all variation in E and H with both x and y can be neglected within this system:

$$\frac{\partial}{\partial x} = \frac{\partial}{\partial y} = 0 \tag{A.18}$$

The system is excited with a low-frequency sinusoidal excitation, which is uniformly distributed between the plates (at $z = -l$), such that a fixed reference

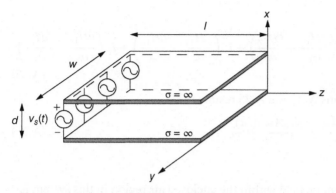

Figure A.2 Parallel-plate capacitor in air. All fringing in the resulting fields can be neglected as $l \gg d$ and $w \gg d$.

voltage between the plates in the $z = 0$ plane is maintained:

$$v_r = A\cos(\omega t) \tag{A.19}$$

Since the reference voltage has an amplitude that is independent of the frequency ω, it follows that the power series expansion consists solely of a zero-order term equal in value to the reference voltage:

$$v_0(z = 0, t) = v_r(t) = A\cos(\omega t)$$
$$v_k(z = 0, t) = 0 \quad \text{for } k \geq 1 \tag{A.20}$$

A.3.1 Quasi-static Solution

The zero-order time-varying electromagnetic fields in this system are identical in form to their static (DC) counterparts:

$$E_0 = -i_x \frac{A\cos(\omega t)}{d} \tag{A.21}$$

i unit vector

$$H_0 = 0 \tag{A.22}$$

The first-order contributions to the magnetic field can be derived from the first-order field laws in Table A.1:

$$\nabla \times H_1 = \varepsilon_0 \frac{\partial E_0}{\partial t} = i_x \frac{\omega \varepsilon_0 A \sin(\omega t)}{d} \tag{A.23}$$

$$\nabla \cdot \mu_0 H_1 = 0 \tag{A.24}$$

with

$$\nabla \times \boldsymbol{H}_1 = \boldsymbol{i}_x \left(\frac{\partial H_{1z}}{\partial y} - \frac{\partial H_{1y}}{\partial z} \right) + \boldsymbol{i}_y \left(\frac{\partial H_{1x}}{\partial z} - \frac{\partial H_{1z}}{\partial x} \right) + \boldsymbol{i}_z \left(\frac{\partial H_{1y}}{\partial x} - \frac{\partial H_{1x}}{\partial y} \right)$$

(A.25)

Combining Equations A.23 and A.25 results in

$$\underbrace{\frac{\partial H_{1z}}{\partial y} - \frac{\partial H_{1y}}{\partial z}}_{0} = \frac{\omega \varepsilon_0 A \sin(\omega t)}{d}$$

(A.26)

The first-order magnetic field within the enclosed air region in this system is

$$\boldsymbol{H}_1 = \boldsymbol{i}_y H_{1y} = -\boldsymbol{i}_y \frac{\omega \varepsilon_0 A \sin(\omega t)}{d} z$$

(A.27)

Note that Equation A.24 is satisfied, as H_{1y} is independent of y (and $H_{1x} = H_{1z} = 0$):

$$\nabla \cdot \mu_0 \boldsymbol{H}_1 = \mu_0 \left(\frac{\partial H_{1x}}{\partial x} + \frac{\partial H_{1y}}{\partial y} + \frac{\partial H_{1z}}{\partial z} \right) = 0$$

(A.28)

The first-order contributions to the electric field can be derived from the first-order field laws in Table A.1:

$$\nabla \times \boldsymbol{E}_1 = -\mu_0 \frac{\partial H_0}{\partial t} = 0$$

(A.29)

$$\nabla \cdot \varepsilon_0 \boldsymbol{E}_1 = \rho_1 = 0$$

(A.30)

Equation A.29 equals zero as H_0 equals zero for all time (Equation A.22) and so does its time derivative. Equation A.30 equals zero as the enclosed air region is free of charge (for all frequencies ω and time t) due to its insulating nature. The first-order electric field within the enclosed air region in this system is

$$\boldsymbol{E}_1 = 0$$

(A.31)

As a result, the expressions for the non-fringing fields between the plates of the capacitor correct up to and including the first-order terms are

$$\boldsymbol{E}_{0,1} = \boldsymbol{E}_0 + \underbrace{\boldsymbol{E}_1}_{0} = -\boldsymbol{i}_x \frac{A \cos(\omega t)}{d}$$

(A.32)

$$\boldsymbol{H}_{0,1} = \underbrace{\boldsymbol{H}_0}_{0} + \boldsymbol{H}_1 = -\boldsymbol{i}_y \frac{\omega \varepsilon_0 A \sin(\omega t)}{d} z$$

(A.33)

For the voltage applied to the parallel-plate capacitor, correct up to and including the first-order contributions, it must hold that

$$[v_s(t)]_{0,1} = -\int_{x=0}^{d} (E_{0,1})_{z=-l}dx = A\cos(\omega t) \tag{A.34}$$

For the source current, correct up to and including the first-order contributions, it yields

$$[i_s(t)]_{0,1} = -\int_{y=0}^{w} (H_{0,1})_{z=-l}dy = -\frac{\varepsilon_0 lw}{d}\omega A\sin(\omega t) \tag{A.35}$$

The input impedance of the parallel-plate capacitor, correct up to and including first-order contributions, can be computed from the ratio between the phasor representations of v_s and i_s:

$$[V_s]_{0,1} = A \tag{A.36}$$

$$[I_s]_{0,1} = j\omega A \frac{\varepsilon_0 lw}{d} \tag{A.37}$$

$$Z_{0,1} = \frac{[V_s]_{0,1}}{[I_s]_{0,1}} = \frac{1}{j\omega \frac{\varepsilon_0 lw}{d}} = \frac{1}{j\omega C} \tag{A.38}$$

The lumped-element representation of the parallel-plate capacitor, correct up to and including first-order contributions, is a lumped capacitor C, where the value of the capacitance equals the DC or static capacitance of the parallel-plate system:

$$C = C_{DC} = \frac{\varepsilon_0 lw}{d} \tag{A.39}$$

A.3.2 Validity of the Quasi-static Approach

In order to check the validity of the quasi-static approximation, we compute the second-order contributions to the power series of the electric and magnetic field. The second-order magnetic field H_2 is generated by the time rate of change of the first-order electric field E_1 (Table A.1). But with E_1 equal to zero at all points in the system, we can expect H_2 to be equal to zero at all points in the system as well:

$$H_2 = 0 \tag{A.40}$$

The second-order contributions to the electric field can be derived from the kth-order ($k=2$) field laws in Table A.1 and the first-order magnetic field in Equation A.27:

$$\nabla \times E_2 = -\mu_0 \frac{\partial H_1}{\partial t} = i_y \frac{\omega^2 \mu_0 \varepsilon_0 A\cos(\omega t)}{d} z \tag{A.41}$$

$$\nabla \cdot \varepsilon_0 E_2 = \rho_2 = 0 \tag{A.42}$$

with

$$\nabla \times \boldsymbol{E}_2 = \boldsymbol{i}_x \left(\frac{\partial E_{2z}}{\partial y} - \frac{\partial E_{2y}}{\partial z} \right) + \boldsymbol{i}_y \left(\frac{\partial E_{2x}}{\partial z} - \frac{\partial E_{2z}}{\partial x} \right) + \boldsymbol{i}_z \left(\frac{\partial E_{2y}}{\partial x} - \frac{\partial E_{2x}}{\partial y} \right)$$

(A.43)

Combining Equations A.41 and A.43 results in

$$\underbrace{\frac{\partial E_{2x}}{\partial z} - \frac{\partial E_{2z}}{\partial x}}_{0} = \frac{\omega^2 \mu_0 \varepsilon_0 A \cos(\omega t)}{d} z$$

(A.44)

The second-order electric field within the enclosed air region in this system is

$$\boldsymbol{E}_2 = \boldsymbol{i}_x E_{2x} = \boldsymbol{i}_x \frac{\omega^2 \mu_0 \varepsilon_0 A \cos(\omega t)}{d} \frac{z^2}{2}$$

(A.45)

Note that Equation A.42 is satisfied, as E_{2x} is independent of x (and $E_{2y} = E_{2z} = 0$):

$$\nabla \cdot \varepsilon_0 \boldsymbol{E}_2 = \varepsilon_0 \left(\frac{\partial E_{2x}}{\partial x} + \frac{\partial E_{2y}}{\partial y} + \frac{\partial E_{2z}}{\partial z} \right) = 0$$

(A.46)

The expressions for the non-fringing fields between the plates of the capacitor correct up to and including the second-order terms are

$$\boldsymbol{E}_{0,1,2} = \boldsymbol{E}_0 + \underbrace{\boldsymbol{E}_1}_{0} + \boldsymbol{E}_2 = -\boldsymbol{i}_x \frac{A \cos(\omega t)}{d} \left(1 - \frac{\omega^2 \mu_0 \varepsilon_0 z^2}{2} \right)$$

(A.47)

$$\boldsymbol{H}_{0,1,2} = \underbrace{\boldsymbol{H}_0}_{0} + \boldsymbol{H}_1 + \underbrace{\boldsymbol{H}_2}_{0} = -\boldsymbol{i}_y \frac{\omega \varepsilon_0 A \sin(\omega t)}{d} z$$

(A.48)

For the voltage applied to the parallel-plate capacitor, correct up to and including the second-order contributions, it must hold that

$$[v_s(t)]_{0,1,2} = -\int_{x=0}^{d} (E_{0,1,2})_{z=-l} dx = A \cos(\omega t) \left(1 - \frac{\omega^2 \mu_0 \varepsilon_0 l^2}{2} \right)$$

(A.49)

For the source current, correct up to and including second-order contributions, it yields

$$[i_s(t)]_{0,1,2} = -\int_{y=0}^{w} (H_{0,1,2})_{z=-l} dy = -\frac{\varepsilon_0 l w}{d} \omega A \sin(\omega t)$$

(A.50)

The input impedance of the parallel-plate capacitor, correct up to and including second-order contributions, can be computed from the ratio between the phasor representations of v_s and i_s:

$$[V_s]_{0,1,2} = A \left(1 - \frac{\omega^2 \mu_0 \varepsilon_0 l^2}{2} \right)$$

(A.51)

$$[I_s]_{0,1,2} = j\omega A \frac{\varepsilon_0 lw}{d} \tag{A.52}$$

$$Z_{0,1,2} = \frac{[V_s]_{0,1,2}}{[I_s]_{0,1,2}} = \frac{1}{j\omega \frac{\varepsilon_0 lw}{d}} \left(1 - \frac{\omega^2 \mu_0 \varepsilon_0 l^2}{2}\right) = \frac{1}{j\omega C}(1 - \omega^2 LC) \tag{A.53}$$

The lumped-element representation of the parallel-plate capacitor, correct up to and including second-order contributions, is a series LC network, where the values of the capacitance and the inductance are specified as follows:

$$C = C_{DC} = \frac{\varepsilon_0 lw}{d} \tag{A.54}$$

$$L = \frac{1}{2}\left(\frac{\mu_0 ld}{w}\right) \tag{A.55}$$

The quasi-static approach is valid only when the second-order contribution is negligible:

$$\frac{\omega^2 \mu_0 \varepsilon_0 l^2}{2} \ll 1 \tag{A.56}$$

The wavelength of an electromagnetic wave is defined as

$$\lambda = \frac{2\pi}{\omega \sqrt{\mu_0 \varepsilon_0}} \tag{A.57}$$

The use of the circuit theory model of a lumped capacitor for the parallel-plate system is justified as long as the values of the wavelength (λ) are much higher than the length of the plates (l):

$$\frac{\omega^2 \mu_0 \varepsilon_0 l^2}{2} = 2\pi^2 \left(\frac{l}{\lambda}\right)^2 \ll 1 \tag{A.58}$$

A.4 Quasi-static Field of a Single-turn Inductor

Consider the single-turn inductor in air as shown in Figure A.3. The plates consist of perfectly conducting material ($\sigma = \infty$), and the length and width of the plates are much larger than their separation distance, so that all fringing in the resulting fields can be neglected. Therefore, all variation in E and H with both x and y can be neglected within this system:

$$\frac{\partial}{\partial x} = \frac{\partial}{\partial y} = 0 \tag{A.59}$$

The system is fed by a low-frequency sinusoidal current source (at $z = -l$), such that the reference surface current density:

$$K_r = A\cos(\omega t) \tag{A.60}$$

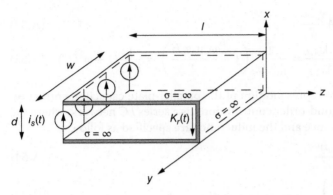

Figure A.3 Single-turn inductor in air. All fringing in the resulting fields can be neglected as $l \gg d$ and $w \gg d$.

A.4.1 Quasi-static Solution

The zero-order time-varying electromagnetic fields in this system are identical in form to their static (DC) counterparts:

$$H_0 = -i_y K_r = -i_y A \cos(\omega t) \tag{A.61}$$

$$E_0 = 0 \tag{A.62}$$

The first-order contributions to the electric field can be derived from the first-order field laws in Table A.1:

$$\nabla \times E_1 = -\mu_0 \frac{\partial H_0}{\partial t} = -i_y \mu_0 \omega A \sin(\omega t) \tag{A.63}$$

$$\nabla \cdot \varepsilon_0 E_1 = \rho_1 = 0 \tag{A.64}$$

with

$$\nabla \times E_1 = i_x \left(\frac{\partial E_{1z}}{\partial y} - \frac{\partial E_{1y}}{\partial z} \right) + i_y \left(\frac{\partial E_{1x}}{\partial z} - \frac{\partial E_{1z}}{\partial x} \right) + i_z \left(\frac{\partial E_{1y}}{\partial x} - \frac{\partial E_{1x}}{\partial y} \right) \tag{A.65}$$

Combining Equations A.63 and A.65 results in

$$\frac{\partial E_{1x}}{\partial z} - \underbrace{\frac{\partial E_{1z}}{\partial x}}_{0} = -\mu_0 \omega A \sin(\omega t) \tag{A.66}$$

The first-order electric field within the enclosed air region in this system is

$$E_1 = i_x E_{1x} = -i_x \mu_0 \omega z A \sin(\omega t) \tag{A.67}$$

Note that Equation A.64 is satisfied, as E_{1x} is independent of x (and $E_{1y} = E_{1z} = 0$):

$$\nabla \cdot \varepsilon_0 E_1 = \varepsilon_0 \left(\frac{\partial E_{1x}}{\partial x} + \frac{\partial E_{1y}}{\partial y} + \frac{\partial E_{1z}}{\partial z} \right) = 0 \tag{A.68}$$

The first-order contributions to the magnetic field can be derived from the first-order field laws in Table A.1:

$$\nabla \times H_1 = \varepsilon_0 \frac{\partial E_0}{\partial t} = 0 \tag{A.69}$$

$$\nabla \cdot \mu_0 H_1 = 0 \tag{A.70}$$

The first-order magnetic field within the enclosed air region in this system is

$$H_1 = 0 \tag{A.71}$$

As a result, the expressions for the non-fringing fields of the inductor correct up to and including first-order terms are

$$E_{0,1} = \underbrace{E_0}_{0} + E_1 = -i_x \mu_0 \omega z A \sin(\omega t) \tag{A.72}$$

$$H_{0,1} = H_0 + \underbrace{H_1}_{0} = -i_y A \cos(\omega t) \tag{A.73}$$

For the voltage across the current source, correct up to and including first-order contributions, it must hold that

$$[v_s(t)]_{0,1} = -\int_{x=0}^{d} (E_{0,1})_{z=-1} dx = -\mu_0 \omega l d A \sin(\omega t) \tag{A.74}$$

For the source current, correct up to and including first-order contributions, it yields

$$[i_s(t)]_{0,1} = -\int_{y=0}^{w} (H_{0,1})_{z=-l} dy = w A \cos(\omega t) \tag{A.75}$$

The input impedance of the single-turn inductor, correct up to and including first-order contributions, can be computed from the ratio between the phasor representations of v_s and i_s:

$$[V_s]_{0,1} = j\mu_0 \omega l d A \tag{A.76}$$

$$[I_s]_{0,1} = w A \tag{A.77}$$

$$Z_{0,1} = \frac{[V_s]_{0,1}}{[I_s]_{0,1}} = \frac{j\mu_0 \omega l d A}{w A} = j\omega \frac{\mu_0 l d}{w} = j\omega L \tag{A.78}$$

The lumped-element representation of the single-turn inductor, correct up to and including first-order contributions, is a lumped inductor L, where the value of the inductance equals the DC or static inductance of the single-turn inductor:

$$L = L_{DC} = \frac{\mu_0 l d}{w} \tag{A.79}$$

A.4.2 Validity of the Quasi-static Approach

In order to check the validity of the quasi-static approximation, we compute the second-order contributions to the power series of the electric and magnetic field. The second-order electric field E_2 is generated by the time rate of change of the first-order magnetic field H_1 (Table A.1). But with H_1 equal to zero at all points in the system, we can expect E_2 to be equal to zero at all points in the system as well:

$$E_2 = 0 \tag{A.80}$$

The second-order contributions to the magnetic field can be derived from the kth-order ($k = 2$) field laws in Table A.1 and the first-order electric field in Equation A.67:

$$\nabla \times H_2 = \varepsilon_0 \frac{\partial E_1}{\partial t} = -i_x \omega^2 \mu_0 \varepsilon_0 z A \cos(\omega t) \tag{A.81}$$

$$\nabla \cdot \mu_0 H_2 = 0 \tag{A.82}$$

with

$$\nabla \times H_2 = i_x \left(\frac{\partial H_{2z}}{\partial y} - \frac{\partial H_{2y}}{\partial z} \right) + i_y \left(\frac{\partial H_{2x}}{\partial z} - \frac{\partial H_{2z}}{\partial x} \right) + i_z \left(\frac{\partial H_{2y}}{\partial x} - \frac{\partial H_{2x}}{\partial y} \right) \tag{A.83}$$

Combining Equations A.81 and A.83 results in

$$\underbrace{\frac{\partial H_{2z}}{\partial y} - \frac{\partial H_{2y}}{\partial z}}_{0} = -\omega^2 \mu_0 \varepsilon_0 z A \cos(\omega t) \tag{A.84}$$

The second-order magnetic field within the enclosed air region in this system is

$$H_2 = i_y H_{2y} = i_y \frac{\omega^2 \mu_0 \varepsilon_0 z^2}{2} A \cos(\omega t) \tag{A.85}$$

Note that Equation A.82 is satisfied, as H_{2y} is independent of y (and $H_{2x} = H_{2z} = 0$):

$$\nabla \cdot \mu_0 H_2 = \mu_0 \left(\frac{\partial H_{2x}}{\partial x} + \frac{\partial H_{2y}}{\partial y} + \frac{\partial H_{2z}}{\partial z} \right) = 0 \tag{A.86}$$

The expressions for the non-fringing fields of the inductor correct up to and including second-order terms are

$$E_{0,1,2} = \underbrace{E_0}_{0} + E_1 + \underbrace{E_2}_{0} = -i_x\mu_0\omega z A \sin(\omega t) \tag{A.87}$$

$$H_{0,1,2} = H_0 + \underbrace{H_1}_{0} + H_2 = -i_y A\cos(\omega t)\left(1 - \frac{\omega^2\mu_0\varepsilon_0 z^2}{2}\right) \tag{A.88}$$

For the voltage across the current source, correct up to and including second-order contributions, it must hold that

$$[v_s(t)]_{0,1,2} = -\int_{x=0}^{d}(E_{0,1,2})_{z=-1}dx = -\mu_0\omega ldA\sin(\omega t) \tag{A.89}$$

For the source current, correct up to and including second-order contributions, it yields

$$[i_s(t)]_{0,1,2} = -\int_{y=0}^{w}(H_{0,1,2})_{z=-1}dy = wA\cos(\omega t)\left(1 - \frac{\omega^2\mu_0\varepsilon_0 l^2}{2}\right) \tag{A.90}$$

The input impedance of the single-turn inductor, correct up to and including second-order contributions, can be computed from the ratio between the phasor representations of v_s and i_s:

$$[V_s]_{0,1,2} = j\mu_0\omega ldA \tag{A.91}$$

$$[I_s]_{0,1,2} = wA\left(1 - \frac{\omega^2\mu_0\varepsilon_0 l^2}{2}\right) \tag{A.92}$$

$$Z_{0,1,2} = \frac{[V_s]_{0,1,2}}{[I_s]_{0,1,2}} = \frac{j\omega\mu_0 ldA}{wA\left(1 - \frac{\omega^2\mu_0\varepsilon_0 l^2}{2}\right)} = \frac{j\omega L}{(1 - \omega^2 LC)} \tag{A.93}$$

The lumped-element representation of the single-turn inductor, correct up to and including second-order contributions, is a parallel LC network, where the values of the capacitance and the inductance are specified as follows:

$$L = L_{DC} = \frac{\mu_0 ld}{w} \tag{A.94}$$

$$C = \frac{1}{2}\left(\frac{\varepsilon_0 lw}{d}\right) \tag{A.95}$$

The quasi-static approach is valid only when the second-order contribution is negligible:

$$\frac{\omega^2\mu_0\varepsilon_0 l^2}{2} = 2\pi^2\left(\frac{l}{\lambda}\right)^2 \ll 1 \tag{A.96}$$

This result is identical to what we found in Section A.3 for the parallel-plate capacitor.

A.5 Quasi-static Field of a Resistor

Consider the resistor in air as shown in Figure A.4, formed by placing a resistive sheet of uniform surface conductivity σ_s at one end of two perfectly conducting plates ($\sigma = \infty$). The length and width of the plates are much larger than their separation distance, so that all fringing in the resulting fields can be neglected. Therefore, all variation in E and H with both x and y can be neglected within this system:

$$\frac{\partial}{\partial x} = \frac{\partial}{\partial y} = 0 \tag{A.97}$$

The system is excited with a low-frequency sinusoidal excitation, which is uniformly distributed between the plates (at $z = -l$), such that a fixed reference voltage across the resistive sheet (in the $z = 0$ plane) is maintained:

$$v_r = A\cos(\omega t) \tag{A.98}$$

A.5.1 Quasi-static Solution

The zero-order time-varying electromagnetic fields in this system are identical in form to their static (DC) counterparts:

$$E_0 = -i_x\frac{v_r}{d} = -i_x\frac{A\cos(\omega t)}{d} \tag{A.99}$$

$$H_0 = -i_y\frac{\sigma_s v_r}{d} = -i_y\frac{\sigma_s A\cos(\omega t)}{d} \tag{A.100}$$

The first-order contributions to the electric field can be derived from the first-order field laws in Table A.1:

$$\nabla \times E_1 = -\mu_0\frac{\partial H_0}{\partial t} = -i_y\frac{\mu_0\sigma_s\omega A\sin(\omega t)}{d} \tag{A.101}$$

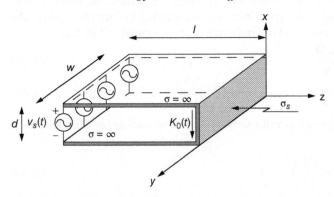

Figure A.4 Resistor in air. All fringing in the resulting fields can be neglected as $l \gg d$ and $w \gg d$.

$$\nabla \cdot \varepsilon_0 E_1 = \rho_1 = 0 \tag{A.102}$$

with

$$\nabla \times E_1 = i_x \left(\frac{\partial E_{1z}}{\partial y} - \frac{\partial E_{1y}}{\partial z} \right) + i_y \left(\frac{\partial E_{1x}}{\partial z} - \frac{\partial E_{1z}}{\partial x} \right) + i_z \left(\frac{\partial E_{1y}}{\partial x} - \frac{\partial E_{1x}}{\partial y} \right) \tag{A.103}$$

Combining Equations A.101 and A.103 results in

$$\frac{\partial E_{1x}}{\partial z} - \underbrace{\frac{\partial E_{1z}}{\partial x}}_{0} = -\frac{\mu_0 \sigma_s \omega A \sin(\omega t)}{d} \tag{A.104}$$

The first-order electric field within the enclosed air region in this system is

$$E_1 = i_x E_{1x} = -i_x \frac{\mu_0 \sigma_s \omega z A \sin(\omega t)}{d} \tag{A.105}$$

Note that Equation A.102 is satisfied, as E_{1x} is independent of x (and $E_{1y} = E_{1z} = 0$):

$$\nabla \cdot \varepsilon_0 E_1 = \varepsilon_0 \left(\frac{\partial E_{1x}}{\partial x} + \frac{\partial E_{1y}}{\partial y} + \frac{\partial E_{1z}}{\partial z} \right) = 0 \tag{A.106}$$

The first-order contributions to the magnetic field can be derived from the first-order field laws in Table A.1:

$$\nabla \times H_1 = \varepsilon_0 \frac{\partial E_0}{\partial t} = i_x \frac{\varepsilon_0 \omega A \sin(\omega t)}{d} \tag{A.107}$$

$$\nabla \cdot \mu_0 H_1 = 0 \tag{A.108}$$

with

$$\nabla \times H_1 = i_x \left(\frac{\partial H_{1z}}{\partial y} - \frac{\partial H_{1y}}{\partial z} \right) + i_y \left(\frac{\partial H_{1x}}{\partial z} - \frac{\partial H_{1z}}{\partial x} \right) + i_z \left(\frac{\partial H_{1y}}{\partial x} - \frac{\partial H_{1x}}{\partial y} \right) \tag{A.109}$$

Combining Equations A.107 and A.109 results in

$$\frac{\partial H_{1z}}{\partial y} - \underbrace{\frac{\partial H_{1y}}{\partial z}}_{0} = \frac{\varepsilon_0 \omega A \sin(\omega t)}{d} \tag{A.110}$$

The first-order magnetic field within the enclosed air region in this system is

$$H_1 = i_y H_{1y} = -i_y \frac{\varepsilon_0 \omega z A \sin(\omega t)}{d} \tag{A.111}$$

Note that Equation A.108 is satisfied, as H_{1y} is independent of y (and $H_{1x} = H_{1z} = 0$):

$$\nabla \cdot \mu_0 H_1 = \mu_0 \left(\frac{\partial H_{1x}}{\partial x} + \frac{\partial H_{1y}}{\partial y} + \frac{\partial H_{1z}}{\partial z} \right) = 0 \tag{A.112}$$

The expressions for the non-fringing fields within the enclosed air space in the system correct up to and including first-order terms are

$$E_{0,1} = E_0 + E_1 = -i_x \left(\frac{v_r(t)}{d} - \frac{\mu_0 \sigma_s}{d} \frac{dv_r(t)}{dt} z \right)$$
$$= -i_x \left(\frac{A\cos(\omega t)}{d} + \frac{\mu_0 \sigma_s}{d} \omega z A \sin(\omega t) \right) \tag{A.113}$$

$$H_{0,1} = H_0 + H_1 = -i_y \left(\frac{\sigma_s v_r(t)}{d} - \frac{\varepsilon_0}{d} \frac{dv_r(t)}{dt} z \right)$$
$$= -i_y \left(\frac{\sigma_s A\cos(\omega t)}{d} + \frac{\varepsilon_0}{d} \omega z A \sin(\omega t) \right) \tag{A.114}$$

For the voltage applied to the resistor, correct up to and including first-order contributions, it must hold that

$$[v_s(t)]_{0,1} = -\int_{x=0}^{d} (E_{0,1})_{z=-l}dx = A\cos(\omega t) - \mu_0 \sigma_s \omega lA \sin(\omega t) \tag{A.115}$$

For the source current, correct up to and including first-order contributions, it yields

$$[i_s(t)]_{0,1} = -\int_{y=0}^{w} (H_{0,1})_{z=-l}dy = \frac{\sigma_s wA\cos(\omega t)}{d} - \frac{\varepsilon_0 w}{d}\omega lA \sin(\omega t) \tag{A.116}$$

The input impedance of the resistor, correct up to and including first-order contributions, can be computed from the ratio between the phasor representations of v_s and i_s:

$$[V_s]_{0,1} = A + j\mu_0 \sigma_s \omega lA \tag{A.117}$$

$$[I_s]_{0,1} = \frac{\sigma_s wA}{d} + j\frac{\varepsilon_0 w}{d}\omega lA \tag{A.118}$$

$$Z_{0,1} = \frac{[V_s]_{0,1}}{[I_s]_{0,1}} = \frac{1 + j\mu_0 \sigma_s \omega l}{\frac{\sigma_s w}{d} + j\frac{\varepsilon_0 w}{d}\omega l} = R\frac{1 + j\omega(L/R)}{1 + j\omega RC} \tag{A.119}$$

The lumped-element representation of the resistor, correct up to and including first-order contributions, is an RLC circuit, where the values of the resistance, inductance, and capacitance are specified as follows:

$$R = R_{DC} = \frac{d}{\sigma_s w} \tag{A.120}$$

$$L = L_{DC} = \frac{\mu_0 l d}{w} \tag{A.121}$$

$$C = C_{DC} = \frac{\varepsilon_0 l w}{d} \tag{A.122}$$

In the case of $\sigma_s = \infty$ ($R = 0$), the system changes into the single-turn inductor as described in Section A.4. In the case of $\sigma_s = 0$ ($R = \infty$), the system changes into the parallel-plate capacitor as described in Section A.3.

The lumped-element representation of the resistor is dependent on the actual value of the resistance R:

- $R \ll \sqrt{L/C}$. The lumped-element representation of the resistor equals a series LR circuit, as the input impedance is approximately

$$Z_{0,1} = R \frac{1 + j\omega(L/R)}{1 + j\omega RC} \approx R + j\omega L \tag{A.123}$$

- $R \gg \sqrt{L/C}$. The lumped-element representation of the resistor equals a parallel RC circuit, as the input admittance is approximately

$$Y_{0,1} = \frac{1}{R} \frac{1 + j\omega RC}{1 + j\omega(L/R)} \approx \frac{1}{R} + j\omega C \tag{A.124}$$

- $R = \sqrt{L/C}$. The lumped-element representation of the resistor reduces to the DC resistance, as the input impedance equals

$$Z_{0,1} = R \frac{1 + j\omega(L/R)}{1 + j\omega RC} = R \tag{A.125}$$

A.6 Circuit Modeling

It is the electrical size of the structure – its size in terms of the minimum wavelength of interest in the bandwidth over which the model must be valid – that dictates the sophistication and complexity of the required model. Although this is only an approximate criterion, which is closely related to Equations A.58 and A.96, an electromagnetic structure is said to be electrically small if its dimensions are smaller than one-tenth of the smallest wavelength under consideration:

$$l < \frac{1}{10}\lambda \tag{A.126}$$

l the physical dimension [m]
λ the wavelength [m]

In the case of a steady-state analysis of the power system, only the power frequency (50 or 60 Hz) is considered, and the wavelength of a 50 Hz sinusoidal voltage or current equals

$$\lambda = \frac{v}{f} = \frac{3 \times 10^5}{50} = 6000 \, \text{km} \tag{A.127}$$

λ the wavelength [km]
v the speed of light ≈ 300000 [km/s]
f the frequency [Hz]

Therefore, a component, for example, a transmission line, is "electrically small" in the steady-state analysis when the dimensions are smaller than $6000/10 = 600 \, \text{km}$. In that case, Maxwell's equations can be approximated by a quasi-static approach, and the component can accurately be modeled by lumped elements. Kirchhoff's laws are then applicable to compute the voltages and currents.

Reference

1 Magid, Leonard M.: *Electromagnetic Fields, Energy and Waves*, John Wiley & Sons, Inc., New York, 1972, ISBN 0-471-56334-X.

B

Power Transformer Model

B.1   Introduction

Generation of electrical power at the synchronous generator level is normally done at voltage levels, ranging from a few kilovolts up to 25 or 30 kV. To minimize the power loss and voltage drop because of the conductor resistance, the transmission of electrical energy is done at rather high voltage levels. Power transformers raise the voltage from the level of generation to the higher level of transmission. They are called step-up transformers. Power transformers also connect the transmission and distribution networks, which operate at different voltage levels. Distribution transformers step down the voltage to levels that are more safe for the consumers to use. Transformers consist essentially of two coils on an iron core. The iron core increases the magnetic coupling between the two coils and ensures that almost all the magnetic flux created by one coil links the other coil, which results in an efficient device. The operation of the transformer is based on Faraday's law of induction. The transformer has no moving parts and is therefore a "relatively simple" piece of equipment. For the analysis of the behavior of the transformer in the power system, a qualitative description alone is not sufficient. The central item of this appendix is the mathematical description of the voltage–current relations of the transformer under different operating conditions.

B.2   The Ideal Transformer

Consider the two-winding transformer as shown in Figure B.1. Two coils are magnetically coupled: the primary winding, for which we use suffix 1, and the secondary winding, for which we use suffix 2. The number of turns in the primary and secondary windings is indicated with N_1 and N_2, respectively. The primary winding is connected to a sinusoidal voltage source v_1, while the secondary winding supplies a load current. The resistance of each winding and the losses in the transformer core are for the moment neglected. Furthermore, we

Electrical Power System Essentials, Second Edition. Pieter Schavemaker and Lou van der Sluis.
© 2017 John Wiley & Sons Ltd. Published 2017 by John Wiley & Sons Ltd.

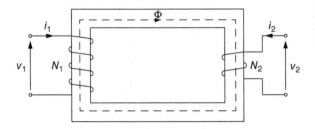

assume the permeability of the core to be infinite and all the flux to be confined to circulate within the transformer core as is shown in Figure B.1 (i.e., there is no leakage flux).

The voltage v_1 supplies the current i_1 in the primary winding, and this current creates a magnetic flux Φ in the transformer core. According to Faraday's law of induction, the magnetic flux in the core induces an EMF in the primary winding that opposes the applied voltage, and it also induces an EMF in the secondary winding. The transformer is connected to a load, and the magnitude of the secondary current i_2 depends on the load impedance. As there is no flux leakage, the flux Φ links both the primary and the secondary windings. Ampère's law states that the line integral of the magnetic field H around a closed path C is equal to the current traversing the surface S bounded by that path. Ampère's law for the closed path of flux, shown by the dashed line in Figure B.1, can be written as

$$\oint_C H \cdot \tau \, ds = N_1 i_1 + N_2 i_2 \tag{B.1}$$

H the magnetic field intensity vector [A/m]
N_1 the number of turns of the primary winding
i_1 the current in the primary winding [A]
N_2 the number of turns of the secondary winding
i_2 the current in the secondary winding [A]

We assumed the permeability of the iron core to be infinite ($\mu_r = \infty$). A nonzero magnetic field intensity H in the iron would imply that the magnetic flux density in the iron ($B = \mu_0 \mu_r H$) is infinite, which is not possible. Therefore, the magnetic field intensity in the iron equals zero, and Equation B.1 can be written as follows:

$$N_1 i_1 + N_2 i_2 = 0 \tag{B.2}$$

From this equation the current ratio for an ideal transformer results:

$$\frac{i_1}{i_2} = -\frac{N_2}{N_1} \tag{B.3}$$

The terminal voltages are given by the following equations:

$$v_1 = \frac{d\psi_1}{dt} = N_1\frac{d\Phi}{dt} \tag{B.4}$$

$$v_2 = \frac{d\psi_2}{dt} = N_2\frac{d\Phi}{dt} \tag{B.5}$$

Therefore, the voltage ratio of the transformer can be derived as follows:

$$\frac{v_1}{v_2} = \frac{N_1}{N_2} \tag{B.6}$$

The circuit representation of the ideal transformer is shown in Figure B.2.

We now switch to the phasor domain (see also Section 1.4) in order to demonstrate the impedance transformation property of the ideal transformer. This means that an impedance on either side of the ideal transformer can be transformed to the other side. In case of a parallel-connected impedance, as shown in Figure B.3, the voltage–current relations of the ideal transformer with an impedance connected at the secondary side are given by

$$V_1 = \frac{N_1}{N_2}V_2 \quad \text{and} \quad I_1 = -\frac{N_2}{N_1}\left(I_2 - \frac{V_2}{Z_2}\right) \tag{B.7}$$

The voltage–current relations when an impedance is connected at the primary side are

$$V_1 = \frac{N_1}{N_2}V_2 \quad \text{and} \quad I_1 - \frac{V_1}{Z_1} = -\frac{N_2}{N_1}I_2 \tag{B.8}$$

Those voltage–current relations are identical if and only if

$$Z_1 = \left(\frac{N_1}{N_2}\right)^2 Z_2 \tag{B.9}$$

Figure B.2 The circuit representation of an ideal transformer.

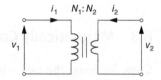

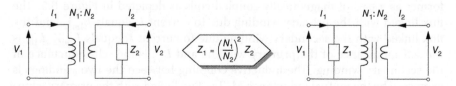

Figure B.3 Transformation of a parallel-connected impedance.

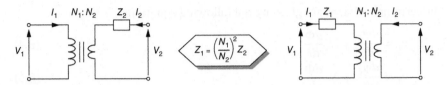

Figure B.4 Transformation of a series-connected impedance.

The two circuits in Figure B.3 are identical when the relation between Z_1 and Z_2 is given by Equation B.9.

In case of a series-connected impedance, as shown in Figure B.4, the voltage–current relations of the ideal transformer with an impedance connected at the secondary side are given by

$$V_1 = \frac{N_1}{N_2}(V_2 - Z_2 I_2) \quad \text{and} \quad I_1 = -\frac{N_2}{N_1} I_2 \tag{B.10}$$

The voltage–current relations when an impedance is connected at the primary side are

$$(V_1 - Z_1 I_1) = \frac{N_1}{N_2} V_2 \quad \text{and} \quad I_1 = -\frac{N_2}{N_1} I_2 \tag{B.11}$$

Again, those voltage–current relations are identical if and only if

$$Z_1 = \left(\frac{N_1}{N_2}\right)^2 Z_2 \tag{B.12}$$

The two circuits in Figure B.4 are identical when the relation between Z_1 and Z_2 is given by Equation B.12.

Note that the factors that are needed to transform a parallel- or a series-connected impedance to the other side of the ideal transformer are identical.

B.3 Magnetically Coupled Coils

Consider again the two-winding transformer as shown in Figure B.1. In this section we follow a more general approach and describe the two-winding transformer as a pair of magnetically coupled coils as depicted in Figure B.5. The flux linked with the primary winding due to current i_1 equals $L_{1m}i_1$, and the flux linked with the secondary winding due to current i_2 equals $L_{2m}i_2$. L_{1m} is the self-inductance of the primary winding and L_{2m} is the self-inductance of the secondary winding. The inductive coupling between the two windings is expressed by the mutual inductance M. The flux linked with the primary winding due to current i_2 equals Mi_2, and the flux linked with the secondary winding

Figure B.5 Two magnetically coupled coils.

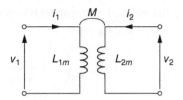

due to current i_1 equals Mi_1. We assume here that the coils have an ideal coupling so that the mutual inductance takes its maximum value (i.e., the geometric mean of the self-inductances L_{1m} and L_{2m}):

$$M = \sqrt{L_{1m}L_{2m}} \tag{B.13}$$

Now, the flux linked with the primary and secondary winding can be written as

$$\psi_1 = L_{1m}i_1 + Mi_2 \tag{B.14}$$
$$\psi_2 = L_{2m}i_2 + Mi_1 \tag{B.15}$$

ψ_1 the flux linked with the primary winding [Wb]
ψ_2 the flux linked with the secondary winding [Wb]
L_{1m} the self-inductance of the primary winding [H]
L_{2m} the self-inductance of the secondary winding [H]
M the mutual inductance [H]

When we apply Equation B.13, we can represent the magnetically coupled coils of Figure B.5 as a combination of an ideal transformer and a coil in parallel as is shown in Figure B.6. We can prove that both representations of the magnetically coupled coils are equivalent by comparing their voltage–current relations. The voltage–current relations of the circuit in Figure B.5, using Equations B.14 and B.15, are

$$v_1 = \frac{d\psi_1}{dt} = L_{1m}\frac{di_1}{dt} + M\frac{di_2}{dt} \tag{B.16}$$
$$v_2 = \frac{d\psi_2}{dt} = M\frac{di_1}{dt} + L_{2m}\frac{di_2}{dt} \tag{B.17}$$

Figure B.6 Magnetically coupled coils as a combination of an ideal transformer and a coil.

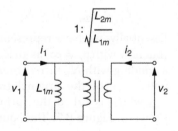

The voltage–current relations of the circuit in Figure B.6, using the relation $M = \sqrt{L_{1m}L_{2m}}$, are

$$v_1 = L_{1m}\frac{d}{dt}\left(i_1 + \sqrt{\frac{L_{2m}}{L_{1m}}}\,i_2\right) = L_{1m}\frac{di_1}{dt} + \sqrt{L_{1m}L_{2m}}\frac{di_2}{dt}$$

$$= L_{1m}\frac{di_1}{dt} + M\frac{di_2}{dt} \tag{B.18}$$

$$v_2 = \sqrt{\frac{L_{2m}}{L_{1m}}}\,v_1 = \sqrt{L_{1m}L_{2m}}\frac{di_1}{dt} + L_{2m}\frac{di_2}{dt} = M\frac{di_1}{dt} + L_{2m}\frac{di_2}{dt} \tag{B.19}$$

The voltage–current relations are equal for both circuits, which proves that they are equivalent.

Division of Equation B.16 by Equation B.17, and using again the relation $M = \sqrt{L_{1m}L_{2m}}$ (i.e., an ideal coupling), gives us the voltage ratio:

$$\frac{v_1}{v_2} = \frac{\frac{M^2}{L_{2m}}\frac{di_1}{dt} + M\frac{di_2}{dt}}{M\frac{di_1}{dt} + L_{2m}\frac{di_2}{dt}} = \frac{M}{L_{2m}} \cdot \frac{M\frac{di_1}{dt} + L_{2m}\frac{di_2}{dt}}{M\frac{di_1}{dt} + L_{2m}\frac{di_2}{dt}} = \frac{M}{L_{2m}} = \frac{L_{1m}}{M} = \sqrt{\frac{L_{1m}}{L_{2m}}}$$

$$\tag{B.20}$$

The terminal voltages can also be described by

$$v_1 = \frac{d\psi_1}{dt} = N_1\frac{d\Phi}{dt} \tag{B.21}$$

$$v_2 = \frac{d\psi_2}{dt} = N_2\frac{d\Phi}{dt} \tag{B.22}$$

N_1 the number of turns of the primary winding
N_2 the number of turns of the secondary winding

Therefore, the voltage ratio equals

$$\frac{v_1}{v_2} = \frac{N_1}{N_2} \tag{B.23}$$

When we compare this with the voltage ratio that we found in Equation B.20, we can see that

$$L_{1m} : L_{2m} : M = N_1^2 : N_2^2 : N_1N_2 \tag{B.24}$$

Essentially, we have represented the two-winding transformer of Figure B.1 (but now with $\mu_r \neq \infty$) as an ideal transformer with a winding ratio $N_1 : N_2$ and a winding with N_1 turns around a magnetic core that is identical to the magnetic core of the original transformer and is magnetized by the current i_{1m}, which is called the (primary) magnetization current. This approach is shown in Figure B.7. The equivalent circuit of the magnetically coupled coils

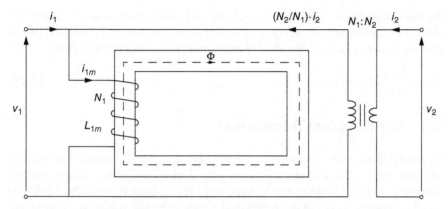

Figure B.7 Approach of representing magnetically coupled coils as a combination of an ideal transformer and a coil.

Figure B.8 Magnetically coupled coils as a combination of an ideal transformer and a coil.

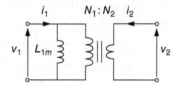

that we derived in Figure B.6 can now be altered to the equivalent circuit in Figure B.8.

B.3.1 Equivalence with the Ideal Transformer

In this section we see that we need an extra assumption in order to make the ideally magnetically coupled coils equivalent to the ideal transformer that we derived in Section B.2: we have to assume that the coupled coils have infinite mutual and self-inductances, which correspond to a situation where the coils have a common core with an infinite permeability (which was indeed one of the assumptions in Section B.2):

$$L_{1m} \to \infty, \quad L_{2m} \to \infty, \quad M \to \infty \tag{B.25}$$

When we divide Equation B.16 by M and Equation B.17 by L_{2m} and use Equation B.24, we get the equations

$$\frac{v_1}{M} = \frac{L_{1m}}{M} \frac{di_1}{dt} + \frac{di_2}{dt} = \frac{d}{dt}\left(\frac{N_1}{N_2} i_1 + i_2\right) \tag{B.26}$$

$$\frac{v_2}{L_{2m}} = \frac{M}{L_{2m}} \frac{di_1}{dt} + \frac{di_2}{dt} = \frac{d}{dt}\left(\frac{N_1}{N_2} i_1 + i_2\right) \tag{B.27}$$

As the coupled coils have infinite mutual and self-inductances, the left-hand sides of those two equations equal zero, and we end up with the current relation that we found earlier for the ideal transformer (Equations B.2 and B.3):

$$\frac{N_1}{N_2}i_1 + i_2 = 0 \tag{B.28}$$

B.4 The Nonideal Transformer

In reality, the currents flowing in the primary and the secondary winding do not only create a flux in the iron core (the ideal situation) but also in the air surrounding the windings: the leakage flux. The leakage flux will be relatively small compared with the main flux, but needs to be taken into account when modeling a nonideal transformer. A transformer with leakage flux is illustrated in Figure B.9. Φ_m is the main flux that links all the turns of the primary and secondary coils. $\Phi_{1\sigma}$ and $\Phi_{2\sigma}$ are the fluxes that link either the primary or secondary coil. The flux linked with the primary and secondary winding can be written as

$$\psi_1 = L_{1\sigma}i_1 + L_{1m}i_1 + Mi_2 \tag{B.29}$$
$$\psi_2 = L_{2\sigma}i_2 + L_{2m}i_2 + Mi_1 \tag{B.30}$$

L_{1m} the coefficient of the self-inductance of the primary winding that is related to the main flux [H]

L_{2m} the coefficient of the self-inductance of the secondary winding that is related to the main flux [H]

$L_{1\sigma}$ the coefficient of the self-inductance of the primary winding that is related to the primary leakage flux [H]

$L_{2\sigma}$ the coefficient of the self-inductance of the secondary winding that is related to the secondary leakage flux [H]

M the mutual inductance [H]

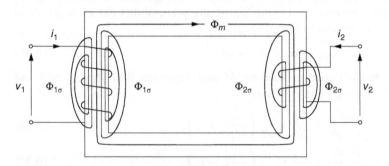

Figure B.9 A transformer with leakage flux.

The coefficients of the self-inductance L_{1m} and L_{2m} have an ideal magnetic coupling ($M = \sqrt{L_{1m}L_{2m}}$), whereas the coefficients $L_{1\sigma}$ and $L_{2\sigma}$ have no magnetic coupling.

The voltage–current relations of the nonideal transformer are

$$v_1 = i_1 R_1 + \frac{d\psi_1}{dt} = i_1 R_1 + L_{1\sigma}\frac{di_1}{dt} + L_{1m}\frac{di_1}{dt} + M\frac{di_2}{dt} \tag{B.31}$$

$$v_2 = i_2 R_2 + \frac{d\psi_2}{dt} = i_2 R_2 + L_{2\sigma}\frac{di_2}{dt} + L_{2m}\frac{di_2}{dt} + M\frac{di_1}{dt} \tag{B.32}$$

R_1 the copper resistance of the primary winding [Ω]
R_2 the copper resistance of the secondary winding [Ω]

When we compare these two equations with Equations B.16 and B.17 and the accompanying graphical representation in Figures B.5 and B.8, we can see that Equations B.31 and B.32 can be depicted as in Figure B.10. The core losses, which we touched upon in Section 3.8, consisting of the hysteresis losses and the eddy current losses, are still missing in this equivalent circuit. The hysteresis losses quantify the energy dissipated in the ferromagnetic material due to the continuous change (50 Hz) of direction of the Weiss particles. The eddy current losses originate from the eddy currents caused by the time-varying magnetic flux in the electrically conductive magnetic core. The hysteresis losses and eddy current losses can by approximation be incorporated in the equivalent circuit by putting a resistance R_m in parallel with the main inductance L_{1m} as shown in Figure B.11.

For system studies the transformer equivalent circuit can be simplified. The secondary circuit elements R_2 and $L_{2\sigma}$ can be referred to the primary side of the ideal transformer as shown in Figure B.12. In practice the values of the series elements $L_{2\sigma}(N_1/N_2)^2$ and $R_2(N_1/N_2)^2$ are small compared with the shunt elements R_m and L_{1m} and may therefore be interchanged. The series resistance $R = R_1 + R_2(N_1/N_2)^2$, accounting for the copper losses, is much smaller than the leakage reactance $\omega L_\sigma = \omega(L_{1\sigma} + L_{2\sigma}(N_1/N_2)^2)$ and can therefore be neglected. Both the shunt elements R_m and ωL_{1m} are much larger than ωL_σ and can often be left out of the circuit model. This results in an, for short-circuit

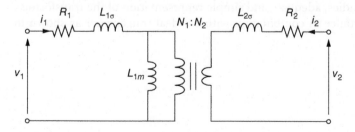

Figure B.10 Transformer equivalent circuit without core losses.

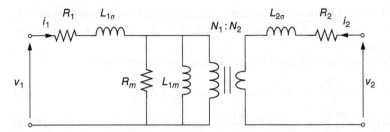

Figure B.11 Transformer equivalent circuit.

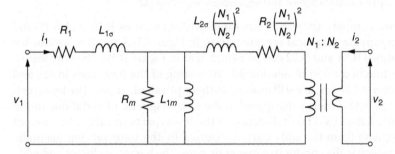

Figure B.12 Transformer equivalent circuit with the secondary circuit elements referred to the primary side of the ideal transformer.

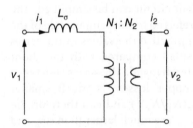

Figure B.13 Simplified transformer equivalent circuit.

and load flow studies, adequate and simple representation of the transformer: the leakage reactance in combination with an ideal transformer as shown in Figure B.13.

B.5 Three-Phase Transformer

In single-phase transformers, only the ratio of the voltage magnitudes (i.e., the lengths of the voltage phasors) between the primary and secondary side can be altered, and the turns ratio is a real number: the number of windings at the primary side divided by the number of windings at the secondary side. In three-phase transformers it is possible to arrange the windings (and terminals) in such a way that not only the voltage magnitudes between the primary and secondary sides are different but also the phase angles (see also Section 3.8). If we make a single-phase model of a three-phase transformer that causes a phase shift of the voltage phasors, the model as shown in Figure B.12 can be easily adjusted by putting an extra ideal transformer in series that takes care of the phase shift, as shown in Figure B.14. The phase shift is established by giving this extra ideal transformer a complex turns ratio. For example, the complex turns ratio $1:1\angle 30°$ means that the voltage phasor at the secondary side is rotated with $30°$ counterclockwise with regard to the voltage phasor at the primary side (its amplitude remains constant!).

The three coils at the primary or the secondary side of a three-phase transformer can be connected in wye (Y) or delta (D). The single-phase equivalent models of an ideal three-phase transformer are shown in Table B.1 for various combinations of these winding connections. Note that we did not mention a clock reference in the column labeled "designation." Possible phase shifts can be dealt with by putting an extra ideal transformer with a complex turns ratio in series.

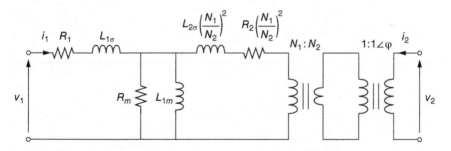

Figure B.14 Single-phase model of a three-phase, phase-shifting, transformer.

Table B.1 Single-phase equivalent models of three-phase transformers.

Designation	Winding Connection	Single-Phase Equivalent
Yy		$N_1 : N_2$
Yd		$N_1 : N_2/\sqrt{3}$
Dy		$N_1/\sqrt{3} : N_2$
Dd		$N_1/\sqrt{3} : N_2/\sqrt{3}$

C

Synchronous Machine Model

C.1 Introduction

The workhorse for the generation of electricity is the synchronous machine. The bulk of electric energy is produced by three-phase synchronous generators. Synchronous generators with power ratings of several hundred MVA are common. Under steady-state conditions, they operate at a speed fixed by the power system frequency, and therefore they are called synchronous machines. Like most rotating machines, synchronous machines can operate both as a motor and as a generator. They are used as motors in constant speed drives. With a power-electronic frequency changer, however, they can be turned into a variable speed drive. As generators, several synchronous machines usually operate in parallel in the larger power stations.

The operation of a synchronous generator is based on Faraday's law of induction. In this appendix we first describe the principle of operation of the synchronous machine and discuss some constructional aspects. For the analysis of the behavior of the synchronous machine in the power system, a qualitative description alone is not sufficient. The central item of this appendix is the mathematical description of the voltage–current relation of the synchronous generator under different operating conditions.

C.2 The Primitive Synchronous Machine

In Chapter 1 (Section 1.3.3) the rotating magnetic field was introduced. In the three-phase coil system, the resulting magnetic field vector rotates with a constant amplitude (see Figure 1.9).

A synchronous generator generates electricity, as we call it, by conversion of mechanical energy into electric energy. The two basic parts of the synchronous machine are the rotor and the armature or stator. The iron rotor is equipped

Electrical Power System Essentials, Second Edition. Pieter Schavemaker and Lou van der Sluis.
© 2017 John Wiley & Sons Ltd. Published 2017 by John Wiley & Sons Ltd.

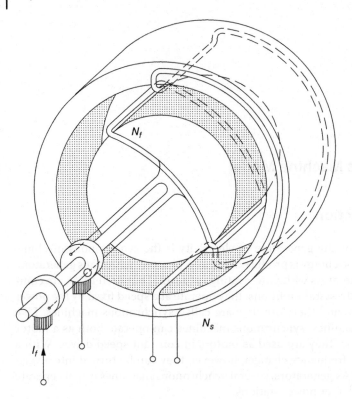

I_f

Figure C.1 Three-dimensional view of a simple synchronous machine.

with a DC-excited winding, which acts as an electromagnet. When the rotor rotates and the rotor winding is excited, a rotating magnetic field is present in the air gap between the rotor and the armature. The armature has a three-phase winding in which the time-varying EMF is generated by the rotating magnetic field. A simple synchronous machine is shown in Figure C.1. The machine has a cylindrical rotor with two poles, which are excited with a DC current. In the armature the winding consists of one single turn. The cross section of this machine is shown in Figure C.2.

The rotor current creates a magnetic field in the air gap between rotor and armature. This magnetic field can be calculated with Ampère's law:

$$\oint_C \boldsymbol{H} \cdot \boldsymbol{\tau} ds = \iint_S \boldsymbol{J} \cdot \boldsymbol{n} dA \tag{C.1}$$

$\boldsymbol{H}$ the magnetic field intensity vector [A/m]
$\boldsymbol{J}$ the current density vector [A/m^2]

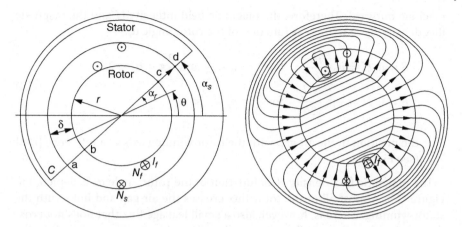

Figure C.2 Cross section of a simple synchronous machine and the field lines of the magnetic flux in the case that only the rotor current is present.

Ampère's law states that the line integral of the magnetic field intensity H around a closed path C is equal to the current traversing the surface S bounded by that path. Let us assume that the rotor winding carrying a current I_f has N_f turns; then Ampère's law, for the contour C in Figure C.2, can be written as

$$\oint_C \boldsymbol{H} \cdot \boldsymbol{\tau} ds = \int_a^b \boldsymbol{H} \cdot \boldsymbol{\tau} ds + \int_b^c \boldsymbol{H} \cdot \boldsymbol{\tau} ds + \int_c^d \boldsymbol{H} \cdot \boldsymbol{\tau} ds + \int_d^a \boldsymbol{H} \cdot \boldsymbol{\tau} ds \qquad \text{(C.2)}$$
$$= N_f I_f$$

N_f the number of turns of the rotor winding
I_f the DC current in the rotor winding [A]

When we assume the permeability of the iron in both the rotor and the armature to be infinite ($\mu_r = \infty$), the magnetic field intensity in the iron is zero.

The size of the air gap is rather small compared to the diameter of the rotor. Because of the symmetry, the magnetic field H for the path from a to b is equal to the magnetic field for the path from c to d. The magnetic field vector points in the air gap in the direction of the contour, and Equation C.2 becomes

$$\boldsymbol{H} \cdot \boldsymbol{\tau} \delta + 0 + \boldsymbol{H} \cdot \boldsymbol{\tau} \delta + 0 = N_f I_f \quad \text{or} \quad 2H\delta = N_f I_f \qquad \text{(C.3)}$$

δ the size of the air gap [m]

The sign of H and B in the air gap is usually taken as positive when the direction of the related vector points from rotor to armature or, in other words, when the associated field lines or flux lines are drawn from the center of the machine outward. Therefore, the magnetic field intensity H and the magnetic flux density B are written as function of the rotor angle α_r:

$$B = \mu_0 H = \mu_0 \frac{N_f I_f}{2\delta} \quad \text{for } 0 < \alpha_r < \frac{\pi}{2} \text{ and } \frac{3}{2}\pi < \alpha_r < 2\pi$$

$$B = \mu_0 H = -\mu_0 \frac{N_f I_f}{2\delta} \quad \text{for } \frac{\pi}{2} < \alpha_r < \frac{3}{2}\pi$$

(C.4)

α_r the position in the air gap with the rotor winding axis as a reference [rad] (see Figure C.2)

The magnetic flux density B as function of the rotor angle α_r is depicted in Figure C.3. Almost all of the rotor flux crosses the air gap and links with the stator winding. There is, however, also a small leakage flux that does not cross the air gap. The magnetic flux surrounding the end part of the turns, where the winding comes outside the stator iron, is, for instance, a part of this leakage flux.

The flux linkage changes with time because the rotor flux in the air gap rotates and in the stator winding an EMF is generated (Faraday's law of induction), which appears at the stator terminals, when the winding is left open, as the so-called open-circuit voltage. For the calculation of the flux linkage with the stator, we choose a surface S, with the shape of a half cylinder, in the air gap between rotor and stator. The surface S is schematically depicted in Figure C.4. The flux crossing the surface S is

$$\Phi = \iint_S \boldsymbol{B} \cdot \boldsymbol{n} dA = \int_{-\pi/2}^{\pi/2} Blr d\alpha_s$$

(C.5)

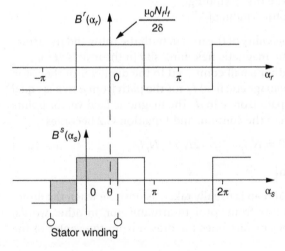

Figure C.3 The distribution of the magnetic flux density in the air gap for the simple synchronous machine of Figure C.2.

Figure C.4 The surface S as used to calculate the flux linkage with the stator winding.

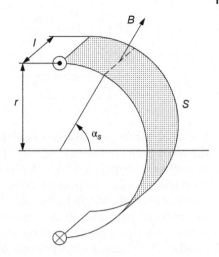

Φ the flux crossing the surface S [Wb]
B the magnetic flux density vector [Wb/m^2]
l the actual length of the machine [m]
r the radius of the rotor [m]
α_s the position in the air gap with the stator winding axis as a reference [rad] (see Figure C.2)

The integral equals the shaded area in Figure C.3 (after multiplication with lr), and the magnetic flux is proportional to the position of the rotor angle (see Figure C.5). The relation between α_s and α_r is given by (see Figure C.2)

$$\alpha_s = \alpha_r + \theta \tag{C.6}$$

Thus the integral of Equation C.5 is maximum for $\theta = 0$ (i.e., $\alpha_s = \alpha_r$) and minimum for $\theta = \pi$. The induced voltage in a single turn equals

$$E = \frac{d\Phi}{dt} \tag{C.7}$$

In the case of a constant rotor speed ($\theta = \omega_m t$), a voltage as shown in Figure C.6 results. The induced voltage in the stator winding has a rectangular shape – the same shape as the magnetic flux density B in the air gap. The voltage induced in

Figure C.5 Flux linkage with a single stator turn.

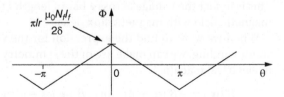

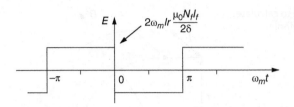

Figure C.6 The induced voltage in a single stator turn.

the stator winding can also be calculated in a different way, in order to show that the relationship between $B^r(\alpha_r)$ and $E(t)$ at a constant rotor speed is also valid for different shapes of $B^r(\alpha_r)$. When the rotor turns in the direction that has been referenced as the positive direction in Figure C.2, the rectangular-shaped distribution of the magnetic flux density in the air gap, as depicted in Figure C.3, moves to the right, and $d\theta/dt$ is positive. As a result, the shaded area of the part below the x-axis increases, and the shaded area of the part above the x-axis gets smaller. The change in flux per unit of time can be expressed for the left-hand side of the considered area as

$$\left.\frac{d\Phi}{dt}\right|_{\text{left}} = lrB^s\left(-\frac{\pi}{2}\right)\frac{d\theta}{dt} \tag{C.8}$$

The change in flux per unit of time for the right-hand side of the considered area is

$$\left.\frac{d\Phi}{dt}\right|_{\text{right}} = -lrB^s\left(\frac{\pi}{2}\right)\frac{d\theta}{dt} \tag{C.9}$$

With $\omega_m = d\theta/dt$, the voltage induced in the stator winding can be written as

$$E = \frac{d\Phi}{dt} = \omega_m lr\left(B^s\left(-\frac{\pi}{2}\right) - B^s\left(\frac{\pi}{2}\right)\right) \tag{C.10}$$

This shows us that the voltage induced in the stator winding depends only on the actual flux density B at the position where the turns of the stator winding lie. This is in general valid when the flux density distribution, seen from the rotor, is constant in time.

When we substitute the rotational speed at the rotor surface ($v = \omega_m r$) in Equation C.10, we recognize the expression

$$E = Blv \tag{C.11}$$

This is in fact the voltage across a bar of length l that moves with a speed v in a magnetic field with magnetic flux density B.

When we want to find the expression for the voltage at the terminals of the stator winding, we can make use of the symmetry that is often so nicely present in electrical machines:

$$B^s(\alpha_s) = -B^s(\alpha_s - \pi) \quad \text{or} \quad B^r(\alpha_r) = -B^r(\alpha_r - \pi) \tag{C.12}$$

and Equation C.10 can be written as

$$E = 2\omega_m lrB^s \left(-\frac{\pi}{2}\right) \tag{C.13}$$

The voltage at the terminals of the stator winding is the multiplication of the voltage of one turn with the number of turns:

$$E_s = 2\omega_m lrN_s B^s \left(-\frac{\pi}{2}\right) \tag{C.14}$$

N_s the number of turns of the stator winding

This equation shows that, at a constant angular rotor speed and for constant magnetic flux density (as seen from the rotor), the voltage induced in the stator winding is proportional to the magnetic flux density at position $\alpha_s = -\pi/2$. The induced voltage as a function of time has therefore the same shape as the magnetic flux density B as a function of α_r.

C.3   The Single-Phase Synchronous Machine

In the previous section, we considered a synchronous machine generating a rectangular-shaped stator voltage instead of the desired sinusoidal-shaped voltage. In this section we discuss the way how this can be achieved. We saw that, under normal circumstances, the voltage induced in a stator winding has the shape of the distribution of the magnetic flux density B in the air gap, as seen from the rotor reference system. This means that we can acquire a sinusoidal-shaped voltage by creating a sinusoidal magnetic flux density in the air gap. This can, for instance, be done by changing the air-gap size as a function of α_r or by redistributing the rotor winding at the rotor surface with a "sinusoidal density." Another way is to distribute the stator winding with a "sinusoidal density," as is shown in Figure C.7.

In Figure C.7 (a), the cross section of a machine is shown. The stator winding is not concentrated at one spot, but is distributed. The rotor winding is connected with a current source supplying a constant rotor current I_f. There is no current flowing in the stator winding. The voltage at the terminals of the stator winding is the sum of the winding voltages E_1, E_2, and E_3. We can already recognize a rudimentary sine wave shape.

In practice a combination of techniques is used to acquire a sine wave-shaped voltage, but in this appendix we restrict ourselves to the sinusoidal stator winding distribution, of which an example is shown in Figure C.8. We can see from this figure that the layout of the armature winding along the stator surface is only a rough approximation of the sine wave shape. For our description of the machine however, we assume an ideal sinusoidal winding distribution

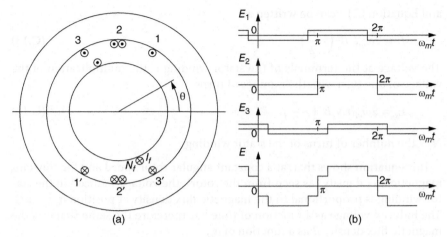

Figure C.7 A distributed stator winding and the induced voltages.

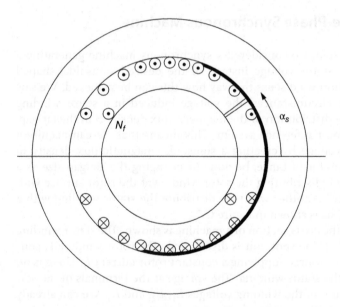

Figure C.8 A sinusoidal distribution of the stator winding.

(the letter "*a*" in the subscript refers to phase *a*, as later on also phases *b* and *c* are introduced):

$$Z_{sa}(\alpha_s) = \hat{Z}_s \sin(\alpha_s) \tag{C.15}$$

Z_{sa} the stator winding distribution [1/m]

The relation between the total number of turns of the stator winding and the amplitude of the winding distribution is

$$N_s = \int_0^\pi Z_{sa}(\alpha_s)rd\alpha_s = \int_0^\pi \hat{Z}_s \sin(\alpha_s)rd\alpha_s = 2r\hat{Z}_s \tag{C.16}$$

N_s the number of turns of the stator winding
r the radius of the inner diameter of the stator [m]

Therefore, the stator winding distribution can be written as

$$Z_{sa}(\alpha_s) = \frac{N_s}{2r} \sin(\alpha_s) \tag{C.17}$$

The open-circuit voltage of the stator is the voltage that appears at the terminals of the stator winding when there is no load connected; the machine supplies no current. For the calculation of the open-circuit voltage, we need to know the flux that is linked with the stator winding. The flux distribution in the air gap $B^s(\alpha_s)$ is a periodic function that repeats itself every 2π and can be expressed as a Fourier series:

$$B^s(\alpha_s) = \sum_{n=1}^\infty \hat{B}_n \cos(n\alpha_s - \varphi_n) \tag{C.18}$$

Both $\hat{B}_n$ and φ_n in this equation can be time dependent. We assume that the stator winding is distributed over the stator surface as depicted in Figure C.8. For the calculation of the flux linked with the stator turns, we choose a surface S, with the shape of a half cylinder as shown in Figure C.4. In Figure C.8 this surface is displayed in bold. An infinitely small amount of flux, which crosses the air gap from the rotor to the stator at angle α_s, can be written as

$$d\Phi(\alpha_s) = B^s(\alpha_s)lrd\alpha_s \tag{C.19}$$

With the expression for the sinusoidal stator winding distribution (Equation C.17), the number of turns that are linked with this flux equals

$$\int_{\alpha_s}^\pi Z_{sa}(\alpha_s)rd\alpha_s = \frac{N_s}{2r} \int_{\alpha_s}^\pi \sin(\alpha_s)rd\alpha_s = \frac{N_s}{2}(1 + \cos(\alpha_s)) \tag{C.20}$$

To find the flux linked with the stator winding, we examine for each part of the air-gap flux how much of it is linked with the stator turns:

$$\psi_{sa} = \int_0^{2\pi} \frac{N_s}{2}(1 + \cos(\alpha_s))B^s(\alpha_s)lrd\alpha_s \tag{C.21}$$
$$= \frac{N_s}{2}lr \int_0^{2\pi} B^s(\alpha_s)d\alpha_s + \frac{N_s}{2}lr \int_0^{2\pi} \cos(\alpha_s)B^s(\alpha_s)d\alpha_s$$

Substitution of Equation C.18 makes the first term of Equation C.21 equal to zero and Equation C.21 reduces to

$$\psi_{sa} = \frac{N_s}{2} lr \sum_{n=1}^{\infty} \int_0^{2\pi} \cos(\alpha_s)\hat{B}_n \cos(n\alpha_s - \varphi_n) d\alpha_s \tag{C.22}$$

$$= \frac{N_s}{4} lr \sum_{n=1}^{\infty} \int_0^{2\pi} \hat{B}_n[\cos((n+1)\alpha_s - \varphi_n) + \cos((n-1)\alpha_s - \varphi_n)] d\alpha_s$$

For $n \neq 1$ this expression is zero, so the only meaningful solution is for $n = 1$

$$\psi_{sa} = \frac{N_s}{4} lr \int_0^{2\pi} \hat{B}_1 \cos(-\varphi_1) d\alpha_s = \frac{\pi}{2} \cdot \hat{B}_1 N_s lr \cos(\varphi_1) \tag{C.23}$$

This is an interesting result because it shows us that for a sinusoidally distributed stator winding, the flux linked with this winding is determined by the first harmonic of the flux density distribution $B^s(\alpha_s)$. The stator winding filters, so to speak, the first harmonic out of the flux distribution of the rotor field, which is rectangular shaped in our case. For the first harmonic of the rectangular-shaped rotor flux, induced by the rotor current I_f in the rotor winding, we can write, by applying Fourier analysis,

$$\hat{B}_1 = \frac{4}{\pi} \cdot \frac{\mu_0 N_f I_f}{2\delta} \quad \text{when } i_{sa} = 0 \tag{C.24}$$

The maximum of the amplitude of the first harmonic is at $\alpha_s = \theta$ ($\alpha_r = 0$) and that means that $\varphi_1 = \theta$ (see Figure C.3).

After substitution of Equation C.24 in Equation C.23, we can write for the total flux linked with stator winding a:

$$\psi_{sa} = \frac{\mu_0 N_f I_f}{\delta} N_s lr \cos(\theta) \quad \text{when } i_{sa} = 0 \tag{C.25}$$

We have now found the mutual inductance between the stator and the rotor. The maximum mutual inductance is

$$\hat{M}_{sf} = N_s N_f \mu_0 \frac{lr}{\delta} \tag{C.26}$$

For now we are interested in the steady-state conditions. The rotor field rotates with a constant angular velocity. The stator winding has a self-inductance L_s. The inductive coupling between the rotor field and the stator winding is expressed by the mutual inductance. This mutual inductance varies with the rotor position:

$$M_{saf} = \hat{M}_{sf} \cos(\theta) = N_s N_f \mu_0 \frac{lr}{\delta} \cos(\theta) \tag{C.27}$$

M_{saf} the mutual inductance between the rotor and stator winding [H] where the subscript "*a*" refers to phase "*a*"

$\hat{M}_{sf}$ the maximum value of the mutual inductance [H]

N_s the number of turns of the stator winding

N_f the number of turns of the rotor winding

l the rotor length [m]

r the rotor radius [m]

δ the air-gap width [m]

The rotor flux linked with the stator winding equals

$$\psi_{sa} = M_{saf}I_f = \hat{M}_{sf}\cos(\theta)I_f \tag{C.28}$$

The mutual inductance M_{saf} varies sinusoidally with the rotor position angle θ. At constant angular rotor speed θ increases linearly with time, and the flux ψ_{sa} varies sinusoidally with time. This results in a sine wave-shaped stator voltage and that is what we want.

So far we have assumed that the machine supplies no current and $i_{sa} = 0$. When the machine supplies a load, the voltage at the stator terminals will differ from the no-load voltage. The general voltage equation for the stator winding is

$$v_{sa} = R_s i_{sa} + \frac{d\psi_{sa}}{dt} \tag{C.29}$$

v_{sa} the stator winding voltage [V]

R_s the resistance of the stator winding [Ω]

i_{sa} the stator winding current [A]

and for the rotor winding

$$v_f = R_f I_f + \frac{d\psi_f}{dt} \tag{C.30}$$

v_f the rotor winding voltage [V]

R_f the resistance of the rotor winding [Ω]

I_f the DC rotor winding current [A]

The flux linked with the stator winding is

$$\psi_{sa} = L_{sa}i_{sa} + M_{saf}I_f \tag{C.31}$$

L_{sa} the self-inductance of the stator winding [H]

The flux linked with the rotor winding is

$$\psi_f = M_{saf}i_{sa} + L_f I_f \tag{C.32}$$

L_f the self-inductance of the rotor winding [H]

We can see from Figure C.8 that the shape of the magnetic circuit is independent of the rotor position, and therefore L_{sa} and L_f are independent of the rotor angle θ (in contrast with the mutual inductance between the rotor and stator winding M_{saf}).

The last four equations (C.29–C.32) can be used to draw an equivalent circuit for the single-phase synchronous machine, as shown in Figure C.9 (see also Appendix B.3).

The rotor position angle can be written as

$$\theta = \omega_m t + \theta_0 \tag{C.33}$$

θ_0 the rotor position angle at $t = 0$ s [rad]

When we substitute Equation C.31 in Equation C.29 and assume a constant rotor speed, the voltage equation for the stator winding becomes

$$v_{sa} = R_s i_{sa} + L_{sa}\frac{di_{sa}}{dt} + \hat{M}_{sf}I_f\frac{d\cos(\omega_m t + \theta_0)}{dt} \tag{C.34}$$

$$= R_s i_{sa} + L_{sa}\frac{di_{sa}}{dt} + \omega_m \hat{M}_{sf}I_f \cos\left(\omega_m t + \theta_0 + \frac{\pi}{2}\right)$$

This equation can be simplified by choosing $\theta_0 = -\pi/2$ rad. Furthermore, we can use the stator radial frequency ω_s instead of the mechanical rotor angle speed ω_m. The voltage equation for the stator winding can then be written as (for a machine with two poles ($p = 1$))

$$v_{sa} = R_s i_{sa} + L_{sa}\frac{di_{sa}}{dt} + \omega_s \hat{M}_{sf}I_f \cos(\omega_s t) \tag{C.35}$$

The lumped-element equivalent circuit for the single-phase synchronous machine is depicted in Figure C.10.

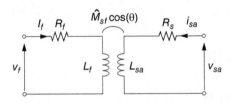

Figure C.9 The single-phase synchronous machine represented as two magnetically coupled coils.

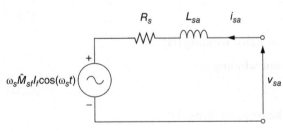

Figure C.10 The lumped-element equivalent circuit for the single-phase synchronous machine.

C.4   The Three-Phase Synchronous Machine

In Section 1.3.3 we discussed the advantages of three-phase systems over single-phase systems. In three-phase systems the electricity is supplied by three-phase synchronous machines. A three-phase synchronous machine has three sinusoidally distributed stator windings that are displaced by 120 electrical degrees along the stator surface; the stator windings are represented schematically as lumped inductances in Figure C.11. The three phases are indicated by the subscripts *a*, *b*, and *c*.

When the synchronous machine supplies a three-phase symmetrical load (generator) or when the machine is supplied by a three-phase symmetrical source (motor), the three sinusoidal phase currents are displaced by 120 electrical degrees in time:

$$i_{sa} = \hat{i}_s \cos(\omega_s t - \varphi)$$

$$i_{sb} = \hat{i}_s \cos\left(\omega_s t - \varphi - \frac{2\pi}{3}\right) \quad\quad\quad (C.36)$$

$$i_{sc} = \hat{i}_s \cos\left(\omega_s t - \varphi - \frac{4\pi}{3}\right)$$

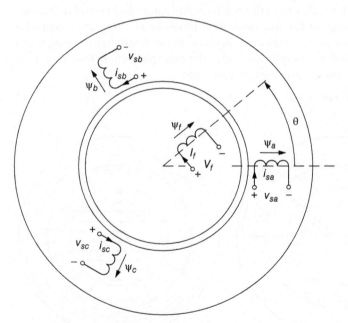

Figure C.11 Cross section of a three-phase synchronous machine.

The current amplitude $\hat{\imath}_s$ and the phase angle φ depend on the actual operating conditions of the machine. To be able to describe the machine's electrical behavior, we need to find an expression for the flux linked with the stator windings. First we consider the air-gap flux density as a result of the stator currents. After that we determine the flux linked with each stator winding and use them to derive the voltage equations.

To determine the magnetic flux density in the air gap, we make use of Figure C.12. We apply Ampère's law on path C in Figure C.12:

$$\oint_C \boldsymbol{H} \cdot \boldsymbol{\tau} ds = \iint_S \boldsymbol{J} \cdot \boldsymbol{n} dA \tag{C.37}$$

By using the expression for the stator winding distribution (Equation C.17), we can write

$$\oint_C \boldsymbol{H} \cdot \boldsymbol{\tau} ds = i_{sa} \int_{\alpha_s}^{\alpha_s - \pi} Z_{sa}(\alpha_s) r d\alpha_s \tag{C.38}$$

$$= i_{sa} \int_{\alpha_s}^{\alpha_s - \pi} \left(\frac{N_s}{2r} \sin(\alpha_s) \right) r d\alpha_s = i_{sa} N_s \cos(\alpha_s)$$

When we assume the permeability of the iron in both the rotor and the stator to be infinite ($\mu_r = \infty$), the magnetic field intensity H in the iron is zero. The size of the air gap is much smaller than the radius of the rotor ($\delta \ll r$), and that implicates that the change in the magnetic field intensity H along the integration path C in the air gap can be neglected. Because of the symmetry the magnitude of the magnetic field intensity H is equal at the two spots where the integration contour C crosses the air gap, and we can write

$$2H\delta = i_{sa} N_s \cos(\alpha_s) \tag{C.39}$$

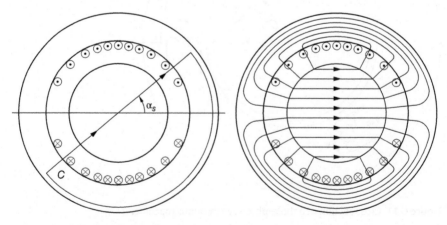

Figure C.12 Cross section of a single-phase machine to calculate the flux density due to the stator current.

And for the magnetic flux density along the integration path C in the air gap, we find the expression

$$B_{sa}^s(\alpha_s) = \mu_0 H = \frac{\mu_0 N_s}{2\delta} i_{sa} \cos(\alpha_s) \tag{C.40}$$

This equation gives us the contribution of the stator winding of phase a to the air-gap flux density. When we take the geometrical displacement of the stator windings b and c into account, the contribution of each stator winding to the air-gap flux is

$$
\begin{aligned}
B_{sa}^s(\alpha_s) &= \frac{\mu_0 N_s}{2\delta} i_{sa} \cos(\alpha_s) \\
B_{sb}^s(\alpha_s) &= \frac{\mu_0 N_s}{2\delta} i_{sb} \cos\left(\alpha_s - \frac{2}{3}\pi\right) \\
B_{sc}^s(\alpha_s) &= \frac{\mu_0 N_s}{2\delta} i_{sc} \cos\left(\alpha_s - \frac{4}{3}\pi\right)
\end{aligned}
\tag{C.41}
$$

When we substitute the stator currents according to Equation C.36, we get

$$
\begin{aligned}
B_s^s(\alpha_s) &= \frac{\mu_0 N_s}{2\delta} \cdot \frac{\hat{i}_s}{2} \cdot [\cos(\omega_s t - \varphi + \alpha_s) + \cos(\omega_s t - \varphi - \alpha_s) \\
&\quad + \cos\left\{ \left(\omega_s t - \varphi - \frac{2}{3}\pi\right) + \left(\alpha_s - \frac{2}{3}\pi\right) \right\} \\
&\quad + \cos\left\{ \left(\omega_s t - \varphi - \frac{2}{3}\pi\right) - \left(\alpha_s - \frac{2}{3}\pi\right) \right\} \\
&\quad + \cos\left\{ \left(\omega_s t - \varphi - \frac{4}{3}\pi\right) + \left(\alpha_s - \frac{4}{3}\pi\right) \right\} \\
&\quad + \cos\left\{ \left(\omega_s t - \varphi - \frac{4}{3}\pi\right) - \left(\alpha_s - \frac{4}{3}\pi\right) \right\} \\
&= \frac{\mu_0 N_s}{2\delta} \cdot \hat{i}_s \cdot \frac{3}{2} \cdot \cos(\omega_s t - \varphi - \alpha_s)
\end{aligned}
\tag{C.42}
$$

We discovered earlier that we only have to take the first harmonic of the flux density distribution into account to find the flux that is linked with the stator windings (Equation C.18):

$$B_1^s(\alpha_s) = \widehat{B}_1 \cos(\alpha_s - \varphi_1) \tag{C.43}$$

And we found out that the flux linked with stator winding a is then given by Equation C.23:

$$\psi_{sma} = \frac{\pi}{2} \cdot \widehat{B}_1 N_s lr \cos(\varphi_1) \tag{C.44}$$

The subscript m indicates the main field or air-gap field. This is not the total flux linked with the stator winding as the leakage flux has not yet been taken into account. When we compare this equation with Equation C.42, we can see what to fill in for $\hat{B}_1$ and φ_1. The main flux linked with stator winding a, due to the stator currents, is

$$\psi_{sma,s} = \frac{\pi}{2} \cdot N_s lr \cdot \frac{\mu_0 N_s}{2\delta} \cdot \hat{i}_s \cdot \frac{3}{2} \cdot \cos(\omega_s t - \varphi) \tag{C.45}$$

In a three-phase configuration, ψ_{smb} lags ψ_{sma} with $2\pi/3$, and ψ_{smc} lags ψ_{smb} with $2\pi/3$. It suffices to consider only one phase: we continue with phase a.

We can define the inductance coefficient L_{sm} as

$$L_{sm} = \frac{\pi}{2} \cdot N_s lr \cdot \frac{\mu_0 N_s}{2\delta} \cdot \frac{3}{2} \tag{C.46}$$

L_{sm} gives the relation between the three-phase currents (the total rotating stator field) and the flux linked with one stator winding; this means that L_{sm} is not a coefficient of self-inductance. With Equation C.46 we can write Equation C.45 as

$$\psi_{sma,s} = L_{sm} \hat{i}_s \cos(\omega_s t - \varphi) \tag{C.47}$$

We now add the contribution of the rotor current to this flux (Equation C.28) and use $\theta = \omega_s t + \theta_0$ as parameter for the rotor angle position:

$$\psi_{sma} = \psi_{sma,s} + \psi_{sma,f} = L_{sm} \hat{i}_s \cos(\omega_s t - \varphi) + \hat{M}_{sf} I_f \cos(\omega_s t + \theta_0) \tag{C.48}$$

For the voltage induced in stator winding a, by the main air-gap field, we can write

$$v_{sma} = \frac{d\psi_{sma}}{dt} \tag{C.49}$$

Because the electrical quantities represent a symmetrical three-phase system of sinusoidal quantities, with angular frequency ω_s, it is sufficient to examine only one phase (phase a), and we can fruitfully use the phasor notation for the voltages and currents in the steady state (see also Section 1.4). For the current in stator winding a, for instance, we can write

$$i_{sa} = \hat{i}_s \cos(\omega_s t - \varphi) = \hat{i}_s \mathrm{Re}(e^{j(\omega_s t - \varphi)}) = \sqrt{2}\mathrm{Re}(I_s e^{j(\omega_s t)}) \tag{C.50}$$

I_s the stator current phasor $I_s = (\hat{i}_s/(\sqrt{2}))e^{-j\varphi} = |I_s| \angle -\varphi$

The stator flux due to the rotor excitation current can be written as

$$\psi_{sma,f} = \hat{M}_{sf} I_f \cos(\omega_s t + \theta_0) = \sqrt{2}\mathrm{Re}(\Psi_{sf} e^{j(\omega_s t)}) \tag{C.51}$$

Ψ_{sf} the stator flux phasor $\Psi_{sf} = (\hat{M}_{sf} I_f/(\sqrt{2}))e^{j\theta_0} = |\Psi_{sf}| \angle \theta_0$

The phasors that we introduced in Equations C.50 and C.51 are now used to write Equations C.48 and C.49 as

$$\begin{aligned} V_{sm} &= j\omega_s \Psi_{sm} \\ \Psi_{sm} &= L_{sm} I_s + \Psi_{sf} \end{aligned} \tag{C.52}$$

Figure C.13 Single-phase model of a three-phase synchronous machine.

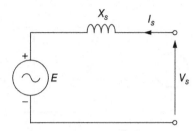

The voltage induced by the stator flux, due to the rotor excitation current, is referred to as the internal EMF of the machine:

$$E = j\omega_s \Psi_{sf} \qquad (C.53)$$

Therefore, Equation C.52 can be written as

$$V_{sm} = j\omega_s L_{sm} I_s + E \qquad (C.54)$$

So far we did not take the leakage flux into account in the total flux linked with the stator winding. This contribution can be substituted in Equation C.54 by adding an extra inductance term $L_s = L_{sm} + L_{s\sigma}$:

$$V_s = j\omega_s L_s I + E = jX_s I + E \qquad (C.55)$$

The single-phase model of a three-phase synchronous machine, in which the resistance of the stator winding is neglected, is shown in Figure C.13.

C.5 Synchronous Generator in the Power System

The generators in the power system are connected to the loads and to each other through the grid. Each individual generator experiences the power system to be an ideal voltage source (i.e., a constant amplitude and a constant frequency). We say that the generator is connected to an infinite bus. The equivalent circuit of a synchronous generator connected to the power system with a constant system voltage V is shown in Figure C.14. We choose for the indication of the positive direction of the current for the generator convention, so that positive current is injected into the infinite bus. Therefore Equation C.55 alters and, when we leave out the subscript s, can be written as

$$E = j\omega L I + V = jXI + V \qquad (C.56)$$

I the current phasor $I = |I| \angle -\varphi$ A
E the machine internal EMF phasor $E = |E| \angle \delta$ V
V the system voltage phasor $V = |V| \angle 0$ V
X the machine reactance $[\Omega]$

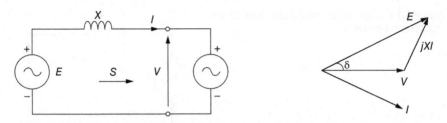

Figure C.14 The equivalent circuit of a synchronous generator connected to an infinite bus with the corresponding phasor diagram; the resistance of the stator winding is neglected.

The expression for the current is

$$I = \frac{E - V}{jX} \tag{C.57}$$

The three-phase complex power supplied to the power system equals

$$S = 3VI^* = 3V\left(\frac{E - V}{jX}\right)^* = \frac{3VE^* - 3VV^*}{-jX} = 3j\frac{|V||E|\angle(-\delta)}{X} - 3j\frac{|V|^2}{X}$$

$$= 3j\frac{|V||E|}{X}(\cos(\delta) - j\sin(\delta)) - 3j\frac{|V|^2}{X} = P + jQ \tag{C.58}$$

The real part of the complex power S is the active power:

$$P = \text{Re}(S) = 3\frac{|V||E|}{X}\sin(\delta) \tag{C.59}$$

The imaginary part of the complex power S is the reactive power:

$$Q = \text{Im}(S) = 3\frac{|V||E|}{X}\cos(\delta) - 3\frac{|V|^2}{X} \tag{C.60}$$

When the excitation current I_f is kept constant, the synchronous internal EMF E remains constant. When the power angle equals $\delta = 0$, we have a no-load situation, and the supplied/consumed active power equals zero. There is no exchange of active power between the synchronous machine and the system. When we increase the mechanical power supplied to the machine, a short-time transient occurs before a new steady state is established. Initially the rotor of the generator slightly accelerates. As the generator is forced to rotate with the same frequency as the other generators in the system, the rotor decelerates to this frequency. The result is a new steady state with a slightly increased (and positive) δ-value. When δ is positive, the internal EMF leads the system voltage, and the active power supplied by the machine to the system is positive; the machine acts as a generator.

When we start again in the no-load situation, that is, the power angle $\delta = 0$, and we slow down the rotor by increasing the mechanical load connected to the shaft, the internal EMF will lag the system voltage. In this case the power

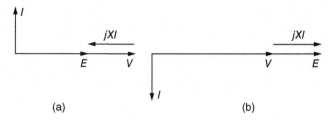

(a)   (b)

Figure C.15 The phasor diagram of an underexcited (a) and an overexcited (b) synchronous machine.

angle δ is negative; the machine consumes active power from the system and acts as a motor.

In the no-load situation of the synchronous machine ($\delta = 0$), the exchange of active power is zero, but the machine can exchange reactive power. From Equation C.60 we see that the machine consumes reactive power ($Q < 0$) when $|E| < |V|$; we call the machine to be underexcited (see Figure C.15 (a)). When the machine is overexcited $|E| > |V|$, $Q > 0$, the machine supplies reactive power (see Figure C.15 (b)).

The exchange of active power between a synchronous machine and the power system is controlled by the torque on the axis. The exchange of reactive power is managed by the excitation.

D

Induction Machine Model

D.1 Introduction

In Appendix C the operation of the synchronous machine is described, and the mathematical description of the voltage–current relation of the synchronous generator under different operating conditions is derived. The rotor of the synchronous machine rotates with the same angular speed (synchronous) as the rotating field in the air gap between the rotor and the stator. This means that there is a direct relation between the frequency of the three-phase voltage supply, to which the machine is connected, and the angular speed of the rotor. When the machine is connected to the grid, it can develop a mean nonzero torque at one speed only. This indicates that when the rotor is at standstill, no mean electromagnetic torque is active, and therefore the machine cannot speed up. For these reasons the synchronous machine is mainly used as a generator and seldom directly as a motor. To be able to use the synchronous machine conveniently as a motor, we are generally obliged to apply a power-electronic converter to create a three-phase source with variable frequency.

The induction machine is an alternating current machine that is very well suited to be used as a motor when it is directly supplied from the grid. The stator of the induction machine has a three-phase winding; the rotor is equipped with a short-circuited rotor winding. When the rotor speed is different from the speed of the rotating magnetic field generated by the stator windings, we call the rotor speed asynchronous (in German-based literature, induction motors are referred to as asynchronous motors). In that case the short-circuited rotor windings are exposed to a varying magnetic field that induces an EMF and currents in the short-circuited rotor windings. The induced rotor currents and the rotating stator field result in an electromagnetic torque that attempts to pull the rotor in the direction of the rotating stator field. For the construction of induction machines, it is common practice to manufacture the short-circuited rotor winding as a robust molded cage (the so-called squirrel cage).

In the next section we analyze the basic principles of the induction machine. The influence of the induced rotor currents on the rotating stator field is

Electrical Power System Essentials, Second Edition. Pieter Schavemaker and Lou van der Sluis.
© 2017 John Wiley & Sons Ltd. Published 2017 by John Wiley & Sons Ltd.

neglected in favor of illustrating the basic principles but is taken into account in Section D.3 when the air-gap field is treated.

D.2 The Basic Principle of the Induction Machine

In Figure D.1, both the cross section of a simple induction machine with a single rotor winding and the field lines of the magnetic flux at time $t = 0$ are shown.

The rotor winding is assumed to be concentrated at one spot rd (rotor direct axis). The stator has three sinusoidally distributed stator windings, represented schematically as lumped inductances. To define a tangential position in the air gap, we make use of two coordinate systems: either the coordinate system of the stator with the angle α_s or the rotor coordinate system with the coordinate α_r. The axis of stator winding a is used as reference for the stator coordinates, and the axis of the rotor winding rd is the reference axis for the rotor coordinates. The rotor position is defined with the same reference as the stator coordinates by means of the angle θ, so that $\alpha_s = \alpha_r + \theta$.

As said, the stator windings are laid out in a sine wave shape so that if the stator windings are connected to a symmetrical three-phase power source, which supplies sine wave-shaped currents, a rotating magnetic field appears in the air gap between stator and rotor, as will be proven. We can write for the currents in the three stator windings

$$
\begin{aligned}
i_{sa} &= \hat{i}_s \cos(\omega_s t) \\
i_{sb} &= \hat{i}_s \cos\left(\omega_s t - \frac{2\pi}{3}\right) \\
i_{sc} &= \hat{i}_s \cos\left(\omega_s t - \frac{4\pi}{3}\right)
\end{aligned}
\tag{D.1}
$$

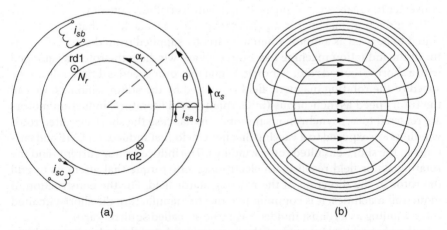

(a)   (b)

Figure D.1 Cross section of a simple induction machine with a single concentrated rotor winding (a) and the field lines of the magnetic flux at time $t = 0$ (b).

If the geometrical position of the stator windings is taken into account, the contribution of each stator winding to the air-gap flux density can be written as

$$B_{sa}^s(\alpha_s) = \frac{\mu_0 N_s}{2\delta} \cdot i_{sa} \cos(\alpha_s)$$

$$B_{sb}^s(\alpha_s) = \frac{\mu_0 N_s}{2\delta} \cdot i_{sb} \cos\left(\alpha_s - \frac{2\pi}{3}\right) \tag{D.2}$$

$$B_{sc}^s(\alpha_s) = \frac{\mu_0 N_s}{2\delta} \cdot i_{sc} \cos\left(\alpha_s - \frac{4\pi}{3}\right)$$

N_s the number of turns of the stator windings
δ the size of the air gap between stator and rotor [m]

After substitution of the stator currents (Equation D.1), this results in

$$B_{sa}^s(\alpha_s) = \frac{\mu_0 N_s}{2\delta} \cdot \hat{i}_s \cdot \frac{1}{2}\{\cos(\omega_s t + \alpha_s) + \cos(\omega_s t - \alpha_s)\}$$

$$B_{sb}^s(\alpha_s) = \frac{\mu_0 N_s}{2\delta} \cdot \hat{i}_s \cdot \frac{1}{2}\left\{\cos\left(\omega_s t + \alpha_s - \frac{4\pi}{3}\right) + \cos(\omega_s t - \alpha_s)\right\} \tag{D.3}$$

$$B_{sc}^s(\alpha_s) = \frac{\mu_0 N_s}{2\delta} \cdot \hat{i}_s \cdot \frac{1}{2}\left\{\cos\left(\omega_s t + \alpha_s - \frac{2\pi}{3}\right) + \cos(\omega_s t - \alpha_s)\right\}$$

The resulting magnetic flux density equals

$$B_s^s(\alpha_s) = B_{sa}^s(\alpha_s) + B_{sb}^s(\alpha_s) + B_{sc}^s(\alpha_s) = \frac{\mu_0 N_s}{2\delta} \cdot \hat{i}_s \cdot \frac{3}{2} \cdot \cos(\omega_s t - \alpha_s)$$

$$= \hat{B}_{s1} \cos(\omega_s t - \alpha_s) \tag{D.4}$$

The superscript s in these formulas indicates that the stator coordinates are used. In Figure D.1 (b) the flux distribution in the induction machine at $t = 0$ is shown (for iron in stator and rotor with $\mu_r = \infty$). If ω_r is positive, the magnetic field rotates counterclockwise at angular speed ω_s.

Because we analyze the behavior of the machine mainly from the rotor position, it is useful to have an expression for the flux density (Equation D.4) in the rotor coordinate system. From Figure D.1 (a) we see that we have to replace angle α_s by an angle α_r such that $\alpha_s = \alpha_r + \theta$:

$$B_s^r(\alpha_r) = \hat{B}_{s1} \cos(\omega_s t - \theta - \alpha_r) \tag{D.5}$$

The superscript r indicates that the rotor coordinates are used.

D.2.1   A Single Rotor Winding

Before we calculate the rotor current, we have to determine the flux linkage between the rotor winding and the stator field. Therefore we choose a surface S with the shape of a half cylinder in the air gap in between the rotor and the stator (we used this approach in Appendix C as well; see Figure C.4). For the flux through a single turn of the rotor winding, we can write (along with Equation D.5)

$$\Phi_{rd,s} = \iint_S \boldsymbol{B} \cdot \boldsymbol{n} dA = \int_{-\pi/2}^{\pi/2} B_s^r(\alpha_r) lr d\alpha_r = \hat{B}_{s1} lr \int_{-\pi/2}^{\pi/2} \cos(\omega_s t - \theta - \alpha_r) d\alpha_r$$

$$= 2\hat{B}_{s1} lr \cos(\omega_s t - \theta) \tag{D.6}$$

l the actual length of the machine [m]
r the radius of the rotor [m]

The flux linkage equals

$$\Psi_{rd,s} = N_r \Phi_{rd,s} = N_r 2 \hat{B}_{s1} lr \cos(\omega_s t - \theta)$$

$$= \hat{\Psi}_{rd,s} \cos(\omega_s t - \theta) \tag{D.7}$$

N_r the number of turns of the rotor winding
$\hat{\Psi}_{rd,s}$ the maximum value of the flux linked with the rotor winding due to the stator currents

For this moment we assume the rotor to be at standstill, so θ is time independent. The short-circuited rotor winding is provisionally supposed to have a rather high ohmic resistance R_{rd}. As a result, the current flowing through the rotor winding is so small that the contribution of this current to the magnetic field in the air gap can be neglected. This is, of course, not completely realistic, but for a first analysis it is very useful. Later on in this appendix we take into account the contribution of the rotor current to the flux.

Using this simplification, we can write for the flux linked with the rotor winding

$$\Psi_{rd} = \Psi_{rd,s} = \hat{\Psi}_{rd,s} \cos(\omega_s t - \theta) \quad (\theta \text{ is constant}) \tag{D.8}$$

The voltage equation for the rotor winding equals

$$0 = R_{rd} i_{rd} + \frac{d\Psi_{rd}}{dt} = R_{rd} i_{rd} - \hat{\Psi}_{rd,s} \omega_s \sin(\omega_s t - \theta) \quad (\theta \text{ is constant}) \tag{D.9}$$

Therefore, the current through the rotor winding equals

$$i_{rd} = \frac{\hat{\Psi}_{rd,s} \omega_s}{R_{rd}} \sin(\omega_s t - \theta) \quad (\theta \text{ is constant}) \tag{D.10}$$

For the calculation of the electromagnetic torque, we make use of the fact that the rotor winding is positioned in the air gap, so we can calculate the mechanical forces on the conductors with $F = Bil$. We pass over the fact that the rotor winding is embedded in a slot in the rotor iron. The electromagnetic torque is the sum of the force on the rotor winding part rd1 and the force on the rotor winding part rd2 (see Figure D.1 (a)):

$$T_{e,rd} = M_{rd1} + M_{rd2} = N_r rli_{rd} B_s^r \left(\frac{\pi}{2}\right) - N_r rli_{rd} B_s^r \left(-\frac{\pi}{2}\right) \tag{D.11}$$

Because of symmetry, $B_s^r(-\pi/2) = -B_s^r(\pi/2)$ (see also Equation D.5), so that Equation D.11 can be written as

$$T_{e,rd} = 2N_r r l i_{rd} B_s^r \left(\frac{\pi}{2}\right)$$

(D.12)

Substituting the expression for $B_s^r(\alpha_r)$ (Equation D.5) and the expression for i_{rd} (Equation D.10) and making use of the expression for $\psi_{rd,s}$ (Equation D.7) gives us

$$T_{e,rd} = 2N_r r l \frac{\widehat{\Psi}_{rd,s} \omega_s}{R_{rd}} \sin(\omega_s t - \theta) \widehat{B}_{s1} \cos\left(\omega_s t - \theta - \frac{\pi}{2}\right)$$

$$= \frac{\widehat{\Psi}_{rd,s}^2 \omega_s}{R_{rd}} \left[\frac{1}{2} - \frac{1}{2}\cos(2\omega_s t - 2\theta)\right] \quad (\theta \text{ is constant})$$

(D.13)

The derived equations are, as a function of time, depicted in Figure D.2. As we can see in Figure D.2, the torque on the rotor has a positive average value: the torque tries to speed up the rotor in the same direction as the rotating stator field. The torque, however, also contains an alternating component with an angular frequency of $2\omega_s$. This alternating component can be avoided by applying more windings, such as is the case for a squirrel-cage rotor. To understand this we now calculate the electromagnetic torque for an induction machine with two rotor windings.

Figure D.2 The derived equations for the primitive induction machine with a single concentrated rotor winding as a function of time.

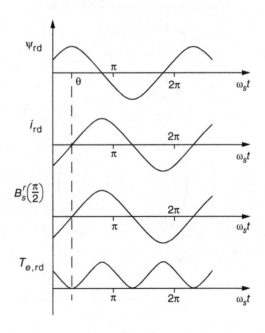

D.2.2 Two Rotor Windings

An additional winding will contribute to the electromagnetic torque in a similar way as the single winding. Because we want to smooth the ripple of the torque, we want the alternating component of the second winding to be out of phase with the alternating component of the first winding. This means that the argument of the cosine function in Equation D.13 should be $2\omega_s t - 2\theta - \pi$ instead of $2\omega_s t - 2\theta$. This can be achieved by positioning the second winding perpendicular (index: rq – rotor quadrature axis) to the first winding (index: rd – rotor direct axis). The axis of the second winding is therefore at $\alpha_r = \pi/2$. The two rotor windings are depicted in the cross section of the primitive induction machine as shown in Figure D.3. If we denote the contribution of the second winding to the electromagnetic torque as $T_{e,rq}$, we find with Equation D.13 for the resulting torque

$$T_e = T_{e,rd} + T_{e,rq} = \frac{\hat{\psi}_{rd,s}^2 \omega_s}{R_{rd}} \quad (\theta \text{ is constant}) \tag{D.14}$$

This is a constant torque, with a positive value: the torque tries to accelerate the rotor.

D.2.3 Rotating Rotor

Until now we assumed that the rotor remained at standstill (θ is constant). We now examine what occurs if the rotor rotates with a constant angular frequency ω_m. The angle position can be expressed as

$$\theta = \omega_m t + \theta_0 \tag{D.15}$$

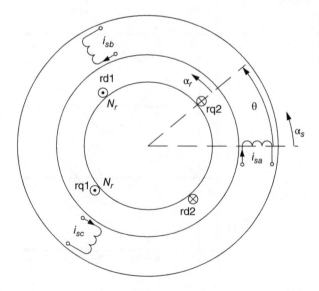

Figure D.3 The cross section of the primitive induction machine with two rotor windings.

Now, the stator field expressed in the rotor coordinate system (Equation D.5) can be written as

$$B_s^r(\alpha_r) = \hat{B}_{s1} \cos(\omega_s t - \omega_m t - \theta_0 - \alpha_r) \tag{D.16}$$

The stator field rotates with an angular velocity $\omega_s - \omega_m$ with respect to the rotor. This "lagging" of the rotor with respect to the rotating stator field is called slip; the angular velocity $\omega_s - \omega_m$ is addressed as the slip angular velocity:

$$\omega_{slip} = \omega_s - \omega_m \tag{D.17}$$

The stator field will induce currents in the rotor with an angular frequency $\omega_{slip} = \omega_s - \omega_m$.

The equation for the flux linked with the rotor winding rd (Equation D.7) can still be applied. With $\theta = \omega_m t + \theta_0$ and $\omega_{slip} = \omega_s - \omega_m$, this equation becomes

$$\begin{aligned} \psi_{rd,s} &= \hat{\psi}_{rd,s} \cos(\omega_s t - \theta) = \hat{\psi}_{rd,s} \cos(\omega_s t - \omega_m t - \theta_0) \\ &= \hat{\psi}_{rd,s} \cos(\omega_{slip} t - \theta_0) \end{aligned} \tag{D.18}$$

When we compare this equation with Equation D.7, we see that we can still apply the equations for the situation that the rotor was at standstill if we realize that the rotor now experiences a rotating field with an angular velocity ω_{slip} instead of ω_s! Furthermore, $\omega_s t - \theta$ should be replaced by $\omega_{slip} t - \theta_0$. The expression of the electromagnetic torque, Equation D.14 with ω_{slip} instead of ω_s and $\omega_{slip} = \omega_s - \omega_m$, becomes

$$T_e = T_{e,rd} + T_{e,rq} = \frac{\hat{\psi}_{rd,s}^2}{R_{rd}} \omega_{slip} = \frac{\hat{\psi}_{rd,s}^2}{R_{rd}} (\omega_s - \omega_m) \tag{D.19}$$

The relation in Equation D.19 is depicted graphically in Figure D.4. It shows us that if $\omega_m < \omega_s (\omega_{slip} > 0)$, the electromagnetic torque is positive and the rotor accelerates in the direction of the rotation of the stator field. If $\omega_m > \omega_s (\omega_{slip} < 0)$, the rotational speed of the rotor is higher than that of the stator field and the electromagnetic torque is negative: the torque forces the rotor to slow down. When the rotor turns at synchronous speed $\omega_m = \omega_s (\omega_{slip} = 0)$, the electromagnetic torque equals zero. Note that in the analysis of the behavior of the induction machine so far, we have neglected the influence of the rotor currents on the air-gap field of the machine.

Figure D.4 The electromagnetic torque of the primitive induction machine as a function of the angular velocity.

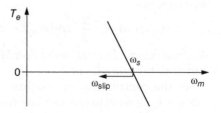

D.3 The Magnetic Field in the Air Gap

The model of the induction machine has to be improved further by taking the influence of the rotor currents on the air-gap field into account. For this exercise we look at the induction machine from the stator and search for a relation between the currents and voltages in the stator windings. Note that the rotor circuit is not electrically accessible in a simple way as the rotor windings are supposed to be short-circuited in the rotor itself. In this paragraph we confine ourselves to the relation between the stator currents and the related flux in the stator. The air-gap flux plays a dominant role. In the next paragraph the derived equations will be transformed into a circuit representation.

The induction machine is assumed to be connected to a symmetrical three-phase current source. This current source supplies the stator currents that contribute to the (rotating) stator field in the air gap. First, we derive the equations for rotor winding rd, and from these equations we can calculate the current in the rotor winding. This current also contributes to the air-gap field. Then, we derive the contribution of rotor winding rq to the air-gap field. Finally, we compute the flux linkage with the stator windings by using the determined expression for the total air-gap field.

D.3.1 Contribution of the Rotor Currents to the Air-Gap Field

Let us first examine the short-circuited winding rd (see Figure D.3). The flux linked with this winding consists of a contribution from the current in this winding (i_{rd}) and from the stator currents. The current i_{rq} does not contribute. Because we apply the same symmetrical three-phase current source as in the previous paragraph, we can make use of Equation D.18 for the contribution of the stator currents to the flux linked with the rotor winding rd. Substitution of the expressions for $\hat{\psi}_{rd,s}$ (Equation D.7) and $\hat{B}_{s1}$ (Equation D.4) gives us

$$\psi_{rd,s} = N_r 2lr \hat{B}_{s1} \cos(\omega_{slip}t - \theta_0) = N_r 2lr \cdot \frac{\mu_0 N_s}{2\delta} \cdot \hat{i}_s \cdot \frac{3}{2} \cdot \cos(\omega_{slip}t - \theta_0)$$

(D.20)

If we define the coefficient of mutual induction M as

$$M = N_r 2lr \cdot \frac{\mu_0 N_s}{2\delta}$$

(D.21)

we can write

$$\psi_{rd,s} = M \cdot \hat{i}_s \cdot \frac{3}{2} \cdot \cos(\omega_{slip}t - \theta_0)$$

(D.22)

The coefficient of mutual induction M is similar to the expression that we found earlier in Equation C.26.

For the contribution of the flux linked with rotor winding rd due to the current i_{rd} flowing in this winding (the self-inductance), we use the knowledge

Figure D.5 The magnetic flux density distribution in the air gap due to the current i_{rd}. The dashed trace is the first harmonic of the flux density.

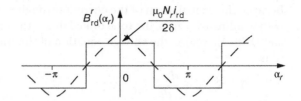

developed in Appendix C.2. In that section we have seen that a current through a single winding results in a rectangular-shaped flux density in the air gap (see Figure C.3). The current i_{rd} brings about the rectangular-shaped flux density in the air gap that is shown in Figure D.5. For the calculation of the flux linkage of i_{rd} with the stator windings, only the first harmonic (the dashed line in Figure D.5) is of importance (see also Appendix C.3). We can see from Figure D.5 that the first harmonic reaches its maximum when $\alpha_r = 0$ ($\alpha_s = 0$). The equation for the first harmonic is

$$B_{rd1}^r(\alpha_r) = \frac{4}{\pi} \cdot \frac{\mu_0 N_r i_{rd}}{2\delta} \cos(\alpha_r) \tag{D.23}$$

To find the contribution of this first harmonic to the flux linked with the rotor winding rd, we follow the same approach as with the derivation of Equations D.6 and D.7:

$$\Psi_{rdm,r} = N_r \Phi_{rdm,r} = N_r \iint_S \boldsymbol{B} \cdot \boldsymbol{n} dA = N_r \int_{-\pi/2}^{\pi/2} B_{rd1,r}^r(\alpha_r) lr d\alpha_r \tag{D.24}$$

l the actual length of the machine [m]
r the radius of the rotor [m]

We use the letter m (main) in the subscript to indicate that this flux contribution is made by the first harmonic of the air-gap field (i.e., the main field). Substitution of Equation D.23 results in

$$\Psi_{rdm,r} = N_r lr \int_{-\pi/2}^{\pi/2} \frac{4}{\pi} \cdot \frac{\mu_0 N_r i_{rd}}{2\delta} \cos(\alpha_r) d\alpha_r = 2N_r lr \cdot \frac{4}{\pi} \cdot \frac{\mu_0 N_r}{2\delta} \cdot i_{rd} \tag{D.25}$$

If we define the induction coefficient as

$$L_{rdm} = 2N_r lr \cdot \frac{4}{\pi} \cdot \frac{\mu_0 N_r}{2\delta} \tag{D.26}$$

Equation D.25 simplifies to

$$\Psi_{rdm,r} = L_{rdm} i_{rd} \tag{D.27}$$

The contribution of the higher harmonics of the flux density distribution (according to Figure D.5) to the flux linked with winding rd is proportional to the current i_{rd} and can be regarded as part of the leakage flux because there is no coupling with the stator windings:

$$\Psi_{rd\sigma} = L_{rd\sigma} i_{rd} \tag{D.28}$$

Because the contribution of the current through winding rq to the flux linkage with winding rd is zero, the total linked flux is the sum of Equations D.22, D.27, and D.28. Therefore, the voltage equation of the short-circuited rotor winding rd is

$$0 = R_{rd}i_{rd} + \frac{d\psi_{rd}}{dt} \tag{D.29}$$

R_{rd} the ohmic resistance of winding rd [Ω]

with

$$\psi_{rd} = \psi_{rd,s} + \psi_{rdm,r} + \psi_{rd\sigma} = M \cdot \hat{i}_s \cdot \frac{3}{2} \cdot \cos(\omega_{slip}t - \theta_0)$$
$$+ L_{rdm}i_{rd} + L_{rd\sigma}i_{rd} \tag{D.30}$$
$$= \psi_{rdm} + L_{rd\sigma}i_{rd}$$

ψ_{rdm} the contribution of the first harmonic of the flux density distribution in the air gap [Wb]

After substitution of Equation D.30 in Equation D.29, we see that the latter turns out to be a first-order linear differential equation. The particular solution of this differential equation is the expression for the current in the short-circuited rotor winding rd:

$$i_{rd} = \hat{i}_{rd} \cos(\omega_{slip}t - \theta_0 - \beta) \tag{D.31}$$

If we substitute this expression for the current in the equation for the first harmonic of the flux density distribution (Equation D.23), we find for the contribution of the current i_{rd} to the first harmonic

$$B_{rd1}^r(\alpha_r) = \frac{4}{\pi} \cdot \frac{\mu_0 N_r}{2\delta} \cdot \hat{i}_{rd} \cos(\omega_{slip}t - \theta_0 - \beta) \cos(\alpha_r)$$
$$= \hat{B}_{r1} \cos(\omega_{slip}t - \theta_0 - \beta) \cos(\alpha_r) \tag{D.32}$$

This is the equation for a standing wave. We can rewrite Equation D.32 as

$$B_{rd1}^r(\alpha_r) = \hat{B}_{r1} \cdot \frac{1}{2} \cdot [\cos(\omega_{slip}t - \theta_0 - \beta + \alpha_r) + \cos(\omega_{slip}t - \theta_0 - \beta - \alpha_r)] \tag{D.33}$$

This equation consists of two traveling waves: a wave related to the first cosine term traveling with an angular velocity $-\omega_{slip}$ with respect to the rotor and a wave related to the second cosine term traveling with an angular velocity ω_{slip} with respect to the rotor.

In a similar way one can derive the equation for the flux density distribution due to the current in the other short-circuited rotor winding (i_{rq}). The geometrical axis of winding rq is at $\alpha_r = \pi/2$ and the current i_{rq} lags i_{rd} by $\pi/2$. So we can write for the contribution of i_{rq}

$$B^r_{rq1}(\alpha_r) = \widehat{B}_{r1} \cos\left(\omega_{slip}t - \frac{\pi}{2} - \theta_0 - \beta\right) \cos\left(\alpha_r - \frac{\pi}{2}\right)$$

$$= \widehat{B}_{r1} \cdot \frac{1}{2} \cdot [\cos(\omega_{slip}t - \theta_0 - \beta + \alpha_r - \pi) \qquad (D.34)$$

$$+ \cos(\omega_{slip}t - \theta_0 - \beta - \alpha_r)]$$

To find the flux density distribution in the air gap, we have to add the contribution of both i_{rd} (Equation D.33) and i_{rq} (Equation D.34):

$$B^r_{r1}(\alpha_r) = B^r_{rd1}(\alpha_r) + B^r_{rq1}(\alpha_r) = \widehat{B}_{r1} \cos(\omega_{slip}t - \theta_0 - \beta - \alpha_r) \qquad (D.35)$$

We see from this equation that only the wave that travels with an angular velocity ω_{slip} with respect to the rotor remains.

To find the flux linked with the stator windings, we have to write the flux density distribution (Equation D.35) in stator coordinates. With $\alpha_r = \alpha_s - \omega_m t - \theta_0$ and $\omega_{slip} = \omega_s - \omega_m$, this is

$$B^s_{r1}(\alpha_s) = \widehat{B}_{r1} \cos(\omega_{slip}t - \beta + \omega_m t - \alpha_s) = \widehat{B}_{r1} \cos(\omega_s t - \beta - \alpha_s) \qquad (D.36)$$

Note that the position θ_0 of the rotor at $t = 0$ has disappeared from the equation. We conclude that the stator field induces currents in the rotor windings, which, in turn, create a rotor field that has the same velocity as the stator field.

D.3.2 The Flux Linkage with the Stator Windings

To derive the flux linkage between the air-gap field and the stator windings, only the contribution of the first harmonic of the magnetic flux density is of importance. We derived the contribution of the stator currents to this first harmonic of the magnetic flux density in Equation D.4 and the contribution of the rotor currents in Equation D.36:

$$B^s_1 = B^s_s + B^s_{r1} = \widehat{B}_{s1} \cos(\omega_s t - \alpha_s) + \widehat{B}_{r1} \cos(\omega_s t - \beta - \alpha_s) \qquad (D.37)$$

To find the flux linked with the stator winding of phase a, we can use Equation C.23. The angle φ_1 in this expression corresponds to an angle α_s where the B-distribution is at its maximum. The maximum of the stator rotating field is at $\alpha_s = \omega_s t$, whereas the maximum of the rotor rotating field is at $\alpha_s = \omega_s t - \beta$. The flux linked with the stator winding of phase a is

$$\psi_{sma} = N_s lr \cdot \frac{\pi}{2} \cdot [\widehat{B}_{s1} \cos(\omega_s t) + \widehat{B}_{r1} \cos(\omega_s t - \beta)] \qquad (D.38)$$

The subscript m indicates the main field, because the leakage flux of the stator winding is not taken into consideration. Because the main field in the air gap rotates with an angular velocity ω_s, the fluxes linked to the three stator windings form a symmetrical three-phase system: ψ_{smb} lags ψ_{sma} with $2\pi/3$ rad and ψ_{smc}

lags ψ_{smb} with $2\pi/3$ rad. By substituting the amplitude in Equation D.4 for $\hat{B}_{s1}$ and the amplitude in Equation D.32 for $\hat{B}_{r1}$, Equation D.38 can be written as

$$\psi_{sma} = N_s lr \cdot \frac{\pi}{2} \cdot \left[\frac{\mu_0 N_s}{2\delta} \cdot \hat{i}_s \cdot \frac{3}{2} \cdot \cos(\omega_s t) + \frac{4}{\pi} \cdot \frac{\mu_0 N_r}{2\delta} \cdot \hat{i}_{rd} \cos(\omega_s t - \beta) \right]$$

$$= L_{sm}\hat{i}_s \cos(\omega_s t) + M\hat{i}_{rd} \cos(\omega_s t - \beta) \tag{D.39}$$

D.4 A Simple Circuit Model for the Induction Machine

For the description of a simple circuit model that describes the system behavior of the induction machine, we assume the stator windings to be ideal: they have no ohmic resistance and generate a sine wave-shaped magnetic air-gap flux (stator leakage fluxes are left out of consideration). Despite these simplifications, the simple circuit model describes the system behavior of the induction machine quite adequately.

As we are dealing with time-dependent sinusoidal quantities, we use the phasor notation (see Section 1.4) in the following sections.

D.4.1 The Stator Voltage Equation

Because we neglect the ohmic resistance and the leakage fluxes, the voltage equation for stator winding a is

$$v_{sma} = \frac{d\psi_{sma}}{dt} \tag{D.40}$$

We must be careful when we apply the phasor notation because the rotor equations have another angular frequency (ω_{slip}) than the stator equations (ω_s). We can write for the current in phase a of the stator

$$i_{sa} = \hat{i}_s \cos(\omega_s t) = \hat{i}_s \text{Re}(e^{j\omega_s t}) = \sqrt{2}\text{Re}(I_s e^{j\omega_s t}) \tag{D.41}$$

Re   the operator that takes the real part of a complex quantity

I_s the current phasor $I_s = (\hat{i}_s/(\sqrt{2}))\angle 0$ (this phasor represents the symmetrical stator currents (see also Equation D.1); that is why the subscript a is left out)

The rotor current i_{rd} can be written in a similar way (Equation D.31) as

$$i_{rd} = \hat{i}_{rd} \cos(\omega_{slip} t - \theta_0 - \beta) = \hat{i}_{rd}\text{Re}(e^{j(\omega_{slip} t - \theta_0 - \beta)})$$

$$= \sqrt{2}\text{Re}(I_{rd}e^{j(\omega_{slip} t - \theta_0)}) \tag{D.42}$$

I_{rd} the rotor current phasor $I_{rd} = (\hat{i}_{rd}/(\sqrt{2}))e^{-j\beta} = |I_{rd}|\angle(-\beta)$

With the phasors according to Equations D.41 and D.42, we can express the flux linked with the stator windings (Equation D.39) and the voltage equation of the stator winding (Equation D.40) as

$$\Psi_{sm}e^{j\omega_s t} = (L_{sm}I_s + MI_{rd})e^{j\omega_s t} \tag{D.43}$$

$$V_{sm}e^{j\omega_s t} = j\omega_s \Psi_{sm}e^{j\omega_s t} \tag{D.44}$$

Using the phasor notation, the rotor equations (Equations D.29 and D.30), after eliminating ψ_{rd}, become

$$0 = R_{rd}I_{rd}e^{j(\omega_{slip}t-\theta_0)} + j\omega_{slip}L_{rd\sigma}I_{rd}e^{j(\omega_{slip}t-\theta_0)} + j\omega_{slip}\Psi_{rdm}e^{j(\omega_{slip}t-\theta_0)} \tag{D.45}$$

$$\Psi_{rdm}e^{j(\omega_{slip}t-\theta_0)} = \left(L_{rdm}I_{rd} + \frac{3}{2}MI_s\right)e^{j(\omega_{slip}t-\theta_0)} \tag{D.46}$$

In the previous equations the exponential functions were preserved only so that we could clearly recognize the stator and rotor angular velocity. When the exponential functions are left out, the equations become fairly simple:

$$\Psi_{sm} = L_{sm}I_s + MI_{rd} \tag{D.47}$$

$$V_{sm} = j\omega_s \Psi_{sm} \tag{D.48}$$

$$0 = R_{rd}I_{rd} + j\omega_{slip}L_{rd\sigma}I_{rd} + j\omega_{slip}\Psi_{rdm} \tag{D.49}$$

$$\Psi_{rdm} = L_{rdm}I_{rd} + \frac{3}{2}MI_s \tag{D.50}$$

D.4.2 The Induction Machine as Two Magnetically Coupled Coils

The induction machine is, similar to a transformer, an apparatus with magnetically linked circuits as can be seen from Equations D.47–D.50. Therefore, it appears to be possible to construct a circuit model based on magnetically coupled coils. The problem that the rotor quantities have an angular frequency (ω_{slip}) different from the stator quantities (ω_s) can be solved by multiplying the rotor voltage equation (Equation D.49) with ω_s/ω_{slip}. We additionally multiply the rotor equations (Equations D.49 and D.50) with 2/3 to eliminate the factor 3/2 at M. This factor 3/2 appears in our equations because of the fact that we have two rotor windings (d and q) and three stator windings. The equations for the induction machine can now be written as

$$\Psi_{sm} = L_{sm}I_s + MI_{rd} \tag{D.51}$$

$$V_{sm} = j\omega_s \Psi_{sm} \tag{D.52}$$

$$0 = \left(\frac{\omega_s}{\omega_{slip}} \cdot \frac{2}{3} \cdot R_{rd} + j\omega_s \cdot \frac{2}{3} \cdot L_{rd\sigma}\right)I_{rd} + j\omega_s \cdot \frac{2}{3} \cdot \Psi_{rdm} \tag{D.53}$$

$$\frac{2}{3} \cdot \Psi_{rdm} = \frac{2}{3} \cdot L_{rdm}I_{rd} + MI_s \tag{D.54}$$

With these equations we can draw the circuit model of Figure D.6. The magnetically coupled coils in the circuit model originate from the first harmonic of

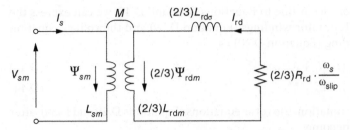

Figure D.6 A circuit model for the induction machine with two magnetically coupled coils.

the flux density distribution in the air gap and have an ideal coupling (see also Appendix B.3):

$$M = \sqrt{\frac{2}{3}L_{rdm}L_{sm}} \tag{D.55}$$

D.4.3 A Practical Model of the Induction Machine

As the rotor windings are short-circuited, they are electrically not accessible. This is not a problem because we are in general not interested in the actual values of the rotor main flux and the rotor current. With the theory developed in Appendix B.3, we can replace the magnetically coupled coils by an ideal transformer and a coil in parallel (see Figure B.6), so that the circuit model as shown in Figure D.7 results. Now we can transform the rotor parameters to the stator side, so that the ideal transformer itself vanishes from the circuit. When we use Equation D.55, the winding ratio $(N_1:N_2)$ can be written as

$$1 : \sqrt{\frac{2}{3}\frac{L_{rdm}}{L_{sm}}} \rightarrow L_{sm} : M \tag{D.56}$$

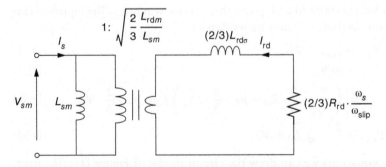

Figure D.7 A circuit model for the induction machine with an ideal transformer.

Figure D.8 A practical circuit model for the induction machine.

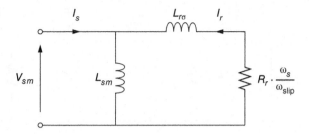

The rotor current phasor I_{rd} transformed to the stator side gives us the phasor I_r:

$$I_r = \frac{M}{L_{sm}} \cdot I_{rd} \tag{D.57}$$

The rotor resistance and inductance transformed to the stator side gives

$$R_r = \left(\frac{L_{sm}}{M}\right)^2 \cdot \frac{2}{3} \cdot R_{rd} \quad \text{and} \quad L_{r\sigma} = \left(\frac{L_{sm}}{M}\right)^2 \cdot \frac{2}{3} \cdot L_{rd\sigma} \tag{D.58}$$

Now the practical circuit model as shown in Figure D.8 is the result.

D.5   Induction Motor in the Power System

When the induction machine is applied as a motor, as is the case in many appliances, we are interested in the mechanical torque produced by the induction machine. So let us consider the mechanical side of the machine. The mechanical torque is equal to the electromechanical torque minus the friction torque. The electromechanical torque is calculated from the power balance of the induction machine in the steady state. We take as a starting point the practical circuit model for the induction machine from Figure D.8, in which the stator resistance and the stator leakage are neglected. In the steady state, the rotor rotates at constant speed, so its kinetic energy is constant. Because the voltages and currents are periodic harmonic functions, the energy stored in the magnetic field will not change over one time period. The only place where power is converted is in the resistance $R_r(\omega_s/\omega_{slip})$. We have to realize that this is not a physical resistance as such, but the quantity has the unit ohm. The rotor consists of two rotor windings, as we saw earlier, and the dissipated power in these two rotor windings equals

$$P_{diss} = 2R_{rd}|I_{rd}|^2 \tag{D.59}$$

Substitution of Equations D.57 and D.58 yields

$$P_{diss} = 2\left(\frac{3}{2} \cdot \frac{M^2}{L_{sm}^2} \cdot R_r\right)\left(\frac{L_{sm}}{M} \cdot |I_r|\right)^2 = 3R_r|I_r|^2 \tag{D.60}$$

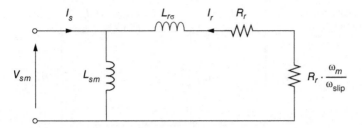

Figure D.9 An induction machine model for power calculations.

The power dissipation in the two rotor windings is equal to three times the dissipated power in resistance R_r. The factor three comes from the fact that the circuit model in Figure D.8 is a single-phase representation of a three-phase induction machine. The resistance in Figure D.8 can be divided in two parts, when we use $\omega_{slip} = \omega_s - \omega_m$ (the slip angular velocity introduced in Equation D.17):

$$\frac{\omega_s}{\omega_{slip}} \cdot R_r = \frac{\omega_{slip}}{\omega_{slip}} \cdot R_r + \frac{\omega_s - \omega_{slip}}{\omega_{slip}} \cdot R_r = R_r + \frac{\omega_m}{\omega_{slip}} \cdot R_r \tag{D.61}$$

The corresponding circuit model is depicted in Figure D.9.

The resistance R_r multiplied with $|I_r|^2$ represents 1/3 of the dissipated power in the machine, and the resistance $R_r(\omega_m/\omega_{slip})$ multiplied with $|I_r|^2$ represents 1/3 of the power P_{em} delivered electromagnetically to the axis, so that we can write for the electromagnetic torque

$$T_e = \frac{P_{em}}{\omega_m} = \frac{3 \cdot \frac{\omega_m}{\omega_{slip}} \cdot R_r |I_r|^2}{\omega_m} = 3 \cdot \frac{R_r}{\omega_{slip}} \cdot |I_r|^2 \tag{D.62}$$

The electric power consumed by the stator of the induction motor equals

$$P_{sm} = \omega_s T_e = 3 \cdot \frac{\omega_s}{\omega_{slip}} \cdot R_r |I_r|^2 \tag{D.63}$$

This electric power P_{sm} is supplied to the three-phase stator terminals of the induction motor. Because there is no power dissipation in the stator (we assumed the stator resistance to be zero), this power is transferred to the air-gap field, which exercises the torque T_e. The air-gap field itself rotates with an angular velocity ω_s, and therefore the power delivered by the stator to the rotor equals $\omega_s T_e$, often called the air-gap power.

E

The Representation of Lines and Cables

E.1 Introduction

When we speak of electricity, we think of current flowing through the conductors of overhead transmission lines and underground cables on its way from generator to load. This approach is valid because the physical dimensions of the power system are generally small compared with the wavelength of the currents and voltages (the wavelength is 6000 km for 50 Hz; see also Appendix A.6). This enables us to apply Kirchhoff's voltage and current laws and use lumped elements in our modeling of overhead transmission lines and underground cables.

E.2 The Long Transmission Line

When we want to make a model of an overhead transmission line, we should distinguish the short transmission line, the medium-length transmission line, and the long transmission line. For the long transmission line, we have, even for 50 or 60 Hz voltages and currents, to take the line length into account. The line parameters are not lumped but distributed uniformly throughout the length of the line. We can distinguish four parameters for a transmission line: the series resistance (due to the resistivity of the conductor), the inductance (due to the magnetic field surrounding the conductors), the capacitance (due to the electric field between the conductors), and the shunt conductance (due to leakage currents in the insulation).

As a first approximation of the series resistance, the DC resistance of the conductor can be calculated, which is affected by the operating temperature of the conductor (i.e., nearly linearly increasing with the temperature). When a conductor is operating on AC, the current density distribution across the conductor cross section becomes nonuniform. This is known as the skin effect, and as a consequence, the AC resistance of a conductor is higher than the DC resistance. The skin effect is frequency dependent.

Electrical Power System Essentials, Second Edition. Pieter Schavemaker and Lou van der Sluis.
© 2017 John Wiley & Sons Ltd. Published 2017 by John Wiley & Sons Ltd.

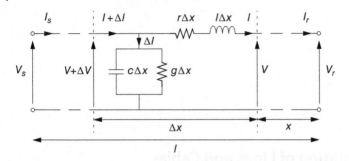

Figure E.1 Incremental length of a transmission line.

The shunt conductance accounts for the losses in the dielectric medium between the conductors when the transmission link is an underground cable. In the case of an overhead line, it accounts for the leakage currents across the insulator strings, the influence of the ground resistance, and the corona losses.

Finding the analytical expressions for the inductance and capacitance of a specific line or cable configuration is rather straightforward. Many classical textbooks give these formulas and show how to derive them. For practical power system analysis, it is in most of the cases sufficient to rely on tables to find the series resistance, the series inductance, the shunt capacitance, and the shunt conductance per unit length.

Once the line parameters are obtained on a per-phase per unit length basis, we are interested in the performance of lines with an arbitrary length l. The parameters are distributed evenly along the line length, and to be able to represent them by lumped elements and to apply Kirchhoff's voltage and current laws, we consider the line to consist of an infinite number of incremental line pieces, each with a differential length. We consider the transmission line in a balanced, sinusoidal steady state. This allows us to use phasors and impedances.

In Figure E.1 an incremental length of a transmission line is depicted. The per-phase terminal voltages and currents are V_s and I_s at the sending end and V_r and I_r at the receiving end. The loads are connected at the receiving end, and the supply is done from the sending end. In the daily operation of the power system, it is of primary importance to keep the voltage at the load at a nearly constant value. Starting from the load, the receiving end of the transmission line is located at $x = 0$ and the sending end is at $x = l$.

When we use lowercase letters for the distributed parameters, the series impedance per unit length can be written as

$$z = r + j\omega l \tag{E.1}$$

z the series impedance per unit length per phase [Ω/m]
r the series resistance per unit length per phase [Ω/m]
l the series inductance per unit length per phase [H/m]

and for the shunt admittance per unit length

$$y = g + j\omega c \tag{E.2}$$

y the shunt admittance per unit length per phase to neutral [S/m]
g the shunt conductance per unit length per phase to neutral [S/m]
c the shunt capacitance per unit length per phase to neutral [F/m]

Applying Kirchhoff's voltage and current law to the incremental section, shown in Figure E.1, results in

$$\Delta V = Iz\Delta x$$
$$\Delta I = (V + \Delta V)y\Delta x \tag{E.3}$$

When we make x infinitely small, the following simplifications can be made. The length of the incremental section Δx can be written as dx. Furthermore, the contribution $\Delta V y \, \Delta x$ to the current (i.e., a product of two differential quantities) can be neglected. By doing so, two first-order linear differential equations are obtained:

$$\frac{dV}{dx} = zI \tag{E.4}$$

$$\frac{dI}{dx} = yV \tag{E.5}$$

Differentiating Equations E.4 and E.5 with respect to x leads to the following expressions:

$$\frac{d^2V}{dx^2} = z\frac{dI}{dx} \tag{E.6}$$

$$\frac{d^2I}{dx^2} = y\frac{dV}{dx} \tag{E.7}$$

Now we can substitute Equation E.4 into Equation E.7 and Equation E.5 into Equation E.6, and we obtain the second-order linear differential equations:

$$\frac{d^2V}{dx^2} = zyV = \gamma^2V \tag{E.8}$$

$$\frac{d^2I}{dx^2} = zyI = \gamma^2I \tag{E.9}$$

γ the propagation constant [1/m]; $\gamma = \sqrt{zy}$. For a lossless line, the series resistance r and the shunt conductance g are zero, and the propagation constant reduces to a pure imaginary number $\gamma = j\omega\sqrt{lc} = j2\pi/\lambda \, \text{rad/m}$, in which λ is the wavelength in meters. When we substitute $\lambda = v/f$ in the equation of the propagation constant of a lossless line ($\gamma = j\omega\sqrt{lc} = j2\pi/\lambda = j2\pi f/v = j\omega/v$), we find a value for the velocity of propagation of the wave v: $v = 1/\sqrt{lc} \, \text{m/s}$. This velocity of propagation will be close to the speed of light in air (approximately 3×10^8 m/s).

The general solution of the second-order differential equation (Equation E.8) is given by

$$V(x) = k_1 e^{\gamma x} + k_2 e^{-\gamma x} \tag{E.10}$$

The expression for the current follows directly from Equations E.10 and E.4:

$$I(x) = \frac{1}{z}\frac{dV}{dx} = \frac{k_1\gamma}{z}e^{\gamma x} - \frac{k_2\gamma}{z}e^{-\gamma x} \tag{E.11}$$

At the receiving end of the line (at $x=0$), $V = V_r$ and $I = I_r$. Substitution in Equations E.10 and E.11 gives $V_r = k_1 + k_2$ and $I_r = (\gamma/z)(k_1 - k_2)$. The constants k_1 and k_2 can be determined to be

$$k_1 = \frac{V_r + I_r(z/\gamma)}{2} \quad \text{and} \quad k_2 = \frac{V_r - I_r(z/\gamma)}{2} \tag{E.12}$$

Substitution of the two constants into Equations E.10 and E.11 gives us the equations of the voltage and the current at any point along the line:

$$V(x) = \frac{V_r + Z_c I_r}{2}e^{\gamma x} + \frac{V_r - Z_c I_r}{2}e^{-\gamma x}$$

$$I(x) = \frac{V_r/Z_c + I_r}{2}e^{\gamma x} - \frac{V_r/Z_c - I_r}{2}e^{-\gamma x} \tag{E.13}$$

Z_c the characteristic impedance [Ω]; $Z_c = \sqrt{z/y}$. For a lossless line, the series resistance r and the shunt conductance g are zero, and the characteristic impedance equals $Z_c = \sqrt{l/c}$.

A more common notation of these equations is by using hyperbolic functions

$$\sinh(\gamma x) = \frac{e^{\gamma x} - e^{-\gamma x}}{2} \quad \cosh(\gamma x) = \frac{e^{\gamma x} + e^{-\gamma x}}{2} \tag{E.14}$$

This leads to the following pair of hyperbolic equations that describe the voltage and current on the line as a function of x:

$$V(x) = V_r \cosh(\gamma x) + Z_c I_r \sinh(\gamma x)$$

$$I(x) = I_r \cosh(\gamma x) + \frac{V_r}{Z_c} \sinh(\gamma x) \tag{E.15}$$

Of particular interest is the relation between the receiving end voltage and current and the sending end voltage and current, which we can obtain by substituting $x = l$ in Equation E.15:

$$V_s = V_r \cosh(\gamma l) + Z_c I_r \sinh(\gamma l)$$

$$I_s = I_r \cosh(\gamma l) + \frac{V_r}{Z_c} \sinh(\gamma l) \tag{E.16}$$

Now we are able to calculate the voltage and line current at the sending end of the transmission line, while the receiving end voltage and current are determined by the load.

When Equation E.16 is written as a matrix expression, the following equation results:

$$\begin{bmatrix} V_s \\ I_s \end{bmatrix} = T \begin{bmatrix} V_r \\ I_r \end{bmatrix} = \begin{bmatrix} \cosh(\gamma l) & Z_c \sinh(\gamma l) \\ \sinh(\gamma l)/Z_c & \cosh(\gamma l) \end{bmatrix} \begin{bmatrix} V_r \\ I_r \end{bmatrix} \tag{E.17}$$

T the transmission matrix or chain matrix

In fact this is nothing more than a general two-port equation. The advantage of this notation is that we can make use of the properties of two-port calculations. When two transmission lines, with different line parameters, are connected in series, the resulting transmission matrix is the product of the individual transmission matrices, as illustrated in Figure E.2.

The equivalent circuit of the long transmission line is depicted in Figure E.3. The expressions for the voltage and the current at the sending end of the equivalent circuit are

$$V_s = \left(\frac{Z'Y'}{2} + 1 \right) V_r + Z'I_r$$

$$I_s = (V_s + V_r)\frac{Y'}{2} + I_r = \left(\frac{Z'Y'}{4} + 1 \right) Y'V_r + \left(\frac{Z'Y'}{2} + 1 \right) I_r \tag{E.18}$$

Or, as a matrix expression,

$$\begin{bmatrix} V_s \\ I_s \end{bmatrix} = \begin{bmatrix} \dfrac{Z'Y'}{2} + 1 & Z' \\ \left(\dfrac{Z'Y'}{4} + 1 \right) Y' & \dfrac{Z'Y'}{2} + 1 \end{bmatrix} \begin{bmatrix} V_r \\ I_r \end{bmatrix} \tag{E.19}$$

The expressions for the elements in the equivalent circuit can be obtained by comparing the elements of the transmission matrix in Equation E.19 with those in Equation E.17. The element in the upper right corner gives

$$Z' = Z_c \sinh(\gamma l) = Z \cdot \frac{\sinh(\gamma l)}{\gamma l} \tag{E.20}$$

Z the total series impedance per phase $[\Omega]$; $Z = z \cdot l$

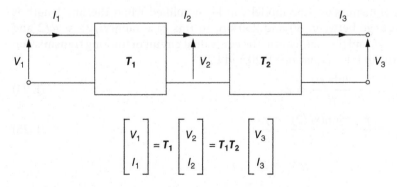

$$\begin{bmatrix} V_1 \\ I_1 \end{bmatrix} = T_1 \begin{bmatrix} V_2 \\ I_2 \end{bmatrix} = T_1 T_2 \begin{bmatrix} V_3 \\ I_3 \end{bmatrix}$$

Figure E.2 Series connection of two transmission lines, represented as two ports.

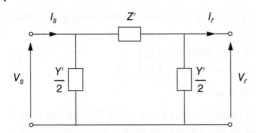

The elements on the main diagonal lead to the following correspondence: $(Z'Y')/2 + 1 = \cosh(\gamma l)$. Substituting Equation E.20 in this relation leads us to

$$\frac{Y'}{2} = \frac{\cosh(\gamma l) - 1}{Z_c \sinh(\gamma l)} = \frac{1}{Z_c} \tanh\left(\frac{\gamma l}{2}\right) = \frac{Y}{2} \cdot \frac{\tanh(\gamma l/2)}{\gamma l/2} \tag{E.21}$$

Y the total shunt admittance per phase to neutral [S]; $Y = y \cdot l$

Now, Equations E.20 and E.21 can be used to verify that also the lower left elements in the transmission matrices correspond to each other:

$$\frac{Z'Y'}{2} + 1 = \cosh(\gamma l) \rightarrow \frac{Z'Y'}{4} = \frac{\cosh(\gamma l) - 1}{2} \rightarrow \frac{Z'Y'}{4} + 1 = \frac{\cosh(\gamma l) + 1}{2} \tag{E.22}$$

$$\left(\frac{Z'Y'}{4} + 1\right) Y' = \frac{\cosh(\gamma l) + 1}{2} \cdot 2\frac{\cosh(\gamma l) - 1}{Z_c \sinh(\gamma l)} = \frac{\cosh^2(\gamma l) - 1}{Z_c \sinh(\gamma l)} \tag{E.23}$$

$$= \frac{\sinh^2(\gamma l)}{Z_c \sinh(\gamma l)} = \frac{\sinh(\gamma l)}{Z_c}$$

E.3 The Medium-Length Transmission Line

The long transmission line model can be simplified when the line length is within a range between 80 and 240 km. In this case $\tanh(\gamma l/2) \approx \gamma l/2$ and $\sinh(\gamma l) \approx \gamma l$, and the elements in the equivalent circuit of the long transmission line transform into (Equations E.20 and E.21)

$$Z' = Z \cdot \frac{\sinh(\gamma l)}{\gamma l} = Z \tag{E.24}$$

$$\frac{Y'}{2} = \frac{Y}{2} \cdot \frac{\tanh(\gamma l/2)}{\gamma l/2} = \frac{Y}{2} \tag{E.25}$$

Figure E.4 Equivalent circuit of a medium-length transmission line.

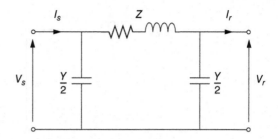

The equivalent circuit of the medium-length transmission line is depicted in Figure E.4. The matrix expression for the voltage and the current at the sending end is

$$\begin{bmatrix} V_s \\ I_s \end{bmatrix} = \begin{bmatrix} \dfrac{ZY}{2}+1 & Z \\ \left(\dfrac{ZY}{4}+1\right)Y & \dfrac{ZY}{2}+1 \end{bmatrix}\begin{bmatrix} V_r \\ I_r \end{bmatrix} \tag{E.26}$$

The equivalent circuit of the medium-length transmission line is made up of the total series impedance of the transmission line with half of the total shunt capacitance at each side of the line.

E.4 The Short Transmission Line

When the line length falls below a value of 80 km, the equivalent circuit of the medium-length transmission line model may be simplified by omitting the shunt capacitances. In this way, only the total series impedance of the transmission line is taken into account.

The equivalent circuit of the medium-length transmission line is depicted in Figure E.5. The matrix expression for the voltage and the current at the sending end is

$$\begin{bmatrix} V_s \\ I_s \end{bmatrix} = \begin{bmatrix} 1 & Z \\ 0 & 1 \end{bmatrix}\begin{bmatrix} V_r \\ I_r \end{bmatrix} \tag{E.27}$$

In the following section, it will be made plausible that the shunt capacitances can be omitted for the shorter line lengths.

E.5 Comparison of the Three Line Models

The line models are put together in Table E.1 for easy comparison.

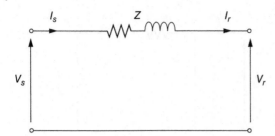

Figure E.5 Equivalent circuit of a short transmission line.

Example E.1 *Comparison of the three line models at a line length of 40 km*
A three-phase transmission line is 40 km long. It has a total series impedance of $Z = 5 + j20 = 20.6\angle76°\ \Omega$ and a total shunt admittance of $Y = j133 \times 10^{-6} = 133 \times 10^{-6}\angle90°$ S. At the receiving end, a three-phase active power of 40 MW is consumed at a voltage of 220 kV with a power factor of 0.9 lagging. The voltage at the sending end will be computed by using the three line models.

First, the current at the receiving end will be computed:

$$|I_r| = \frac{40 \times 10^6}{\sqrt{3} \cdot 220 \times 10^3 \cdot 0.9} = 116.6\,\text{A}, \quad I_r = 116.6\angle{-25.8°}\,\text{A} \qquad (\text{E.28})$$

The voltage at the sending end when using the short line model equals

$$V_s = V_r + I_r Z = 128.55 \times 10^3 \angle 0.82°\,\text{V} \qquad (\text{E.29})$$

The voltage at the sending end when using the medium-length line model equals

$$V_s = \left(\frac{ZY}{2} + 1\right) V_r + I_r Z = 128.38 \times 10^3 \angle 0.84°\,\text{V} \qquad (\text{E.30})$$

The voltage at the sending end when using the long line model equals

$$V_s = V_r \cosh(\gamma l) + Z_c I_r \sinh(\gamma l) = 128.38 \times 10^3 \angle 0.84°\,\text{V} \qquad (\text{E.31})$$

with $\gamma l = \sqrt{ZY} = 0.0064 + j0.0520$ and $Z_c = \sqrt{Z/Y} = 390.76 - j48.104\ \Omega$.

A comparison of the sending end voltages obtained with the three line models when applied at three different line lengths is shown in Table E.2 (the results are based on the (line) data in Example E.1 (p. 372)).

In this table the results obtained with the long line model are exact; deviations in the results obtained with the other models are caused by the simplifications in the line modeling.

From the results in Table E.2, it is evident that for the short line length, even the simplest model (i.e., the short line model) gives acceptable results: the

Table E.1 Line models for various line lengths.

Line model	Two-Port Equation	Equivalent Circuit
Short $l < 80\,\text{km}$	$\begin{bmatrix} V_s \\ I_s \end{bmatrix} = \begin{bmatrix} 1 & Z \\ 0 & 1 \end{bmatrix} \begin{bmatrix} V_r \\ I_r \end{bmatrix}$ with $Z = z \cdot l$	
Medium length $80\,\text{km} < l < 240\,\text{km}$	$\begin{bmatrix} V_s \\ I_s \end{bmatrix} = \begin{bmatrix} \dfrac{ZY}{2}+1 & Z \\ \left(\dfrac{ZY}{4}+1\right)Y & \dfrac{ZY}{2}+1 \end{bmatrix} \begin{bmatrix} V_r \\ I_r \end{bmatrix}$ with $Z = z \cdot l$ $Y = y \cdot l$	
Long $l > 240\,\text{km}$	$\begin{bmatrix} V_s \\ I_s \end{bmatrix} = \begin{bmatrix} \dfrac{Z'Y'}{2}+1 & Z' \\ \left(\dfrac{Z'Y'}{4}+1\right)Y' & \dfrac{Z'Y'}{2}+1 \end{bmatrix} \begin{bmatrix} V_r \\ I_r \end{bmatrix}$ with $Z' = Z \cdot \dfrac{\sinh(\gamma l)}{\gamma l}$ $\dfrac{Y'}{2} = \dfrac{Y}{2} \cdot \dfrac{\tanh(\gamma l/2)}{\gamma l/2}$	

Table E.2 Sending end voltages computed with the three line models at various line lengths.

Line Model	$|V_s|$ [kV]		
	$l = 40$ km	$l = 150$ km	$l = 400$ km
Short	128.55	132.96	143.59
Medium	128.38	130.62	127.54
Long	128.38	130.57	126.88

difference is less than 200 V. For the medium line length, the short line model already gives an unacceptable deviation: more than 2 kV. For the long line length, the medium-length line model starts to give inaccurate results too: the difference almost equals 700 V.

E.6 The Underground Cable

In the previous sections the focus was on the modeling of overhead transmission lines. The models that are developed and tabulated in Table E.1 are equally valid for underground cables. One should, however, bear in mind that the capacitance of an underground cable is much higher than it is for an overhead transmission line (e.g., see also Section 3.9.2 and Table 3.2). Therefore the use of the short line model to represent a short underground cable, thereby neglecting the influence of the capacitance, is not recommended.

Solutions

CHAPTER 1

Solution 1.1

a. $v(t) = 325.27 \sin(\omega t + \varphi)$ V and the injected current $i(t) = 141.42 \sin(\omega t - \varphi)$ A reach their maximum when the sinusoidal term equals 1. Hence $\hat{V} = 325.27$ V and $\hat{I} = 141.42$ A. The RMS value of the voltage can be calculated with Equation 1.12

$$|V| = \hat{V}/\sqrt{2} = 230 \text{ V}$$

The RMS value of the current can be calculated in the same way: $|I| = 100$ A.

b. $v(t) = \hat{V} \sin(\omega t + 30) = \hat{V} \cos(\omega t - 60)$. The phasors are $V = 230\angle -60°$ V and $I = 100\angle -120°$ A

c. The complex power is given by

$$S = P + jQ = VI^* = 11.5 + j19.9 \text{ kVA}$$

$\text{Im}(S) = 19.9$ kvar, and has a positive sign. As such, the connected circuit absorbs reactive power, and the circuit is inductive.

Solution 1.2

$$V_{ba} = V_{bn} - V_{an} = 90 \left(\cos(-15) + j\sin(-15) \right)$$
$$-140 \left(\cos(45) + j\sin(45) \right)$$
$$= (86.93 - 98.99) + j(-23.29 - 98.99) = -12.1 - j122.3 \text{ V}$$

Solution 1.3

a. Please refer to Figure 1.22. When you put your book upside down (so that V_{an} is on the negative real axis), you can see that the phasor V_{bc} would be located on the positive imaginary axis. This is the situation as described in this problem. We can see from the figure that $V_{cn} = 400/\sqrt{3}\angle -60°$ V.

Electrical Power System Essentials, Second Edition. Pieter Schavemaker and Lou van der Sluis.
© 2017 John Wiley & Sons Ltd. Published 2017 by John Wiley & Sons Ltd.

b. $I_{cn} = \dfrac{V_{cn}}{Z} = \dfrac{400\angle-60°}{\sqrt{3}\cdot10\angle30°} = 40/\sqrt{3}\angle-90°$ A.

c. $S_{3\phi} = 3V_{cn}I_{cn}{}^* = 3\dfrac{400}{\sqrt{3}}\angle-60°\cdot\dfrac{40}{\sqrt{3}}\angle90° = 16000\angle30° = 13.85 + 8j$ kVA

Solution 1.4

a. Because of the generator convention $I_2 = -I_1$ and this gives us $E_2 = E_1 - I_1 Z \Rightarrow I_1 = \dfrac{E_1 - E_2}{Z}$. The currents are $I_1 = \dfrac{100 - 100\left(\cos(30) + j\sin(30)\right)}{j5} = \dfrac{(100 - 86.6) - j50}{j5} = -10 - j2.68 = 10.4\angle195°$ A and $I_2 = -I_1 = 10.4\angle15°$ A.

b. The complex power drawn by source 1 equals $S_1 = -E_1I_1^* = 100(10 - j2.68) = 1000 - j268$ VA and for source 2 the complex power equals $S_2 = -E_2I_2^* = -1000 - j268$ VA.

c. Source 2 draws negative active power and thus operates in generator mode. Source 1 operates as a load and is a motor.

d. The impedance has no resistive part and therefore the system is lossless. The line, however, has reactive losses which amount to $Q = |I_1|^2X = 10.4^2 \times 5 = 540$ var.

Solution 1.5

a. From Equation (1.99), which is also valid for three phase systems: $\cos(\varphi) = \dfrac{P}{|S|} = \dfrac{P}{\sqrt{P^2 + Q^2}} = 0.86$

b. $P_{3\phi} = \sqrt{3}|V_{LL}||I|\cos(\varphi) \rightarrow |I| = \dfrac{P_{3\phi}}{\sqrt{3}|V_{LL}|\cos(\varphi)} = 177.3$ A

Solution 1.6

a. Please refer to Figure 1.22. When you rotate the phasors such that V_{ab} is on the positive real axis (rotate clockwise with 30 degrees), you can see that the phasor $V_{an} = 173.2/\sqrt{3}\angle-30°$ V, $V_{bn} = 173.2/\sqrt{3}\angle-150°$ V, $V_{cn} = 173.2/\sqrt{3}\angle90°$ V.

b. $I_a = \dfrac{V_{an}}{Z} = \dfrac{100\angle-30°}{10} = 10\angle-30°$ A, $I_b = 10\angle-150°$ A and $I_c = 10\angle90°$ A

Solution 1.7

a. $S_{3\phi} = P_{3\phi} + jQ_{3\phi} = P_{load} + j(Q_{load} + Q_{capbank}) = 250 \times 10^3 + j(250 \times 10^3$
$\tan(\cos^{-1}0.8) - 60 \times 10^3) = 250 + 127.5j$ kVA $= 280.6\angle27°$ kVA
$V_{LL} = 400\angle0°$ V, $S_{3\phi} = \sqrt{3}\,V_{LL}I^* = 280.6\angle27°$ kVA
$\rightarrow I = \dfrac{280.6}{400\sqrt{3}} \cdot 10^3 \angle -27° = 405\angle -27°$ A

b. Power factor $= \dfrac{P_{3\phi}}{|S_{3\phi}|} = \dfrac{250}{280.6} = 0.89$

Solution 1.8

a. $I_a = \dfrac{V_{an}}{Z} = \dfrac{100 \times 10^3}{\sqrt{3}} \cdot \dfrac{1}{85.4}\angle -20.6° = 676\angle -20.6°$ A,
$I_b = 676\angle -140.6°$ A and $I_c = 676\angle99.4°$ A.

b. $S_{3\phi} = 3V_{an}I_a^* = 3 \cdot \dfrac{100 \times 10^3}{\sqrt{3}} \cdot 676\angle20.6° = 117.1\angle20.6°$ MVA,
$P = \text{Re}\{S\} = 109.6$ MW and $Q = \text{Im}\{S\} = 41.2$ Mvar.

Solution 1.9

a. $S_1 = 3V_{LN1}I^* = \sqrt{3}V_{LL1}I^* \Rightarrow I^* = \dfrac{S_1}{\sqrt{3}V_{LL1}} = 551.1 + 78.7j = 556.7\angle8.13°$ A
and $I = 556.7\angle -8.13°$ A. $V_{LN1} = IZ + V_{LN2} \Rightarrow V_{LN2} = V_{LN1} - IZ = 123.9\angle -15.3°$ kV, $S_2 = 3V_{LN2}I^* = 205.3 - 25.8j$ MVA.

b. $P_{loss} = \text{Re}\{S_1 - S_2\} = 4.7$ MW or $P_{loss} = 3|I|^2R = 4.7$ MW, $Q_{loss} = \text{Im}\{S_1 - S_2\} = 55.8$ Mvar, or $Q_{loss} = 3|I|^2X = 55.8$ Mvar.

Solution 1.10

- $S_2 = 0.68 - 0.086j$ pu, $V_{LN2} = 0.95\angle -15.3°$ pu.
- $P_{loss} = 0.016$ pu and $Q_{loss} = 0.186$ pu.

Solution 1.11

$|Z_{b,1}| = \dfrac{|V_{b,1}|^2}{|S_{b,1}|}$, $|Z_{b,2}| = \dfrac{|V_{b,2}|^2}{|S_{b,2}|}$ and by assessing $\dfrac{|Z_{b,2}|}{|Z_{b,1}|}$ we can derive $\Rightarrow$

$|Z_{b,2}| = |Z_{b,1}|\dfrac{|S_{b,1}|}{|S_{b,2}|}\dfrac{|V_{b,2}|^2}{|V_{b,1}|^2}$

CHAPTER 2

Solution 2.1

a. First step: chemical energy of the fossil fuels is converted into thermal energy by means of combustion. In a nuclear fission reactor the energy of the fission process produces thermal energy.
b. Second step: Thermal energy is used to heat water and to produce steam at high temperatures and high pressures.
c. Third step: The steam expands adiabatically in a steam turbine and the thermal energy of the steam is converted into mechanical energy when the steam passes the turbine blades and makes the turbine rotor rotate.
d. Fourth step: the shaft of the turbine is connected with the shaft of the synchronous generator and the mechanical energy of the rotor is converted into electrical energy, in the air-gap of the machine, by electrical induction.

Solution 2.2

a. Thermal power plants.
b. Nuclear power plants.
c. Hydro power plants.
d. Geothermal power plants.

Solution 2.3

a. Wind energy: wind turbines convert the kinetic energy from the wind into mechanical energy by means of a rotor.
b. Solar energy: photovoltaic panels convert the energy of the photons into direct current by means of semiconducting material.

Solution 2.4

a. A single thermal well usually ranges from 4 MW to 10 MW.
b. It is not practical to transport high-temperature steam over long distances by pipelines, because of the heat losses. For this reason, geothermal power plants are built close to the resources.
c. The efficiency of geothermal power plants is usually in the range of 7–10%. The low efficiency is due to the low operating temperature.
d. Because they can operate 24 hours a day, geothermal power plants supply base load capacity.

Solution 2.5

a. Dry steam power plants: these plants use hydrothermal fluids in the form of steam, which goes directly to the turbine that drives the generator that produces electricity; see also Figure 2.21.
b. Flash steam power plants: these plants use hydrothermal fluids above 175 °C. A part of the fluid vaporizes rapidly to steam and it drives the turbine; see also Figure 2.22.
c. Binary-cycle power plants: these plants use the hot hydrothermal fluid, of a temperature below 175 °C and a secondary fluid with a much lower boiling point than water; see also Figure 2.23.

Solution 2.6
When the generator is connected to a large grid, its output voltage and frequency are locked to the system values and cannot be changed by any action on the generator. We say that the generator is connected to an infinite bus: an ideal voltage source with a fixed voltage amplitude and frequency.

Solution 2.7

a. Diagram: Figure 2.27a
b. Equivalent circuit: Figure 2.27b
c. Phasor diagram: Figure 2.27c

Solution 2.8
The generator voltage must:

Have the same phase sequence as the grid voltages
Have the same frequency as the grid
Have the same amplitude at its terminals as the one of the grid voltage
Be in phase with the grid voltage

Solution 2.9
Calculate the equivalent reactance: $X = 0.4 + 0.8$ pu.
The power angle can be calculated from $P = 3\dfrac{|V||E|\sin(\delta)}{X}$:

$$\delta = \sin^{-1}\left(\frac{XP}{3|V||E|}\right) = \sin^{-1}\left(\frac{1.2 \cdot 0.8}{3 \cdot 1 \cdot 1.3}\right) = 14.25°$$

Solution 2.10

In order to calculate the excitation voltage of the machine it is necessary to know the total reactance: $X = 1.2 + 0.3 = 1.5$ pu.

The excitation voltage can be obtained from $P = 3\dfrac{|V||E|\sin(\delta)}{X}$

For $\sin(\delta) = 1$, $|E| = \dfrac{P_{max} \cdot X}{3|V|} = \dfrac{1.1 \cdot 1.5}{3 \cdot 1} = 0.55$ pu.

Solution 2.11

a. $\delta > 0$: In this case is $P > 0$, thus the machine supplies active power to the grid and acts as a generator.
b. $\delta = 0$: In this case there is no active power exchange with the grid.
c. $\delta < 0$: In this case is $P < 0$, thus the machine absorbs active power from the grid and acts as a motor.

Solution 2.12

The theoretical power angle is $\delta = 90°$, see also Figure 2.31.

CHAPTER 3

Solution 3.1

a. A single phase AC circuit requires two conductors. A three-phase AC circuit, using the same size conductors as the single-phase circuit, can carry three times the power which can be carried by a single-phase circuit and uses three conductors for the three phases and one conductor for the neutral. Thus a three-phase circuit is more economical than a single-phase circuit in terms of initial cost as well as the losses. All transmission and distribution systems are therefore three-phase systems. In fact a balanced three-phase circuit does not require the neutral conductor, as the instantaneous sum of the three line currents is zero.
b. An overhead line is less costly than an underground cable. Not only because of the cost of the cable but also because of the fact that for an underground cable one has to dig a trench.
c. When we apply aluminum conductors as overhead transmission lines (instead of copper), our tower construction and insulator strings can be designed lighter and are therefore cheaper. Additionally, aluminum is a less expensive material than copper, and therefore overhead transmission lines are usually laid out with aluminum instead of copper conductors.

The rather low tensile strength of aluminum is a disadvantage to use it as conductor material for overhead transmission lines, and that is why steel is used as core material of the conductor (Figure 3.42). Aluminum provides the necessary conductivity, while steel provides the necessary mechanical strength.

d. We see in Figure 3.42 that the conductor itself is not solid but is composed of strands (26 aluminum strands and 7 steel strands in this case), which gives the conductor the necessary flexibility. The strands are spiraled – each layer in opposite direction to avoid unwinding.

Solution 3.2

a. A balanced three-phase circuit does not require the neutral conductor, as the instantaneous sum of the three line currents is zero. Therefore the transmission lines and feeders are three-phase three-wire circuits. Distribution systems are three-phase four wire circuits because a neutral wire is necessary to supply the single-phase load of domestic and commercial consumers.

b. In a hydro-electric, thermal and nuclear power plant the generation level is in the range from 10–25 kV. These stations are generally situated far from the load centres. The losses in a circuit are proportional with the square of the current and when we transform the electric energy to a higher voltage level, the current is reduced and so are the losses.

c. An increase in line voltage increases the transmission efficiency. At higher voltages corona causes a significant power loss and interference with communication circuits, if the circuit has one conductor per phase. The use of multiple conductors decreases the voltage gradient in the vicinity of the line and thus reduces the possibility of corona discharge.

d. Aluminium has a larger resistivity than copper and therefore an aluminium conductor has a larger diameter than a copper conductor of the same resistance. A large diameter, for the same voltage, leads to a lower voltage gradient at the conductor surface with a tendency of reduced ionization level of air and corona.

Solution 3.3

An interconnected power system leads to a better overall system efficiency, because the total installed power can be less than the sum of the loads and it brings an improved system reliability, because, when one of the partners in the power pool has a problem, e.g. unforeseen loss of generation power, the other partners can supply the missing generation. Interconnection also results in smaller frequency deviations. In a large interconnected system, a great number of synchronous generators run in parallel. When there is a mismatch in balancing the production with the consumption during a short time, only a small frequency deviation in the system will occur because there is enough rotating mass in the system. An interconnected system covers a large geographical area. The European grid and the North American power pool, for example, span an area with multiple time zones: the peak load in the morning will start at different times in the various time zones. Hydropower can be connected, nuclear and thermal power plants can be built where cooling water is present or where the fuel is available. Wind mills can be erected at a location with a steady wind throughout the year (on-shore or offshore). A geographical spread of the wind production will reduce the chance that no wind power is produced at all: if there is no wind in one area, it is not likely that at the same time this is also the case in other areas. Interconnection also facilitates the dealing and wheeling of power. It becomes possible to exchange the power from the connected power producers and to create in this way a market for electrical energy.

Solution 3.4

See Figure 3.40.

During the short-circuit test $|V_2| = 0$ and the primary voltage is the short-circuit voltage: $|V_1| = |V_k|$.

Because $R_1 \approx R_2 \left(\dfrac{N_1}{N_2}\right)^2$ and $L_{1\sigma} \approx L_{2\sigma} \left(\dfrac{N_1}{N_2}\right)^2$, the resistance R_m sees half the short-circuit voltage. The iron losses per phase during the short-circuit test are $\dfrac{1}{4}\dfrac{|V_k|^2}{R_m} = \dfrac{1}{4}\dfrac{|V_k|^2}{|V_1|^2}\dfrac{|V_1|^2}{R_m} = \dfrac{1}{4}\left(\dfrac{V_k(\%)}{100}\right)^2\dfrac{|V_1|^2}{R_m}$ with $\dfrac{|V_1|^2}{R_m}$ being the iron losses, at no-load, per phase, i.e. 200/3 W. The hysteresis and eddy current losses during the short-circuit test are 0.023 W per phase, so 0.07 W in total and that can be neglected with respect to the copper losses of 1200 W.

Solution 3.5

a. The short-circuit voltage is the percentage of the nominal voltage to be applied to make, in the high-voltage laboratory, the nominal primary current flow when the transformer is short-circuited at its secondary terminals. When the transformer feeds into a short-circuit during normal operation, the primary voltage is the nominal voltage. The nominal power of a transformer is therefore $|S_{nom}| = \dfrac{|S_k||V_k(\%)|}{100} = 0.12 \cdot 1200 \text{ MVA} = 144 \text{ MVA}$.

b. Copper losses and core losses are ohmic losses; the copper losses because of the winding resistance and the core losses due to the losses from the eddy currents and the hysteresis losses. In the high-voltage laboratory the no-load test is used to determine the core losses. When we look at the transformer equivalent circuit (Figure 3.40) we see that $R_2(N_1/N_2)^2$ and $L_{2\sigma}(N_1/N_2)^2$ do not play a role. The dominant component is R_m representing the eddy current and hysteresis losses. $P_{core} = |S_{no-load}| \cos(\varphi) = 0.3 \cdot 2\,\text{MVA} = 0.6\,\text{MW}$.

Solution 3.6

a. Total apparent power: $|S| = \sqrt{P^2 + Q^2}$, active power $P = 100 + 50 \cdot 0.6 = 130\,\text{MW}$, reactive power $Q = 100 \cdot 0 + 50 \cdot 0.8 = 40\,\text{Mvar}$, $|S| = \sqrt{P^2 + Q^2} = 136\,\text{MVA}$.

b. $|I_1|$: $|S| = |V||I| \rightarrow 136 \times 10^6 = |I_1| \cdot 150 \times 10^3 \sqrt{3} \rightarrow |I_1| = 523.5\,\text{A}$.
$|I_2|$: $1 \times 10^8 = |I_2| \cdot 10 \times 10^3 \sqrt{3} \rightarrow |I_2| = 5773.5\,\text{A}$.
$|I_3|$: $50 \times 10^6 = |I_3| \cdot 10 \times 10^3 \sqrt{3} \rightarrow |I_3| = 2886.8\,\text{A}$.

CHAPTER 4

Solution 4.1

a. The peak load is the highest load of the day
b. The valley load is the lowest load of the day. It occurs usually between 4 and 5 a.m.
c. The distribution substation serves several feeder circuits that supply residential and commercial customers.
d. Vertically operated power systems are systems in which large power plants feed bulk power into the high-voltage transmission network that supplies the distribution substations.

Solution 4.2

The stalled rotor current per phase (that is the current drawn from the grid when the rotor is at standstill) can be obtained by using Equation 4.4 considering $\omega_m = 0$: $I_r = \dfrac{-V_{sm}}{j\omega_s L_{r\sigma} + R_r} = \dfrac{-230}{0.7j + 0.1} = 325.3\angle 98.1°\,\text{A}$.

Solution 4.3

a. Incandescent lamp. This lamp consists of a thin filament that conducts current. The current heats the thin filament that radiates light. The metal filament slowly evaporates and deposits on the inside of the glass bulb.

The incandescent lamp is not very efficient because of the heat produced. The lamp is resistive, thus the current is in phase with the applied voltage.

b. Halogen lamp. This lamp can be considered as an improved version of the incandescent lamp. The thin metal filament evaporates but will redeposit on the filament after a chemical reaction with the halogen gas in the bulb and has therefore a longer life-time than the incandescent lamp.

c. Fluorescent lamp. In this lamp an electrical discharge takes place when electrons collide with mercury ions, resulting in ultraviolet radiation that is turned into visible light by the fluorescent coating at the inside of the glass tube. The lamps have an inductive nature and thus they affect the power factor.

d. LED-lamps. These lamps use semiconductors as light source. When a current flows through the semiconductors, they emit light. The energy consumption of LED-lamps is low.

Solution 4.4

A resistance converts electricity for 100% into heat. We consider this in the transmission and distribution of electricity to be unwanted and raise, when possible, the voltage to minimise the losses. But when we heat water in a water cooker or boiler and heat the air in our electrical stove to cook food it is a rather useful phenomenon!

Solution 4.5

Rectifier circuits are made with diodes. Diodes conduct or isolate the current, depending on the direction of the current. As soon as the voltage across the diode becomes positive the diode gets in conducting mode and it isolates when the current passes through zero. For the circuit diagram, and output voltages, see Figure 4.8.

Solution 4.6

a. A battery is built up from one or more voltaic cells. Each voltaic cell has a positive and a negative terminal that are immersed in a solid or liquid electrolyte.

b. The EMF of a battery is equal to the electric potential between the terminals of a battery, that is neither charging or discharging. This voltage can be measured with a voltmeter and is called the open circuit voltage. Lead-acid cells have an EMF of 2 V and lithium cells have an EMF of 3 V.

c. The capacity of a battery is affected by discharge conditions like the duration of the current, the current magnitude and the ambient temperature.

d. Battery manufacturers rate their batteries with a voltage and an ampere-hour rating.

Solution 4.7

Common voltage ratings for distribution grids are 12 kV, 11 kV, 10 kV and 7.2 kV at the primary transformer winding and 400 V at the secondary winding. The power goes from the step-down transformer to the distribution bus; smaller distribution transformers bring the power to lower voltage levels.

Solution 4.8

a. • Residential loads are generally single phase loads
   • Residential loads are connected to the secondary winding of the distribution transformer
b. In order to have a three-phase balanced load during normal system operation, residences are connected alternately between one of the three phases and the neutral.

Solution 4.9

a. The advantages of a medium-voltage distribution network over a low-voltage distribution network are the much lower transmission losses and the lower voltage drops.
b. Industrial loads at the end of a feeder are in most of the cases inductive because of the electrical machines that convert electrical energy into a mechanical torque.
c. Capacitor banks are installed to compensate the lagging power factor.

Solution 4.10

The phase voltage is $|V_{LN}| = |V_{LL}|/\sqrt{3}$. The phase current is calculated with Equation 4.10: $|I| = \dfrac{P}{3|V_{LN}|\cos(\varphi)} = \dfrac{10000}{3 \cdot \left(\dfrac{400}{\sqrt{3}}\right) \cdot 0.8} = 18.04$ A.

Solution 4.11

a. DC systems have a relatively low voltage and require conductors with a large diameter to supply the demanded power.

Typical voltage ratings are 600 V and 750 V for DC overhead line and conductor rail systems (for trams and metros) and 1500 V and 3000 V for DC overhead line systems for trains.

b. AC systems are suited for high intensity railway traffic.

The standard voltages for AC traction overhead line systems for trains are 15 kV, 16 2/3 Hz and 25 kV, 50 Hz. These AC systems are single-phase systems.

CHAPTER 5

Solution 5.1

a. $\Delta f = -R \cdot f_r \cdot \left(\dfrac{\Delta P}{P_r} \right) = -0.02 \cdot 50 \cdot \left(\dfrac{100}{400} \right) = -0.25\,\text{Hz}$

b. $\Delta P = \left(\dfrac{-\Delta f / f_r}{R} \right) \cdot P_r = \left(\dfrac{(0.5/50)}{0.06} \right) \cdot 400 = 66.6\,\text{MW}$

Solution 5.2

a. When the load suddenly consumes more active power ΔP, the frequency drops with an amount Δf. To restore the active power balance, the speed governors increase the prime mover power in accordance with the speed governor characteristics:

$$\left. \begin{array}{l} \Delta P_{g1}/P_{g1,r} = -\dfrac{\Delta f / f_r}{R_{g1}} \\[4mm] \Delta P_{g2}/P_{g2,r} = -\dfrac{\Delta f / f_r}{R_{g2}} \end{array} \right\} \quad \dfrac{\Delta P_{g1}/P_{g1,r}}{\Delta P_{g2}/P_{g2,r}} = \dfrac{R_{g2}}{R_{g1}}$$

$$\dfrac{\Delta P_{g1}/400}{\Delta P_{g2}/400} = \dfrac{0.06}{0.02} = 3 \rightarrow \dfrac{\Delta P_{g1}}{\Delta P_{g2}} = 3$$

The active power increase of the load will be distributed over the two generators ($\Delta P = \Delta P_{g1} + \Delta P_{g2}$) with a ratio 3:1, i.e. generator 1 takes 75 % of the load increase and generator 2 takes 25 %.

b. $\lambda = \dfrac{1}{R_{g1}} \cdot \dfrac{P_{g1,r}}{f_r} + \dfrac{1}{R_{g2}} \cdot \dfrac{P_{g2,r}}{f_r} = \dfrac{1}{0.02} \cdot \dfrac{400}{50} + \dfrac{1}{0.06} \cdot \dfrac{400}{50} = 533.3\,\text{MW/Hz}$

Solution 5.3

a. Hydropower generators are salient pole machines. A generator has to be chosen with the right number of pole pairs. In this case 6 pole pairs. Because 6 times 500 divided by 60 gives 50 Hz (Equation 2.21).

b. Before a synchronous generator can be connected to the grid four conditions must be satisfied. The generator voltage must:
   • have the same phase sequence as the grid voltages;
   • have the same frequency as the grid;
   • have the same amplitude at its terminals as the grid voltage;
   • be in phase with the grid voltage.

c. The active power output of the generator is controlled by adjusting the water flow through the water turbine. The reactive power output is controlled by the excitation of the generator.

d. For the single phase equivalent circuit of the synchronous generator see Figure 2.27 (b).

$Z = 2j = 2\angle 90° \ \Omega$; $\quad V = \dfrac{10.000}{\sqrt{3}}\angle 0° = 5773.5\angle 0°$ V. The current is $I = |I|\angle\varphi$ (φ has a positive sign because the current is leading). $P = \sqrt{3}|V_{LL}||I|\cos(\varphi) = \sqrt{3} \cdot 10000 \cdot |I| \cdot 0.9 = 5 \times 10^6$ W. This gives us for the current amplitude $|I| = 320.7$ A and φ can be calculated from $\cos(\varphi) = 0.9$. The current is therefore $I = 320.7\angle 25.8°$ A. Now the internal EMF E can be calculated: $E = V + ZI = 5773.5\angle 0 + 320.7\angle 25.8 \cdot 2\angle 90 = 5773.5\angle 0 + 641.4\angle 115.8 = (5773.5 - 279.2) + 577.5j = 5524.6\angle 6°$ V.

e. The power factor is 0.9 leading, so the infinite bus can be represented as a capacitor in parallel with a resistor. The three phase active power is $P = 3|V|^2/R$ and for the resistor it holds: $R = 3\dfrac{(10k/\sqrt{3})^2}{5M} = 20 \ \Omega$. The three phase reactive power supplied by the infinite bus can be calculated from: $Q = |S|\sin(\varphi) = \dfrac{P}{\cos(\varphi)}\sin(\varphi) = P\tan(\varphi)$. The three phase reactive power equals $Q = 2.42$ Mvar.

The three phase reactive power supplied by the infinite bus: $Q = -3\dfrac{|V|^2}{X}$ with $X = \dfrac{-1}{\omega C} \rightarrow Q = 3|V|^2\omega C = 3|V|^2 \cdot 2\pi f C$ and this gives us $C = 77 \ \mu$F.

Solution 5.4

a. Figure E.5 in appendix E shows the equivalent circuit of a short transmission line. $Z = 3.3 + 5j \ \Omega = 6\angle 56.6° \ \Omega$, and $V_r = \dfrac{10000}{\sqrt{3}}\angle 0°$ V. The current

$I = |I|\angle\varphi$ (φ has a positive sign because the current is leading) and can be calculated from $P = \sqrt{3}|V||I|\cos(\varphi) = \sqrt{3} \cdot 10000 \cdot |I| \cdot 0.8 = 2500\,\text{kW}$ and this gives us $|I| = 180.4\,\text{A}$. The angle can be calculated from $\cos(\varphi) = 0.8$, and $\varphi = 36.9°$. The complete expression for the current is $I = 180.4\angle36.9°$ A. Now the voltage can be calculated: $V_s = V_r + ZI = 5773.5\angle0 + 180.4\angle36.9 \cdot 6\angle56.6 = 5707.4 + 1080.4j = 5808.8\angle10.7°$ V.

b. The single phase power at the sending end is: $S_{1\phi} = V_s I^* = 5808.8\angle10.7 \cdot 180.4\angle-36.9 = 1\angle-26.2°$ MVA. The three phase active power is $P_{3\phi} = 3 \cdot \text{Re}\{S_{1\phi}\} = 2.8\,\text{MW}$ and the three phase reactive power $Q_{3\phi} = 3 \cdot \text{Im}\{S_{1\phi}\} = -1.4\,\text{Mvar}$.

c. The three phase active power loss in the line: $P = 3|I|^2R = 3(180.4)^2 \cdot 3.3 = 322\,\text{kW}$.

d.

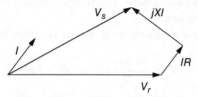

Note that no exact angles and phasor magnitudes have been used in the diagram. The voltage drop across the resistance is drawn in parallel with the current phasor and the voltage drop across the reactance is perpendicular to the current phasor.

Solution 5.5
answer d): None of the answers a), b) and c) is correct.

Solution 5.6
answer c): $\sqrt{3}\angle-30°$

Solution 5.7
answer a): 12.5 MVA; The apparent power is defined as $|S| = \sqrt{P^2 + Q^2}$

Solution 5.8
answer c): The ground wires shield the phase wires from a direct hit by lightning.

Solution 5.9
answer b): 100 Ω.

Solution 5.10
answer c): The leakage reactance.

CHAPTER 6

Solution 6.1

a. The load flow is the most important network computation, as it allows insight into the steady-state behavior of the power system. The purpose of the loadflow computation is to know the voltage magnitude and angle for each node in a three-phase power system. From this, the currents, power flows, and losses can be assessed, and a steady-state insight in the power system is obtained.
b. The loadflow equations, describing the mathematical relation between voltages and power injections are non-linear. It is not so easy to solve the non-linear loadflow equations by means of straightforward arithmetic (see example 6.4) and therefore an iterative method is applied (see example 6.5). With the help of a computer this works rather easy.

Solution 6.2
answer b): The excitation current has to be reduced.

Solution 6.3
answer d): 0.866

Solution 6.4
answer c): A capacitor and a resistor.

Solution 6.5
answer b): Calculations in the phasor domain can only be done for a constant frequency.

Solution 6.6
answer b): $-j15$ S.

Solution 6.7
answer b): 9

Solution 6.8
answer d): The shunt capacitances can be omitted for a short transmission line, but not for a medium-length transmission line.

Solution 6.9
answer b): an underexcited motor.

Solution 6.10
answer c): $0.1\,\Omega$.

Further Reading

In Dutch

- Dommelen, Daniël van: *Productie, transport en distributie van elektriciteit*, Acco, Leuven, 2001, ISBN 90-334-4595-6.
- EnergieNed: *Elektriciteitsdistributienetten*, Kluwer Techniek, Deventer, 1996, ISBN 90-557-6069-2.
- Hoeijmakers, Martin J.: *Elektrische omzettingen*, Delft University Press, Delft, 2003, ISBN 90-407-2455-5.

In German

- Happoldt, Hans, and Oeding, Dietrich: *Elektrische Kraftwerke und Netze*, Springer-Verlag, Berlin, 1978, ISBN 3-540-08305-7.

In English

- Bergen, Arthur R., and Vittal, Vijay: *Power System Analysis*, Prentice Hall, New Jersey, 2000, ISBN 0-13-691990-1.
- El-Hawary, Mohamed E.: *Electrical Power Systems – Design and Analysis*, IEEE Press, New York, 1995, ISBN 0-7803-1140-X.
- El-Sharkawi, Mohamed A.: *Electric Energy – An Introduction*, CRC Press, Boca Raton, 2005, ISBN 0-8493-3078-5.
- Glover, J. Duncan, and Sarma, Mulukutla: *Power System Analysis and Design*, PWSKENT Publishing Company, Boston, 1987, ISBN 0-543-07860-5.
- Grainger, John J., and Stevenson, William D., Jr.: *Power System Analysis*, McGraw-Hill, Singapore, 1994, ISBN 0-07-113338-0.
- Gross, Charles A.: *Power System Analysis – Second Edition*, John Wiley & Sons, Inc., New York, 1986, ISBN 0-471-86206-1.
- Kreuger, Fred H.: *Industrial High Voltage Vol. I*, Delft University Press, Delft, 1991, ISBN 90-6275-561-5.
- Kreuger, Fred H.: *Industrial High Voltage Vol. II*, Delft University Press, Delft, 1992, ISBN 90-6275-562-3.

Electrical Power System Essentials, Second Edition. Pieter Schavemaker and Lou van der Sluis.
© 2017 John Wiley & Sons Ltd. Published 2017 by John Wiley & Sons Ltd.

- Laithwaite, Eric R., and Freris, Leon L.: *Electric Energy: Its Generation, Transmission and Use*, McGraw-Hill, Maidenhead, 1980, ISBN 0-07-084109-8.
- Magid, Leonard M.: *Electromagnetic Fields, Energy and Waves*, John Wiley & Sons, Inc., New York, 1972, ISBN 0-471-56334-X.
- Mathur, R. Mohan, and Varma, Rajiv K.: *Thyristor-Based FACTS Controllers for Electrical Transmission Systems*, John Wiley & Sons, Inc., New York, 2002, ISBN 0-471-20643-1.
- Mohan, Ned, Undeland, Tore M., and Robbins, William P.: *Power Electronics – Converters, Applications and Design*, John Wiley & Sons, Inc., New York, 1995, ISBN 0-471-58408-8.
- Weedy, Birron M.: *Electric Power Systems*, John Wiley & Sons, Ltd, Chichester, 1987, ISBN 0-471-91659-5.

Index

Electrical Power System Essentials, Second Edition. Pieter Schavemaker and Lou van der Sluis.
© 2017 John Wiley & Sons Ltd. Published 2017 by John Wiley & Sons Ltd.